PEAR PRODUCTION AND HANDLING MANUAL

Technical Editors
ELIZABETH J. MITCHAM
RACHEL B. ELKINS

University of California
Agriculture and Natural Resources
Publication 3483
2007

For information about ordering this publication, contact

UC Agriculture and Natural Resources
Publishing
2801 Second Street
Davis, California 95618

E-mail: anrcatalog@ucanr.edu
Visit the ANR Communication Services Web site at
anrpublications.org

Publication 3483

UC PEER REVIEWED This publication has been anonymously peer reviewed for technical accuracy by University of California scientists and other qualified professionals. This review process was managed by the ANR Associate Editor for Pomology, Viticulture, and Subtropical Horticulture, and the Associate Editor for Pest Management.

To simplify information, trade names of products have been used. No endorsement of named or illustrated products is intended, nor is criticism implied of similar products that are not mentioned or illustrated.

Photo credits are given in the captions.

ISBN-13: 978-1-879906-65-5
ISBN-10: 1-879906-65-1
Library of Congress Control Number: 2006926974

1-5m-pr-5/07-SB/CM

WARNING ON THE USE OF CHEMICALS

Pesticides are poisonous. Always read and carefully follow all precautions and safety recommendations given on the container label. Store all chemicals in their original labeled containers in a locked cabinet or shed, away from foods or feeds, and out of the reach of children, unauthorized persons, pets, and livestock.

Recommendations are based on the best information currently available, and treatments based on them should not leave residues exceeding the tolerance established for any particular chemical. Confine chemicals to the area being treated. THE GROWER IS LEGALLY RESPONSIBLE for residues on the grower's crops as well as for problems caused by drift from the grower's property to other properties or crops.

Consult your county agricultural commissioner for correct methods of disposing of leftover spray materials and empty containers. Never burn pesticide containers.

PHYTOTOXICITY: Certain chemicals may cause plant injury if used at the wrong stage of plant development or when temperatures are too high. Injury may also result from excessive amounts or the wrong formulation or from mixing incompatible materials. Inert ingredients, such as wetters, spreaders, emulsifiers, diluents, and solvents, can cause plant injury. Since formulations are often changed by manufacturers, it is possible that plant injury may occur, even though no injury was noted in previous seasons.

Contents

Preface and Acknowledgments

The ***Pear Production and Handling Manual*** provides a basic understanding of the pear tree, its growth, and the development of fruit, as well as practical considerations on pear culture, pest management, and fruit harvesting, storage, and handling. An overview and history of the California pear industry is also provided. The focus of the information is on growing of pears in California, but the information will be useful to pear growers throughout the world. Thirty-six authors contributed to this publication, many of them affiliated with the University of California as Cooperative Extension Farm Advisors and Specialists, and as Agriculture Experiment Station faculty in various departments on the Davis, Berkeley, and Riverside campuses. In addition, several experts on pear production and handling from the United States and abroad were invited contributors. The editors would like to thank the authors for their dedication and perseverance through the long process of developing this manual.

In addition, we would like to thank the California Pear Advisory Board and the Pear Pest Management Research Fund for supporting research that developed much of the information contained in this manual. We view our relationship as symbiotic, each enabling the other to do more. We express our appreciation to the ANR Communication Services staff for assistance and guidance in this process.

Elizabeth J. Mitcham
Rachel B. Elkins

Authors

TECHNICAL EDITORS

Elizabeth J. Mitcham, Extension Pomologist, UC Davis Department of Plant Sciences

Rachel B. Elkins, UC Cooperative Extension Farm Advisor, Lake, Mendocino, and Sutter/Yuba Counties

CONTRIBUTING AUTHORS

James Beutel, retired Emeritus Extension Specialist, UC Davis Department of Pomology

Steven C. Blank, Extension Specialist, UC Davis Department of Agricultural and Resource Economics

Patrick H. Brown, Professor, UC Davis Department of Plant Sciences

David J. Burkhart, Sales Representative, Willow Drive Nursery, Ephrata, WA

Theodore M. DeJong, Professor, UC Davis Department of Pomology

Roger A. Duncan, UC Cooperative Extension Farm Advisor, Stanislaus County

Rachel B. Elkins, UC Cooperative Extension Farm Advisor, Lake, Mendocino, and Sutter/Yuba Counties

Clyde L. Elmore, Extension Weed Specialist Emeritus, UC Davis Department of Vegetable Crops/Weed Science

W. Douglas Gubler, Extension Plant Pathologist, UC Davis Department of Plant Pathology

Chuck A. Ingels, UC Cooperative Extension Farm Advisor, Sacramento County

John Ireland, Product Development and Research Director, Fowler Nurseries, Newcastle, California

Adel A. Kader, Professor, UC Davis Department of Plant Sciences

Karen M. Klonsky, Extension Farm Management Specialist, UC Davis Department of Agricultural and Resource Economics

Steven E. Lindow, Professor, UC Berkeley Department of Microbial Biology

Robert McClain, Field and Research Coordinator, California Pear Advisory Board, Sacramento

Glenn McGourty, UC Cooperative Extension Viticulture and Plant Science Advisor, Mendocino County

Warren C. Micke, Extension Pomologist Emeritus, UC Davis Department of Pomology

Elizabeth J. Mitcham, Extension Pomologist, UC Davis Department of Plant Sciences

F. Gordon Mitchell, Extension Pomologist Emeritus, UC Davis Department of Pomology

Mario S. Moratorio, UC Cooperative Extension Small Farm Advisor, Solano County

Franz J. A. Niederholzer, UC Cooperative Extension Farm Advisor, Sutter-Yuba Counties

Wilbur O. Reil, UC Cooperative Extension Farm Advisor Emeritus, Yolo/Solano Counties

Lawrence J. Schwankl, Extension Irrigation Specialist, UC Davis Department of Land, Air, and Water Resources

Kenneth A. Shackel, Associate Professor, UC Davis Department of Plant Sciences

Jerome B. Siebert, Extension Economist, UC Berkeley Department of Agricultural and Resource Economics

Richard L. Snyder, Extension Meteorologist, Land Air and Water Resources, UC Davis

Stephen M. Southwick, former Extension Pomologist, UC Davis Department of Pomology

Robert Stebbins, Extension Pomologist Emeritus, Oregon State University

David Sugar, Professor, Oregon State University Southern Oregon Research and Extension Center, Medford

James F. Thompson, Extension Agricultural Engineer, UC Davis Department of Biological and Agricultural Engineering

Bas van den Ende, Private Consultant, Shepparton, Victoria, Australia

Robert A. Van Steenwyk, Extension Entomologist, UC Berkeley Department of Entomology and Plant and Soil Microbiology

Lucia G. Varela, UC Cooperative Extension IPM Advisor, Sonoma County

Steven A. Weinbaum, Professor, UC Davis Department of Pomology

Chris Zanobini, Executive Director, California Pear Advisory Board, Sacramento

Broc Zoller, President, The Pear Doctor, Inc., Kelsyville, California

PEAR PRODUCTION AND HANDLING MANUAL

PART 1

INTRODUCTION AND PEAR INDUSTRY OVERVIEW

1

Introduction

RACHEL B. ELKINS

The purpose of the *Pear Production and Handling Manual* is to provide a comprehensive overview of contemporary commercial production and handling practices. While emphasis is on growing Bartlett (Williams Bon Cretien) pears in California, much of the content is also relevant to other varieties and production areas in the western United States.

California pear production has changed dramatically over the past several decades. In 1968, 322,000 tons (292,000 t) of Bartlett pears were produced on almost 35,000 acres (14,165 ha) in California, or just over 9 tons per acre (20 t/ha). At that time, U.C. Cooperative Extension Farm Advisor C. L. Hemstreet of Lake County stated, "Orchards of 40 to 60 acres can economically justify the larger capital investment in equipment and presently appear to be feasible economic units" (Hemstreet n.d.).

In 2004, 242,000 tons (210,000 t) of Bartletts were harvested from 13,500 acres (5,460 ha), or just over 18.5 tons per acre (41.4 t/ha). The number of counties with significant pear acreage has shrunk from 10 or 11 to 4 or 5. The average economic unit is more likely 100 acres (404 ha) for most growers who make a primary living from growing pears. Although comparative absolute numbers are unavailable, the number of active pear growers in California has drastically declined, leaving a small but highly organized, efficient, and dedicated contingent. Changes in the nature of pear production in California are doubtless similar in other western U.S. regions, albeit on varying scales.

The amazingly stable total production in the face of over a 50 percent decline in acreage is testimony to the partnership between the research community and pear industry. Those involved in solving its problems have become highly specialized and professional. Growers are fortunate to have access to progressive and knowledgeable pest control advisers, cannery field personnel, industry organization personnel, and others who are widely respected throughout the industry. These individuals have, in many cases, filled the vacuum created by shrinking public sector resources. They have also been the primary leaders in implementing new and risky production practices.

One sees a much different industry when traveling through California pear country today than in 1968. Poor-producing orchards have, for the most part, disappeared. Many orchards produce 20 or more tons per acre (9 t/ha). The traditional spacing of 24 feet by 24 feet (7.3 m by 7.3 m) has been filled in to as close as 12 feet by 12 feet (3.6 m by 3.6 m). Although new plantings are rare, thousands of trees have been interplanted, which will continue to translate into per-acre tonnage increases in the coming years. New plantings are in the range of 11 feet by 16 feet (3.4 m by 4.9 m) or 9 feet by 15 feet (2.7 m by 4.6 m), in "hedgerow" fashion. Flood irrigation has largely been replaced by solid-set sprinklers, which are used for frost protection as well as for irrigation.

The long-term consolidation and streamlining of the pear industry is the result of several factors: urbanization, economic instability due to continuing declines in the canning sector, retail consolidation and global competition, the devastating diseases pear decline and fire blight, and production cost increases that have steadily outpaced returns. Today the pear grower must produce at least 20 tons per acre (8 t/ha) to break even. Fruit must average size 100 (74 mm, about 3 in), rather than the 120 to 150 (64.5 to 70 mm, about 2½ to 2¾ in) acceptable years ago. Trees must come into bearing early; no longer is a pear orchard planted for the children or grandchildren.

Many of the costs associated with pear production in California, as well as other West Coast production regions, reflect unsolved research problems. Pear trees are still too tall and too late coming into bearing; size-controlling rootstocks and training systems amenable to machine-facilitated pruning and harvest must be found to replace the increasingly expensive and scarce labor. Pest management costs have skyrocketed, and the key pests of pear, codling moth, pear psylla, and mites, have been joined by newer pests such as leafroller as the target of ever-increasing amounts of research. Nonuniform postharvest quality continues to limit consumer acceptance in the fresh market, and new uses for pear fruit are needed to replace diminishing cannery outlets.

The pear industry must continue to find solutions to old problems and anticipate new problems in order to remain viable. Indeed, a review of historical documents, both industry- and research-generated, reveals startlingly similar themes to modern times; many of the problems of yesterday are the problems of today. Yet, upon closer scrutiny, the success stories of research, extension, and industry innovation can be sifted out. Black end was eliminated by improved rootstocks, as was pear decline. Fire blight is now managed using antibiotics and biological control. Per-acre yields have nearly doubled due to better irrigation, fertilization, pruning, and pest control practices. Preharvest fruit drop is prevented by sprays of NAA (1-naphthaleneacetic acid), and rapid precooling has improved storage life and shipping quality.

Past and current advances have dramatically altered the nature of western U.S. pear production in the twenty-first century. While economic, regulatory, and production challenges will continue to evolve, there is no doubt that the West Coast pear industry will respond to the rapidly changing global marketplace by continuing to improve production and handling efficiency, as well as develop innovative market niches for high-quality pears.

REFERENCES

Hemstreet, C. L. N.d. Lake County Bartlett pear industry, 1884 to 1968. Unpublished manuscript in possession of the author.

2

History and Overview of the California Pear Industry

RACHEL B. ELKINS, MARIO S. MORATORIO, ROBERT McCLAIN, AND JEROME B. SIEBERT

EARLY U.S. HISTORY

Pear trees arrived on the East Coast of the United States from Europe with English and French settlers. They were first grown exclusively as backyard trees, attaining heights up to 50 feet (15 m) and living over 200 years in some cases. The French also brought seed and seedlings down from French Canada to New York and Pennsylvania, and from Nova Scotia via the St. Lawrence River to Michigan, Illinois, and Indiana. The first authentic recorded pear variety was the Endicott of Salem, Massachusetts, around 1630. Amateur East Coast breeders active in the late 1700s included William Cox of New Jersey, whose followers bred Seckel, and Robert Manning of Massachusetts, who recognized Bartlett as "Williams" Bon Cretian, its modern name in Europe and the Southern Hemisphere.

Commercial pear growing began mostly in New York and Pennsylvania in the early 1800s. Major varieties included Bartlett, Clapp Favorite, Seckel, Kieffer, Beurré Bosc, Duchess d'Angouline, LeConte, and Beurré d'Anjou. These varieties, some of them European or Asian hybrids, were grown where large bodies of water, such as the Great Lakes and Finger Lakes, moderated the climate and lessened tree loss due to frost, freeze, drought, and fire blight. New York dominated the commercial U.S. pear market until the Gold Rush and improved transportation to and from the western United States. By the early 1900s, the largest plantings were in California, Oregon, and Washington.

THE MOVE TO THE WEST

Fire blight made profitable pear growing in the eastern and midwestern U.S. very difficult. Except for New York, Pennsylvania, and Michigan, where Bartlett was the main variety, growers often had to plant poorer-quality fire blight—resistant varieties such as Kieffer and LeConte, and they struggled with other diseases and pests that thrived in the cool, moist climate. There were also the chronic problems of frost, freeze, and drought, which limited where favored varieties could be grown on a large scale.

The three West Coast states offered ideal conditions. Winters were cold enough for adequate chilling but seldom killed trees. Trees bloomed late enough to avoid major frost and, where needed, could be protected by orchard heaters. Irrigation from local rivers supplied water during the warm, dry summers. Abundant sunshine and long growing seasons produced a large tonnage of high-quality large fruit. Fire blight, while still the major limiting factor, was more easily controlled in the dry climate. Besides fire blight, insect and disease pests were fewer and were kept in check by sprays and cultural practices. Varieties grown in the early western orchards were most of the same ones grown today: Bartlett, Beurré d'Anjou, Easter Beurré, and Beurré Bosc. Comparative seasonal temperature patterns for the major historical and current pear-growing areas are shown in table 2.1.

California began shipping pears by rail in 1870 and displaced New York as the number one pear-producing state by 1930. The Oregon and Washington pear industries began shortly after 1900. Small industries also grew up in Colorado, Idaho, and Utah. Total U.S. acreage of pears was 62,150 (25,152 ha) in 2005. The distribution of acreage in 2005 versus 1935 for major per-growing areas is shown in figure 2.1.

THE EARLY CALIFORNIA INDUSTRY

The Franciscan Fathers, led by Junipero Serra, brought pear seed to plant in the orchards of the California missions, including Mission Santa Clara, in the mid-1700s. As it did in the eastern United States, commercial production lagged behind amateur culture for about 100 years.

The Gold Rush provided impetus for the first commercial orchards. Seeds of *Pyrus communis* were planted in the Sierra foothills to provide fresh fruit to the mining community. Eastern U.S. and European varieties were first brought around the Horn of Africa by ship; Bartlett was brought by this route to the Napa Valley in

Table 2.1. Average growing-season temperatures, by month, in historically important U.S. pear-producing areas

		Average temperatures													
		Mar		Apr		May		June		July		Aug		Sep	
State and area	**Station**	°F	°C	°F	°C	°F	°C	°F	°C	°F	°C	°F	°C	°F	°C
California															
Central	Auburn	51.2	10.7	56.1	13.4	62.4	16.9	71.4	21.9	77.0	25.0	76.0	24.4	69.2	20.7
	Marysville	54.3	12.4	59.4	15.2	64.8	18.2	72.8	22.7	77.8	25.4	76.2	24.6	71.9	22.2
	Sacramento	54.3	12.4	58.1	14.5	63.3	17.4	69.4	20.8	73.2	22.9	72.9	22.7	69.3	20.7
	Upper Lake	49.9	9.9	54.8	12.7	59.6	15.3	66.7.	19.3	73.8	23.2	72.7	22.6	66.7	19.3
Central, coastal	San Jose	53.1	11.9	56.3	13.5	58.5	14.7	62.7	17.0	66.5	19.2	66.1	18.9	64.2	17.9
	Santa Rosa	51.2	10.7	54.4	12.4	57.6	14.2	63.0	17.2	65.2	18.4	64.4	18.0	63.8	17.7
Michigan															
Southwestern	Benton Harbor	34.7	1.5	47.1	8.4	57.9	14.4	68.1	20.0	72.7	22.6	71.3	21.8	64.0	17.8
New York															
Hudson River Valley	Poughkeepsie	37.3	2.9	49.5	9.7	60.7	15.9	69.6	20.9	74.6	23.7	72.4	22.4	64.6	18.1
Western	Rochester	32.3	0.2	45.1	7.3	56.7	13.7	66.9	19.4	71.6	22.0	69.7	20.9	62.4	16.9
Oregon															
Hood River Valley	Hood River	43.3	6.3	49.9	9.9	56.1	114	61.6.	16.4	67.4	19.7	66.6	19.2	59.5	15.3
Rogue River Valley	Medford	46.7	8.2	51.6	10.9	57.7	14.3	65.2	18.4	71.8	22.1	70.8	21.6	63.1	17.3
Washington															
Wenatchee River Valley	Wenatchee	42.8	6.0	51.5	10.8	58.8	14.9	66.2	19.0	73.2	22.9	71.6	22.0	61.6	16.4
Yakima River Valley	Yakima	4.1	6.7	52.5	11.4	59.0	15.0	66.4	19.1	71.4	21.9	69.5	20.8	61.1	16.2

Source: Williams et al. 1978, p. 2.

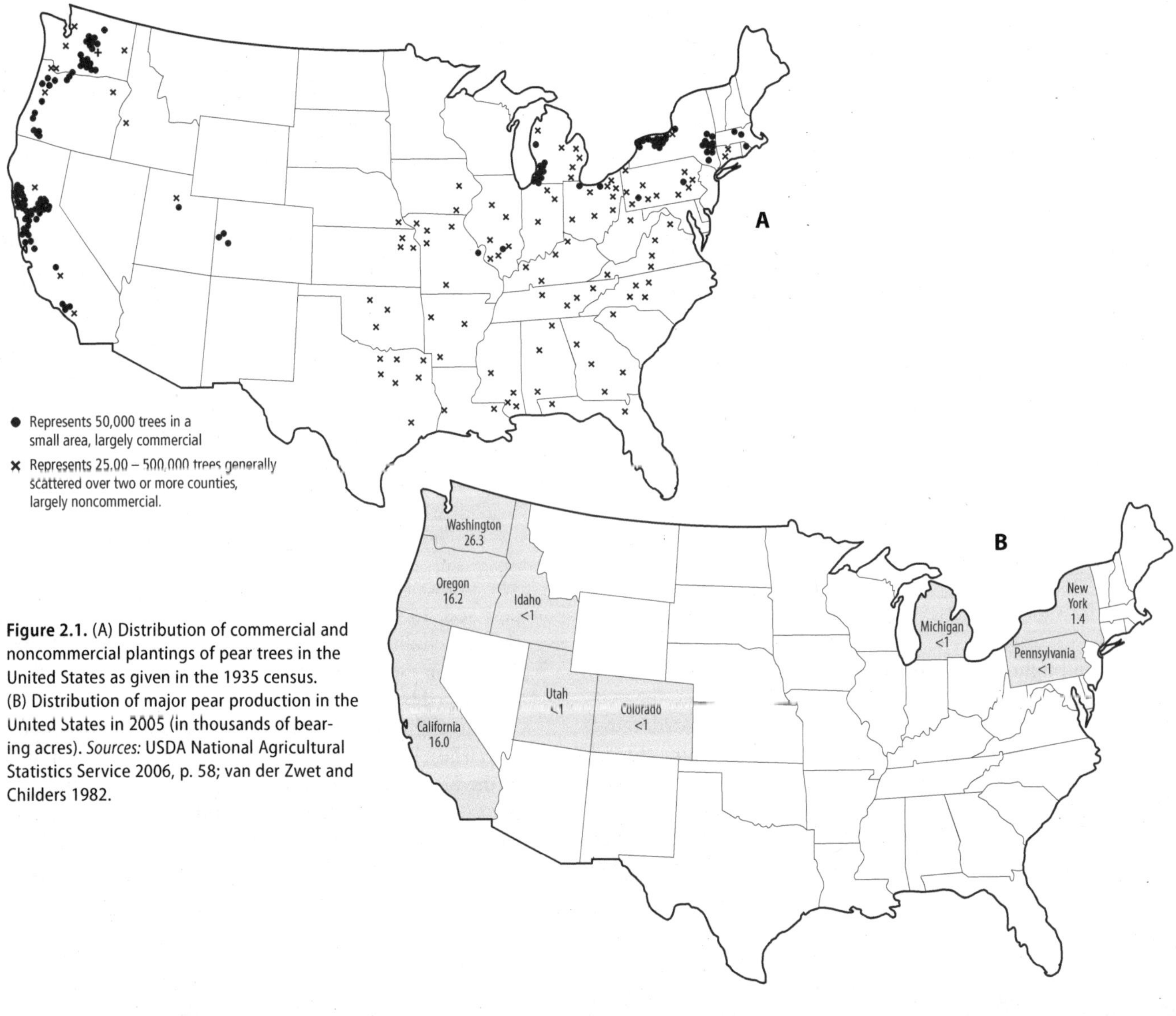

Figure 2.1. (A) Distribution of commercial and noncommercial plantings of pear trees in the United States as given in the 1935 census. (B) Distribution of major pear production in the United States in 2005 (in thousands of bearing acres). *Sources:* USDA National Agricultural Statistics Service 2006, p. 58; van der Zwet and Childers 1982.

1850. The completion of the transcontinental railroad in 1869 led to the rapid increase in western production because pears could now be shipped to eastern markets by rail (Davis and Tufts 1941, p. 3).

In California, pears grew in many counties and climatic areas in the early years. There were two reasons for this. First, only the need for adequate winter chilling limited southern acreage to the higher elevations, and frost limited northern orchards. Second, pears tolerated a wider range of soil conditions than other fruit crops, so they were planted where others failed to thrive. In particular, pear trees could survive in heavy soils with excessive moisture, unlike stone fruits. George Weldon, writing in 1918, declared that "available land for profitable pear culture may be found in practically every county of the state" (Weldon 1918, p. 6) The best areas, however, and those eventually responsible for most of the state's production, had common characteristics. Though tolerant of a wide range of soils, pear trees prefer moist, adequately drained clay loam. Warm summer temperatures, particularly 6 to 8 weeks before harvest, produce the best-quality fruit. Cool springs reduce the hazard of fire blight to manageable levels. Adequate irrigation water or rainfall must be available through the growing season.

By the late 1920s, major acreage existed in Placer, Sacramento, Santa Clara, Solano, and Sonoma Counties, with increasing acreages in El Dorado, Lake, Mendocino, and Nevada Counties. Fire blight quickly took out orchards in much of the upper Sacramento and southern San Joaquin Valleys due to the warm spring and early-summer temperatures during bloom. In other areas, excessively cool summers caused uneven ripening, making fruit unsuitable for canning or shipping.

As the California pear industry evolved in the 1920s, each growing area became known for particular uses, and in some cases, varieties, though Bartlett has always dominated as the only "all-purpose" pear, being suitable for shipping, canning, and drying. Most California fruit was being shipped to the fresh market by the 1930s, while the Pacific Northwest, especially the Yakima Valley in north-central Washington, supplied canneries. There were two main reasons for this. First, after World War I the booming cannery industry developed a preference for Northwest pears because the northern climate produced an elongated neck. The rounder California pears were cut up for fruit salad and cocktail, which was also dominated by California because several major ingredients—pears, cling peaches, and Thompson Seedless grapes—were grown in the state. Much of the California cannery fruit came from the Santa Clara Valley, where cool weather prior to harvest reduced shipping quality. Second, California could ship fruit over a long period. The Sacramento Delta district in those days received high prices for 6 weeks before other pears became available. The later-ripening districts of the Sierra Foothills and North Coast received similar high prices after shipping in other areas had been completed.

According to accounts of those early years, California received a substantial premium over Northwest Bartletts in eastern U.S. markets. As cold storage capacity increased in the Northwest, Washington began shipping some fresh Bartletts after shipping in California was finished, and Northwest growers could pack and ship fruit on demand. The Northwest also established firm domination of the winter pear market, mainly with Anjou, but also Bosc, some Comice, Seckel, and Winter Nelis. Only the Santa Clara Valley grew substantial quantities of winter pears in California, primarily Comice and Winter Nelis.

The high prices and acreage boom of the 1920s came to an end in 1929 with the Depression. Overproduction combined with lack of consumer demand reduced prices by half, to less than production and living costs. World War II cut off exports to Europe, so fresh fruit prices remained low through the war, although War Food Order No. 108 ensured good prices for cannery fruit. By the end of the war, production had declined from a peak of about 60,000 acres (24,300 ha) to 37,000 acres (15,000 ha). Fruit prices further decreased after the war, and acreage continued to decrease gradually to about 34,000 (13,800 ha) in the 1950s.

THE INDUSTRY AFTER PEAR DECLINE

In 1958, a mysterious malady, which was named "pear decline," began killing pear orchards in the Sierra Nevada foothills. After a concerted effort by University of California researchers and extensionists, it was discovered that the disease was caused by a mycoplasma vectored by pear psylla, which had arrived in California from Oregon several years earlier. The pear decline epidemic caused the permanent removal of thousands of acres, mainly in the Sierra foothills. However, many new acres were planted, and many orchards were interplanted. Bartlett acreage had surpassed 36,000 (14,600 ha) by 1967, the highest since the end of World War II.

Several negative factors lead to more acreage removal in the mid-1970s. Fresh fruit was beginning to displace canned fruit consumption, reducing cannery capacity. Irrigated orchards in the Central Valley and north-central Washington produced more popular competing fruits such as apples, cherries, nectarines, peaches, and plums. Production costs increased while returns were flat or negative. Fire blight became impossible to control economically in some districts. Growing population centers made farming difficult and drove up land values and taxes in other districts. The result was a steady decline in acreage from 1974 to the present. Less than 18,000 bearing acres (7,300 ha) of pears remained in California as of 2003, of which 12,500 acres (5,000 ha) were Bartlett. The majority of bearing acres were in three counties: Sacramento, Lake, and Mendocino. Production, however, remained relatively steady as per-acre tonnage has more than doubled since 1972 (table 2.2).

CURRENT INDUSTRY STATUS

World and U.S. Production

A discussion of current pear production trends must include at least a cursory overview of world production. Approximately 19.5 million tons (17.7 million t) of pears were produced worldwide in 2005, including over 11 million tons (10 million t), 380,000 tons (345,000 t), and 441,000 tons (400,000 t) of mainly Asian varieties produced in the People's Republic of China, Republic of Korea, and Japan, respectively (FAO 2005).

Other than China and Japan, most of the Northern Hemisphere production is in the United States and Canada and the European Union. Combined Northern Hemisphere production in 2005 was about 4 million tons (3.6 million t). Other than the United States, northern tier countries that produced over 120,000 tons (109,000 t) were, in descending order, Italy, Spain, France, Netherlands, Germany, Turkey, Iran, Ukraine, and Portugal. Most Northern Hemisphere production is channeled to the fresh market, though France, Italy, the People's Republic of China, Russian Federation, Spain, the United States, and to a lesser extent Turkey, are major producers of Bartlett and other pears for canning.

Table 2.2. California Bartlett pear bearing acreage and production, 1932–2005

Year	Acreage	Production (tons)	Production (tons/acre)
1932	58,209	217,000	3.6
1942	38,881	212,000	5.6
1952	34,452	335,000	9.7
1962	30,229	348,000	11.5
1972	36,093	286,000	7.9
1982	28,705	314,000	10.9
1992	20,200	315,000	15.6
2002	13,500	264,000	19.6
2005	12,000	195,000	16.3

Sources: California Tree Fruit Agreement 1962 p. 3; 1976 p. 11; 1993 p. 8. *California Fruit and Nut Review* 26(10): 1.

The Southern Hemisphere produced just over 1.4 million tons (1.3 million t) of pears in 2005. Production is increasing to meet expanding export opportunities for winter fruit in Europe and, more importantly, North America. Countries producing more than 120,000 tons (109,000 t) included Argentina, the Republic of South Africa, Chile, and Australia. New Zealand is also a relatively small but growing producer. While fresh market exports have driven increased production, Argentina, Australia, and South Africa produce a substantial volume of Bartletts for canning.

Other countries with relatively small but growing pear industries are those of Eastern Europe, particularly Hungary, and Serbia and Montenegro. The People's Republic of China is reportedly expected to export Bartlett pears within the present decade.

Ninety-seven percent of the U.S. pear crop was produced in the three Pacific Coast states of California, Oregon, and Washington as of 2005. Production averaged nearly 854,000 tons (775,000 t) from 2003 to 2005. This was comprised of approximately 51 percent Bartlett, 36 percent Anjou, 11 percent Bosc, and 2 percent other varieties.

California Production

The majority of pear acreage in California lies in the northern interior valleys. Up to 30 to 35 tons per acre (13.4 to 15.6 t/ha) are sometimes produced, although the average Bartlett production from 1996 to 2005 was about 18 tons per acre (8 t/ha). Growers have planted more acres of alternative varieties to compliment or replace Bartletts in recent years (table 2.3).

The major production areas in California (fig. 2.2), in descending order of current acreage, include

- the Delta lands of the Sacramento River Valley between the cities of Sacramento and Isleton
- the Clear Lake district of Lake County
- the Ukiah and Potter Valley districts of Mendocino County
- the Suisun Valley in Solano County
- the Upper Sacramento River Valley
- the Sierra foothills of Placer, Nevada, and El Dorado Counties

Other minor pear-producing areas include

- the central coastal plains adjoining the San Francisco Bay in Santa Clara, San Benito, Alameda, and Contra Costa Counties
- the Napa and Sonoma Valleys and the Pearblossom area in the Tehachapi Mountains of Los Angeles County

Table 2.3. California pear-bearing acreage for Bartlett and other varieties, 1960–2005

	Bearing acreage		Nonbearing acreage	
Year	**Bartlett**	**Other***	**Bartlett**	**Other**
1960	32,489	3,771	9,916	400
1965	32,630	2,460	11,964	1,136
1970	36,040	2,030	9,770	1,380
1975	36,101	1,395	3,607	509
1980	33,900	1,700	1,200	500
1985	23,400	1,300	400	1,300
1990	20,200	3,100	200	2,100
1995	19,100	4,800	N/A	N/A
2000	15,000	4,300	N/A	N/A
2005	12,000	4,000	N/A	N/A

Sources: Pradhan and Moulton 1996, pp. 42–43; California Fruit and Nut Review 26(10): 1.

Note: * Mainly Bosc, but also includes Asian varieties, Red Anjou, Comice, Red Clapp (Starkrimson, Super Red), and others.

Sacramento River Delta

Situated in the lower Sacramento Valley east of the San Francisco Bay Area between the cities of Sacramento and Isleton, the Sacramento Delta is home to the largest pear acreage and tonnage in California. Pears are grown along the banks of the Sacramento River, along various sloughs feeding it, and on islands where sloughs meet each other. Orchards producing 25 tons or more per acre (11.1 t/ha) grow in a narrow band of dark, friable sedimentary soil deposited by river overflows. Subsoils are well-drained peat. Until levees were constructed in the early 1900s by Chinese immigrants, silt was regularly deposited on existing orchards. Away from the levees and to the south of Isleton, the soil becomes shallow and underlain by hardpan; there pears are replaced mainly by field crops. Irrigation water is pumped from the river or sloughs. Drainage is needed where peat is lacking to avoid waterlogging by the high water table. The growing season is warm, but it is moderated by cool westerly breezes from the San Francisco Bay, which may be strong enough to cause some problems with limb rub. Frost protection is unnecessary. The first "River Bartletts" are shipped soon after July 4 and continue for about 6 weeks until fruit softens and the North Coast districts begin to ship. Sacramento also ships the earliest U.S. winter pears. Historically, 80 to 85 percent of total Bartlett production in this area is canned; however, declining cannery capacity since the 1970s has shifted more fruit from this district to the fresh market. Bartletts from the Delta have the shortest length-to-width ratio, about 1:2 to 1:1.

Lake and Mendocino Counties

Located at elevations of 1,200 to 1,400 feet (365 to 425 m) in valleys in the Mayacamas Mountains of the North Coast, about 2 hours north of San Francisco, the Lake and Mendocino County districts are known for "mountain fruit." In recent years, poorer orchards here have been replaced with vineyards.

Until the 1920s, Lake County was famous mainly for dried pears, which were harvested onto beds of straw and dried in the open. It still supplies pears for almost all of the U.S. domestic dried fruit market. In the mid-1920s, a state highway linked the rail-less county to fresh market shipping points in Mendocino, and acre-

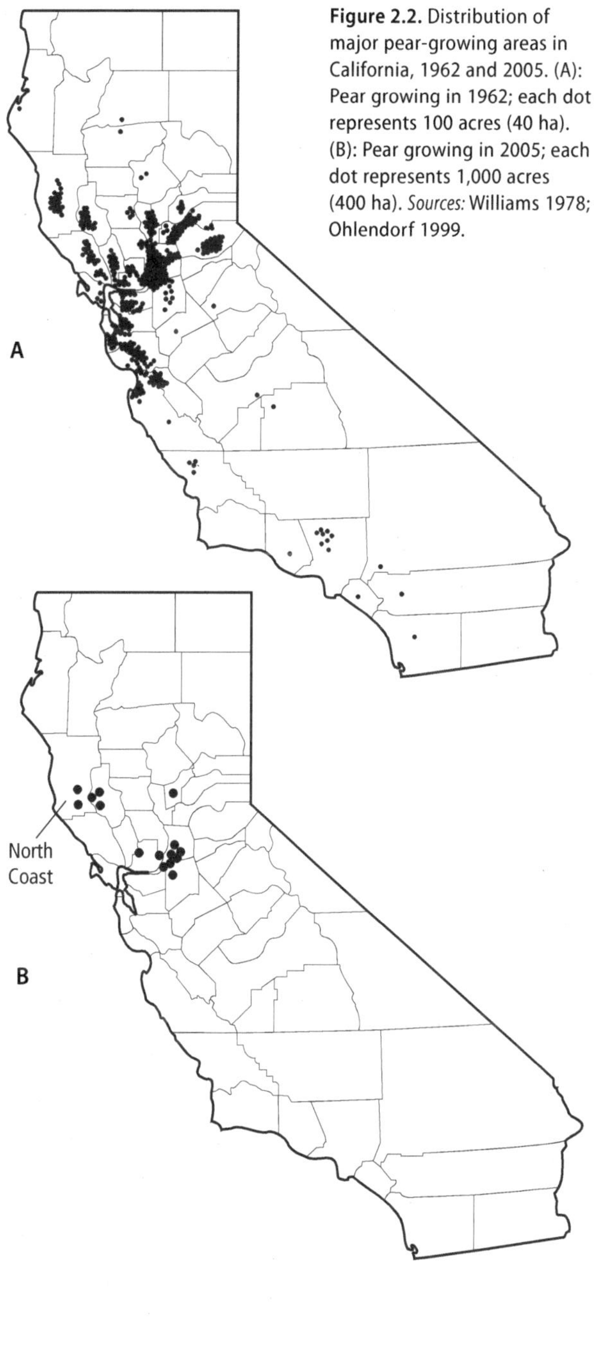

Figure 2.2. Distribution of major pear-growing areas in California, 1962 and 2005. (A): Pear growing in 1962; each dot represents 100 acres (40 ha). (B): Pear growing in 2005; each dot represents 1,000 acres (400 ha). *Sources:* Williams 1978; Ohlendorf 1999.

age increased rapidly in the Big Valley, near the town of Kelseyville. Most orchards were planted near the shores of Clear Lake, the largest natural lake completely within California, which moderates cold spring and hot summer temperatures. Other Lake County districts include Scotts Valley, just west of Lakeport, and Upper Lake at the north end of Clear Lake. Lake County soils are variable, ranging from sandy and silt loams along creeks to clay loam to heavy black adobe clay closer to Clear Lake. Many orchards were nonirrigated until the 1940s, when wells were dug. Spring frost protection is needed in most orchards. Once accomplished by orchard heaters, frost protection is now mainly done by under-tree sprinklers and wind machines. Harvest generally begins the first week of August and continues for about 3 weeks. Pears from this northern district have a longer length-to-diameter ratio than fruit from other California pear districts. For these reasons, close to one-half of Lake County's crop is sold fresh, with the rest mainly canned and dried.

Orchards in Mendocino County are planted mostly along the banks of the Russian River in the Ukiah and Potter Valleys. Soils are sandy loam and loam, and trees grow large, producing tonnages that are among the highest in the state. Irrigation is from wells and water pumped from the Russian River. Fruit shape is intermediate between that of the Sacramento Delta and Lake County. About 25 to 30 percent is shipped fresh, and the rest supplies canneries. Harvest begins about 7 to 10 days after the Delta and an equal number of days before Lake County.

Upper Sacramento Valley

Very few pear orchards survive in the northern Sacramento Valley due to fire blight. Those remaining are large operations under the control of less than a half dozen growers accustomed to battling the disease. Long-lived, high-tonnage orchards are planted on loam soils along the Feather and Yuba Rivers in Sutter and Yuba Counties. Bartlett and Bosc are the main varieties. Harvest begins in late July and continues to mid-August. The fruit shape here is similar to that in the Sacramento Delta. Most growers in this area utilize bees and pollinizer varieties, as fruit reportedly has less tendency to set parthenocarpically.

Suisun Valley, Solano County

This area northeast of San Francisco once produced pears for the fresh market, but today it is almost exclusively a cannery district. Soils are deep loam but water for irrigation is scarce, limiting tree growth and tonnage. As in the Delta, wind can damage fruit. Fire blight, while less problematic than in the Valley, has accelerated tree removal in recent years. Acreage has declined steadily since the 1970s, mainly due to low tonnage, urbanization, and more recently, to vineyard conversion. The demise in 2000 of the district's main fruit purchaser, Tri Valley Growers, Inc., has hastened orchard removal.

Sierra Foothills

In 1949 there were over 10,000 acres (4,000 ha) of pears in El Dorado, Nevada, and Placer Counties. In those days, this late-harvested, high-quality fruit was sold at premium prices in eastern markets. Today less than 400 acres (160 ha) remain, almost exclusively in El Dorado County. The demise was mainly due to pear decline, the mycoplasma disease vectored by pear psylla, which was found in El Dorado County in 1958. Less than 8,500 tons (7,700 t) were produced on only 2,350 acres (950 ha) by 1962. Replanting of pears was limited as a result of pear decline; instead, apples were the crop of choice. Pears are produced on slopes on well-drained granitic soils. Fruit from orchards on the lower slopes at about 1,100 feet (335 m) ripens with fruit in the Sacramento Delta and Suisun Valley. Hot summer days and cool nights produce fine-quality fruit with a characteristic blush at higher elevations to about 3,500 feet (1,070 m). These pears bloom late and are harvested in late August. Most of this mountain fruit is now shipped by a cooperative to midwestern and eastern markets, with some going to canneries. Tonnages are moderate to low. The fruit shape is similar to that grown on the North Coast.

Central Coast

Several pear-growing areas are located on the coastal plains surrounding the San Francisco Bay, between the Coast Ranges and the Pacific Ocean, and in the Napa and Sonoma Valleys. This area is one of the oldest centers of commercial pear growing in the state. In 1945, Santa Clara County hosted the largest producing acreage. Today very little commercial production remains; urbanization is the main factor. Virtually all productive orchard land in the Bay Area region was converted to urban uses in the late 1950s through the 1970s. Many pear growers from this area sold their land and moved to the Central Valley and North Coast, where they planted or bought orchards and continue to farm today. Also, orchards in the Napa and Sonoma Valleys have been replaced by wine grapes, which are generally more profitable. Pear growing is now restricted to about 100 acres (40 ha) in eastern Contra Costa and southern Santa Clara Counties.

Pear trees in cool Central California grew less vigorously than in other areas and were thus less prone to fire blight. However, this climate also reduced shipping quality. Bartletts here were grown mainly for the cannery market, but many other varieties were grown as well, including Anjou, Bosc, Clairgeau, Comice, Easter Beurré, Hardy (a canning variety now sold as French Butter Pear), and Winter Nelis. Pears were grown mainly on the heavy, dark cracking clay near the Bay, while other deciduous fruits were planted on loam soils away from the Bay. Orchards were irrigated and long-lived, but tonnage was only moderate.

Southern California

George Weldon, writing in 1918, stated that pear trees could be found growing below sea level in Southern California. However, pears' high chilling requirement eventually narrowed commercial production to the Tehachapi Mountains north of Los Angeles. Only a few acres exist today, with the industry memorialized by town names such as Pearblossom. This area supplied fruit to Los Angeles metropolitan markets and was also one of the few areas with known boron and copper deficiencies.

INDUSTRY ORGANIZATIONS

Growers and others involved in the pear industry have, over the years, formed various organizations to address short- and long-term challenges. Fresh and processed fruit are generally represented by different organizations, though there has been crossover for certain purposes.

There have been several federal and state marketing orders for Bartlett and winter pears. The first for fresh fruit was the California Fresh Fruit Deciduous Marketing Order and Agreement No. 1. This was established in 1933, revised under the Agricultural Marketing Agreement Act of 1937 as Federal Marketing Order 917, and later known as the California Tree Fruit Agreement (CTFA) No. 917. R. W. Gray was the first manager of CTFA, followed by Galen Geller (1949 to 1983), who provided great vision and leadership for the industry. Through 1991, regulation, promotion, and research for Bartlett pears in California were handled by CTFA, which also managed marketing orders for peaches, plums, and nectarines. Beginning in 1992, CTFA handled only promotion for Bartletts, as regulation and research were overseen by the newly-formed California Pear Advisory Board (CPAB).

Until 1997, non-Bartlett and winter pear varieties in California, Oregon, and Washington were covered under Federal Marketing Order No. 927, managed by the Winter Pear Control Committee (WPCC) based in Portland, Oregon. Promotion of West Coast winter pears, as well as northwest Bartlett pears, was handled by the Oregon-Washington-California Pear Bureau, operated by WPCC staff. California growers withdrew from the WPCC in 1997, at which time the Bureau changed its name to the Northwest Pear Bureau. Herb Diede was president of the Northwest Pear Bureau and also manager of WPCC until 2000, at which time Kevin Moffitt was appointed to these posts. In 2005, Oregon and Washington growers voted to amend Order No. 977 and incorporate summer/fall varieties and to extend the order to processing pears.

Processed pears were originally regulated under the auspices of the California Pear Program Committee, established in 1938 under the enabling California Agricultural Producers Marketing Law. This law allowed for proration "zones" to be established in which the volume of fruit shipped could be restricted. Pear Zone No. 1 was established to control processed pear shipment in all of California. Management of Pear Zone No. 1 was handled by CTFA from 1967 until 1992, when it was replaced by the California Pear Marketing Program. From the late 1980s until 1996, the promotion of processed fruit was handled by the Canned Fruit Promotion Service, based in San Francisco, in conjunction with Oregon and Washington pear producers and the California Cling Peach Board. Prior to that time, pear half promotion was handled by the Pacific Coast Canned Pear Service, and cocktail and mixed fruits were handled by the Cling Peach Advisory Board.

The California pear industry, under the leadership of Alex R. "Tom" Thomas, a Ukiah pear grower, and Jonathan Field, manager of CTFA from 1987 to 2001, consolidated its state processing pear program and federal fresh pear program and voted in 1992 to form the California Pear Marketing Program, a state order with the authority to regulate, promote, and conduct research for both fresh and processed pears. Although replaced by the state order, Federal Order No. 917 still exists, but it remains dormant by vote of the industry. The state order is managed by CPAB, which collects assessments on both fresh and processed tonnage. Jean-Marie Peltier was CPAB's first executive director, and the organization is currently managed by Chris Zanobini.

California Fall Pear Shippers is a voluntary organization created in 1998 to replace some of the services previously supplied by the WPCC and the Oregon-Washington-California Pear Bureau, mainly to compile statistics and promote California winter pears. CPAB manages the business of the California Fall Pear Shippers.

Numerous voluntary organizations have existed since the early 1900s in parallel with federal and state marketing orders. These associations were initially formed on a district basis with very little coordination. The first statewide organization was the California

Pear Growers Association, which existed from 1917 to 1938. Its purpose was to establish good prices and efficient marketing for its members. The California Canning Pear Association was established in 1953 under the Capper Volstead Act of 1922 as a bargaining cooperative to negotiate with canneries for the base price each year. The association changed its name to California Pear Growers (CPG) in 1987, and under the leadership of President Jean-Marie Peltier, it assumed a wider role in the pear industry, including legislative advocacy (especially regarding pest management), new product development, marketing statistics, and field services. Notable was the establishment of the Pear Pest Management Research Fund (PPMRF), which collects monies from growers and canneries to fund research on pest resistance and alternatives to pesticides facing regulatory restrictions.

Today, CPG now focuses largely on cannery negotiations, and in conjunction with CPAB continues efforts in new product development and pest management research through the PPMRF. CPAB has increased its role in advocating industry positions concerning pesticide regulations.

ROLE OF THE UNIVERSITY OF CALIFORNIA

As the land grant university in California, the University of California (UC) has worked closely with the pear industry and other entities to improve production and postharvest technology.

The land grant system began in 1862 with the Morrill Act, which established new agricultural and mechanical arts colleges by granting them tracts of land. This Act was followed by the Hatch Act in the 1870s, which established agricultural experiment stations. The Smith-Lever Act was passed in 1914 to provide agricultural extension programs, the public-service component of university agricultural research, ensuring that research reached the people, firms, and institutions who could use it.

The land grant system in the University of California today provides for agricultural teaching, research, and extension at its Berkeley, Davis, and Riverside campuses. It also provides for field research and extension through nine field stations and the county-based farm advisor offices. The campus-based faculty, who hold teaching and experiment station research appointments, undertake research projects directed at agricultural problems and issues. Specialists, who are located at Berkeley, Davis, Riverside, and the nine field stations, provide a link with farm advisors and campus-based research programs as well as leadership within their program areas. Farm advisors, located in county offices, provide field-related research, often in cooperation with campus-based faculty and specialists, and direct outreach to producers.

Although there have been many contributions to the UC pear-related effort, some can be considered milestones. Dr. A. E. Davey of UC Davis verified the ability of the hormone 1-napthaleneacetic acid (NAA) to delay fruit abscission. This greatly reduced pear losses due to premature drop and lowered labor costs because fruit was picked fewer times. Using NAA increased pear yields dramatically and led to more efficient harvest and storage management. Dr. William Griggs of UC Davis determined that most California Bartlett pears set via vegetative parthenocarpy, without the need of pollinizers or bees. Dr. Peter Ark of UC Berkeley first tested streptomycin to control fire blight. Dr. Larry Claypool of UC Davis established harvest maturity criteria based on firmness and soluble solids. Dr. Steven Lindow of UC Berkeley pioneered and facilitated the first successful commercial application of an antagonistic bacteria to control fire blight as well to as reduce russeting and frost damage. Extension Postharvest Specialist Gordon Mitchell of UC Davis developed tight-fill packing and forced-air cooling, technology currently used widely in the industry.

Two major efforts involved interdisciplinary teams of UC faculty, extensionists, and industry personnel. The first was the UC Research Committee on Pear Decline, a 19-member statewide group launched in 1960 and chaired by Dr. T. A. Shalla of UC Davis. Its goals were to determine the cause of pear decline, factors affecting its development, and methods of control. Funding was provided annually through the early 1970s by the state legislature. The Berkeley, Davis, and Riverside campuses were all represented on the committee, which included economists, entomologists, nematologists, plant pathologists, and pomologists. Extension specialists, farm advisors, and research associates conducted detailed orchard surveys and applied research projects once the cause was identified. Major lasting contributions include pear psylla control, injection of antibiotics into trees to reduce infection, introduction of *Pyrus betulaefolia* and Winter Nelis as resistant rootstocks, and the high-density orchard systems that ultimately replaced the traditional 24- by 24-foot (7.3- by 7.3-m) plantings and increased yield efficiency. After state funding of pear decline research and extension ended, a small amount of research was funded through CTFA, but it would be many years before research once again attained a high level of priority in the industry.

The second notable cooperative project was the Pear Pest Management Project, funded by UC and the U.S. Department of Agriculture (USDA) to demonstrate integrated pest management. The California Pear Pest

Management Steering Committee was headed by UC Berkeley Extension Entomologist Clarence Davis. It included UC faculty and UC Cooperative Extension economists, entomologists, plant pathologists, and farm advisors; California Department of Food and Agriculture and USDA scientists; independent pest management consultants; chemical company representatives; and growers. The project culminated in the 1978 UC Agricultural Experiment Station publication *Pear Pest Management,* the first book of its kind.

Industry-funded pear research continued at a fairly low level through the mid-1980s, with focus on codling moth, fire blight, and postharvest handling and technology. As pest control problems, as well as regulatory pressure to reduce broad-spectrum pesticide use, increased, the level of research funding also gradually increased. Currently, about $200,000 to $250,000 per year is expended by CPAB and PPMRF combined.

FUTURE CHALLENGES

Pear acreage, and to a much lesser extent, tonnage, continues to decline throughout California, though declines are less in Mendocino and Sacramento Counties (tables 2.4 and 2.5). The bankruptcy during the 2000 growing season of the largest purchaser of canning fruit in the United States, the cooperative Tri Valley Growers, Inc., of Modesto, dealt a major blow to the pear industry in California. For the first time, both CPG and CPAB sponsored formal tree pull programs in which growers were offered a monetary incentive to remove trees in order to reduce the risk of pest buildup, catalyzing the removal of nearly 2,000 acres (810 ha) from 2000 to 2002, mainly in Lake County. While the rate of decline has diminished slightly since 2002, this trend is expected to continue.

The North Coast (Lake and Mendocino Counties combined) and Sacramento Delta districts will, however, continue to be viable pear-growing regions as long as they can produce large yields of high-quality fruit, control the major pests (codling moth, pear psylla, and fire blight) economically, and succeed in marketing pears profitably. Pear trees can produce excellent crops even when quite old. Many orchards currently in production in the three major pear-producing areas are growing on rootstocks that resist soil problems and also resist pear decline and can sustain the current production level for several decades. Urban encroachment is also less severe in these districts. It is expected that economics will determine the long-term fate of pear production in these districts, and therefore, in California. Urban encroachment, increasing land values, high cultural costs to grow pears, low profit margins, and the resurgence of wine grapes as a more profitable crop all contribute to placing greater constraints on an industry that is already challenged by changes in customer preferences, a shrinking domestic market, and increasing competition in the international markets. Continuing investment in research, new technologies, and innovative marketing will be necessary to sustain a viable pear industry in the state.

REFERENCES

Batier, L. P., H. A. Schomer, E. J. Newcomer, and D. L. Coyier. 1967. Commercial pear growing. Washington, DC: USDA/ARS Agricultural Handbook No. 330.

Bethell, R. S., ed. 1978. Pear pest management. Oakland: University of California Agricultural Experiment Station Publication 4086.

Table 2.4. Commercial pear-bearing acreage in California for all varieties, in selected counties and years

	Bearing acreage						Nonbearing acreage					
County	**1917**	**1945**	**1974**	**1992**	**2002**	**2005**	**1917**	**1945**	**1974**	**1992**	**2002**	**2005**
Contra Costa	850	2,773	352	261	53	25	1,250	21	58	1	0	10
El Dorado	500	4,233	2,238	966	366	83	1,000	3	238	3	0	0
Lake	600	3,973	7,585	5,061	3,133	2557	2,000	202	1,125	139	16	44
Mendocino	450	3,610	4,085	3,099	2,424	2160	1,100	89	411	32	6	13
Placer/Nevada	2,700	5,080*	1,194	228	38	49	3,010	269*	3	1	0	0
Sacramento	3,300	3,883	6,235	6,318	6,647	6,000	1,700	92	255	112	14	190
Santa Clara	1,530	6,690	1,895	140	45	0	223	128	212	0	0	0
Solano†	1,000	2,605	3,223	1,855	1,005	813	230	106	453	0.	164	113
Sonoma/Napa	1,650	3,154	1,805	188	84	87	1,055	0	114	2	1	1
Sutter/Yuba	604	1,014	2,946	2,567	1,719	1,283	1,183	NA	505	119	13	157

Sources: California Crop and Livestock Reporting Service 1975, 1993; Davis and Tufts 1941, p. 227; Kirman and Magness 1949, p. 5; California County Agricultural Commissioners annual crop reports, various years.

Notes: *Placer County only.

†Includes Ryer Island.

Table 2.5. Commercial pear tonnage in California by county for Bartlett variety, selected years*

County	1945	1974	1992	2000	2002	2005
Contra Costa	12,061	1,645	1,728	574	547	NA
El Dorado	19,826	11,349	2,954	804	420	91
Lake	25,953	58,911	75,911	62,749	55,674	28,732
Los Angeles†	3,392	4,430	151	212	N/A	N/A
Mendocino	20,516	39,719	53,336	47,353	34,493	26,069
Placer/Nevada	28,561	3,189	0	76	0	15
Sacramento/ San Joaquin	44,817	104,884	146,058	137,227	128,416	98,240
Santa Clara	50,380	15,391	460	9	5	0
Sonoma/Napa	9,713	5,764	977	19	47	<1
Solano‡	16,994	19,015	14,083	3,887	1,491	996
Sutter/Yuba§	9,343	30,210	18,097	12,178	10,230	10,781

Sources: California Tree Fruit Agreement 1946, 1976, 1993; California Pear Advisory Board 2003, 2006.

Notes: * Excludes unharvested fruit.

† Includes Tehachapi and Littlerock.

‡ Excludes Ryer Island.

§ Includes Butte, Glenn, and Shasta Counties.

California Agricultural Statistics Service. 2003. County agriculture commissioners' data, calendar year 2002. Sacramento: CASS.

California County Agricultural Commissioners annual crop reports. Various years. Butte, Contra Costa, El Dorado, Glenn, Lake, Mendocino, Napa, Nevada, Placer, Sacramento, Santa Clara, Shasta, Solano, Sonoma, Sutter, and Yuba counties. See the California Department of Agriculture County Liaison Office Web site, http://www.cdfa.ca.gov/exec/cl/countyagmap.htm.

California Crop and Livestock Reporting Service. 1975. 1974 California fruit and nut acreage. Sacramento: California Crop and Livestock Reporting Service.

———. 1993. 1992 California fruit and nut acreage. Sacramento: California Crop and Livestock Reporting Service.

California Pear Advisory Board (CPAB). Various years. Annual reports. CPAB Web site, http://www.calpear.com/brd/_ann_00.cfm.

California Tree Fruit Agreement. Various years. Annual reports. Sacramento: CTFA.

Davis, L. D., and W. P. Tufts. 1941. Pear growing in California. Berkeley: California Agricultural Extension Service Circular 122.

Food and Agriculture Organization of the United Nations (FAO). 2005. Major food and agricultural commodities and producers. Countries by commodity: pears. FAO Economic and Social Department Web site, http://www.fao.org/es/ess/top/commodity.html;jsessionid=98F57DEB00BC27813AC921ED02C3C3AB?item=521&lang=en&year=2005.

Geller, G. 1983. California tree fruit agreement 50th anniversary report. Sacramento: California Tree Fruit Agreement.

Kinman, C. F., and J. R. Magness. 1949. Pear growing in the Pacific Coast states. Washington, DC: USDA Farmers Bulletin No. 1739.

Marcus, G. D. 1994. Farm bargaining cooperatives: Group action, greater gain. Washington, DC: USDA Agricultural Cooperative Service Research Report 130.

Ohlendorf, B. L. 1999. Integrated pest management for apples and pears. 2nd ed. Oakland: University of California Division of Agriculture and Natural Resources Publication 3340.

Pradhan, V., and K. Moulton. 1996. Economic briefs: Time series data relating to California crop production and prices in the United States and California. Berkeley: University of California, Berkeley, Department of Agricultural and Resource Economics.

Shultis, A. 1933. Bartlett pear orchard management. Berkeley: University of California Agricultural Extension Service Circular 78.

Simoons, F. J. 1953. The development of transportation routes in the Clear Lake area. California History 32(4): 363–371.

Tukey, H. B. 1928. The pear and its culture: Designed for both the amateur and the commercial grower. New York: Orange Judd Publishing.

U.S. Department of Agriculture, Foreign Agricultural Service (USDA/FAS). 1998. Canned deciduous fruit situation for selected countries. FAS Online Web site, http://www.fas.usda.gov/htp/Hort_Circular/2001/01-11/Statistics/cndfrt.pdf.

———. 2003. Pears: Production, supply and distribution in selected countries. FAS Online Web site http://www.fas.usda.gov/htp/Hort_Circular/2006/03-06/World%20Pear%20Sitation%202005-06.pdf.

———. 2006. World pear situation. FAS Online Web site, http://www.fas.usda.gov/htp2/circular/1998/98-03/pearfea.html.

U.S. Department of Agriculture, National Agricultural Statistics Service (USDA/NASS). 2006. Noncitrus fruits and nuts 2005 summary. USDA/NASS Crops and Plants Web site, http://usda.mannlib.cornell.edu/usda/nass/NoncFruiNu//2000s/2006/NoncFruiNu-01-24-2006.pdf.

van der Zwet, T., and N. F. Childers, eds. 1982. The pear: Cultivars to marketing. Gainesville, FL: Horticultural Publications.

Weldon, G. P. 1918. Pear growing in California. Monthly Bulletin of the California State Commission of Horticulture 8(5).

Williams, M. W., H. M. Covey, and D. L. Coyier. 1978. Pear production. Washington, DC: USDA Agricultural Handbook No. 526.

3

Marketing

CHRIS ZANOBINI

California pears have been collectively promoted by state or federal marketing boards since the early 1900s. The California Pear Advisory Board (CPAB) was established in 1992, and promoting the entire California Bartlett pear crop has been the focus of this organization. In 1999, $1.7 million, or approximately 70 percent, of the CPAB budget was spent on generic promotion of California pears.

The California pear industry saw a dramatic decline in acreage during the 1990s (see chapter 2, "History and Overview of the California Pear Industry"). Since 1988, pear acreage decreased from approximately 24,000 acres (9,700 ha) to approximately 13,500 acres (5,400 ha) in 2002. Despite this 43 percent decrease, overall production has not dropped at the same rate, averaging 260,000 tons (235,800 t) over the last 5 years. Within that time period, production varied from a high of 363,000 tons (329,200 t) in 1994 to a low of 216,000 tons (195,900 t) in 2003. Although production has declined, the consumption of California pears has remained stable. Ten-year averages indicate that 58 percent of total pear production is used for canning, 28 percent packed for the fresh market, and an average of 14 percent is used for baby food, juice, and dried pears. Utilization data for 2001–02 are shown in figure 3.1.

The majority of California's production is the Bartlett variety, accounting for 93 percent of the total crop. The remaining 7 percent is made up of winter varieties, which include Bosc, Seckel, Comice, and Red pears, and is used primarily in the fresh market.

California is the nation's leading producer of Bartlett pears for both the fresh and processed markets. Washington is the second-largest producer, followed by Oregon (fig. 3.2). Due to the substantial production of winter pears in the Pacific Northwest, the Northwest's fresh pears are promoted year-round by Pear Bureau Northwest.

California produces 57 percent of the canned pears and fruit cocktail in the United States, followed by

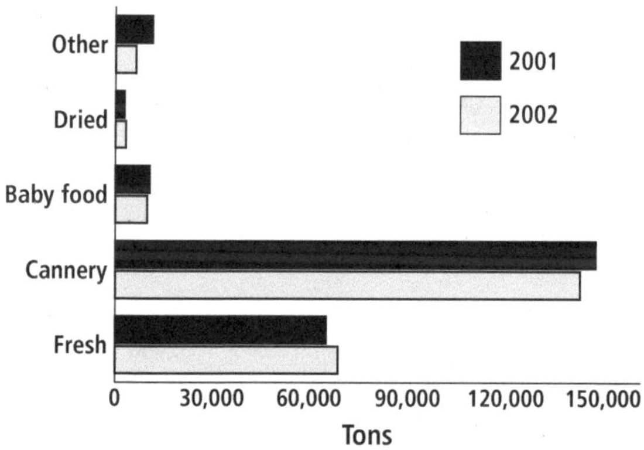

Figure 3.1. Utilization of 2001–02 California pear crop.

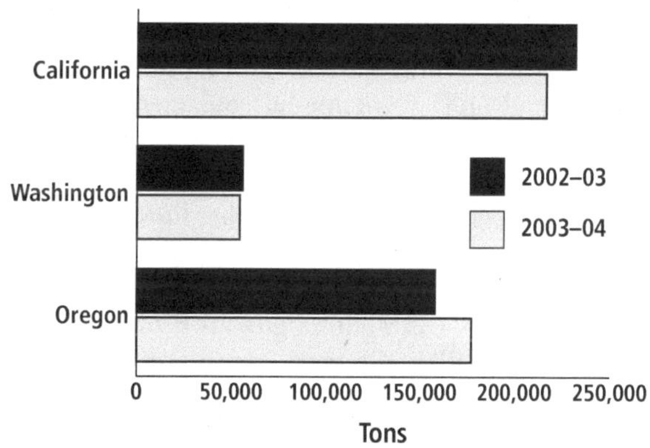

Figure 3.2. U.S. Bartlett production by state, 2001–02.

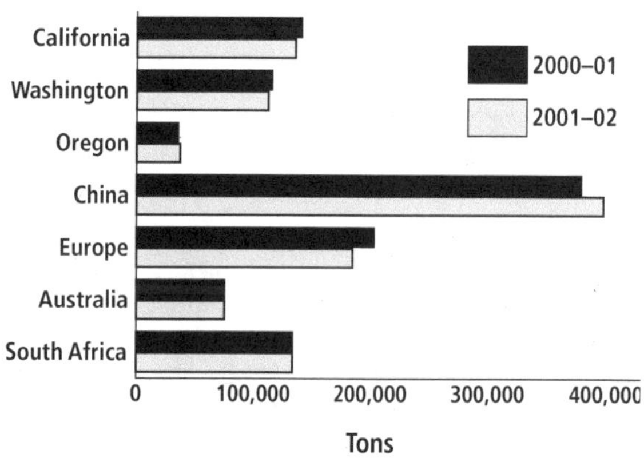

Figure 3.3. World canned pear production, 2001–02.

Washington and Oregon, each producing 31 and 12 percent, respectively. Combined, the United States accounts for 57 percent of the canned pears (fig. 3.3) and fruit cocktail (fig. 3.4; data given for most recent year available) produced around the world. Its major competitors are Europe (chiefly Italy), Australia, South Africa, and more recently, China. The canned fruit industry is suffering a worldwide crisis due to an overall decline in canned fruit consumption, low prices, and an increased number of competitors among principal exporters.

Per capita consumption of fresh pears in the United States has been trending upward slightly over the last 10 years. In 1999, fresh pear consumption averaged 3.4 pounds (1.5 kg) per capita. Canned pear consumption has trended downward along with other canned fruits, with fresh and canned consumption now approximately equal. Per capita consumption of canned pears in the United States averaged 3.5 pounds (1.6 kg) per capita in 1999, down from 3.9 pounds (1.8 kg) in 1988. Per capita consumption of processed pears, including dried and juiced, totaled 4.7 pounds (2.1 kg) in 1999, up from 3.5 pounds (1.6 kg) in 1988.

Currently, California has 22 fresh market packers and 14 fresh market sales agents marketing over 40 different labels and selling an average of 4 million boxes (36-pound tight-fill or volume-fill containers) of pears per year. Fresh product is marketed by these sales agents as well as by packers that also maintain a marketing staff. Some packers perform only the packing function, while others are vertically integrated into marketing.

On the processing side, there are three major California canners: Del Monte, Signature Fruit Company (formerly Tri Valley Growers), and Pacific Coast Producers, which is a cooperative canner. California fruit for processing is also sold to several Northwest- or Washington-based canners. The balance of the crop is purchased by Gerber, Sabroso, Naumes, and other juice and dried processors. California Pear Growers, a voluntary bargaining association, represents California pear growers in negotiating contract prices for canning pears with processors.

Fresh California pears are sorted by size and primarily packed in a 36-pound (16.3-kg) tight fill container or a 40- or 44-pound (18.2- or 20-kg) standard or tissue wrap pack. Specialty containers include tray pack, canning pack, and Los Angeles lug. The fresh market season for California pears starts in early- to mid-July and extends through November, depending on the crop year. Most fresh pears are packed and stored in conventional cold storage. California has one facility that uses controlled atmosphere storage, which extends the marketability of the pears into December.

When canned, the majority of the California pears are packed into fruit cocktail, with a limited amount of grade pack pears canned as halves or diced product. California canners typically pack in three different can sizes. Retail packs are the 15-ounce (425-ml) can, or commonly known as a 2½, and a 29-ounce (857-ml) family-size can. Foodservice product is mainly packed in a 105-ounce (3.1-l), or a number 10, can. Recent produce innovations include pears and mixed fruit in glass jars, clear plastic jars, and individual fruit cups. Dried pears, purees, and juices are packaged and sorted into a variety of types and grades.

MARKETING FRESH PEARS

Domestic

Fresh California Pears are promoted generically across the United States by the California Pear Advisory Board (CPAB). Since 1994, the board has developed a nationwide program that uses a variety of marketing, merchandising, and educational media to aggressively promote the purchase and consumption of California pears. The board works with the top supermarket retail-volume leaders in 18 leading markets and uses a strategy that helps shippers move 80 percent of the fruit.

The first California Bartletts are harvested from the Sacramento Delta district sometime in July. By late August, fresh Bartletts from Lake and Mendocino Counties are competing with Pacific Northwest Bartletts in markets nationwide. The season for California's fresh Bartletts usually ends in October, while Northwest Bartletts are marketed longer because more of the fruit is held in controlled atmosphere storage.

Through a dedicated, contracted regional merchandising staff, the pear industry in California remains in constant contact with produce industry decision

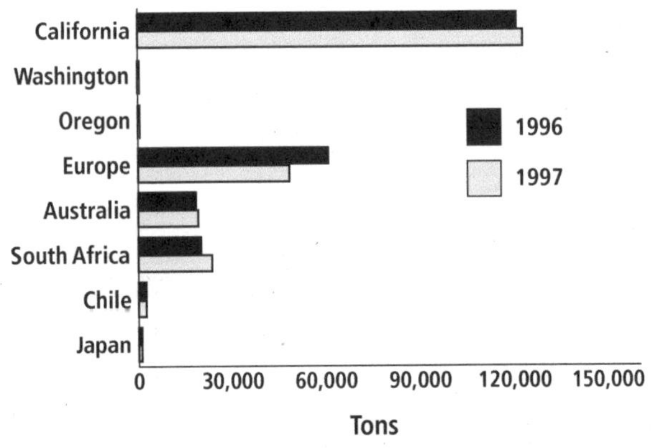

Figure 3.4. World canned fruit cocktail production, 1996–97. *Source:* USDA/FAS 2006.

makers to maintain purchasing and promotion loyalty. Most of the large chains buy directly from shippers. The board works closely with the retail trade to develop ripening and handling protocols to sustain product quality, improve consumer perception, and increase fresh pear sales. A national ripening advisor facilitates customized education seminars and programs on ripening and handling. Consumer outreach is conducted through a variety of public relations activities and supported through retail promotion and media efforts. The board continues to develop new programs that expand the market base for retail purchases.

Periodically the board conducts surveys to assist in the development and direction of the marketing programs. Through this research, the industry is able to concentrate on the key consumers in the best markets for pears.

Export

Through the assistance of the federal government's Market Access Program (MAP) and strong industry support, California exports an average 20 percent of its fresh pear crop. The MAP assists the pear industry to gain access into export markets and provides financial support and in-country resources to execute promotion programs. The primary export markets are Canada and Mexico, and shipments to Central America have been increasing (fig. 3.5).

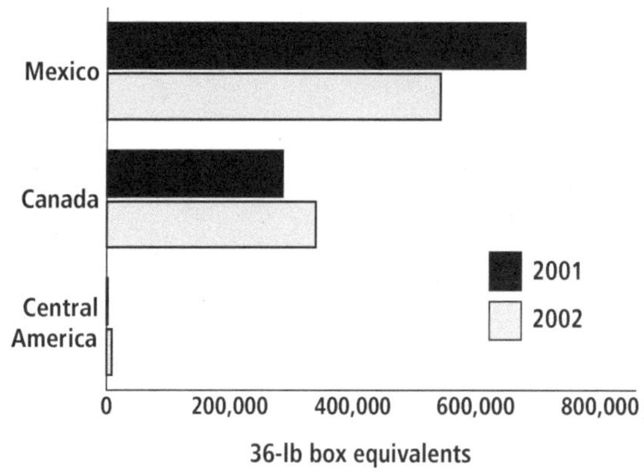

Figure 3.5. California fresh pear exports, 2001–2002. *Source:* CPAB 2003.

MARKETING PROCESSED PEARS

Prior to 1995, the domestic and international promotion of canned fruit was handled by the Canned Fruit Promotion Service and the Pacific Coast Canned Pear Service, a joint effort by California pear and peach producers and Northwest pear producers. The pear industry in California decided to pursue another direction in 1995 and withdrew funding from the organization. The organization was disbanded after the 1995 season.

Processed pears in California are marketed primarily through the three major California canners. Through the Pear Advisory Board and under the discretion of the California canners, the industry has supplemented marketing efforts through consumer-directed food media that highlights the convenience and numerous usage ideas for canned pears.

Export promotions of canned pears and fruit cocktail are handled by a joint pear and peach program managed by the California Cling Peach industry. With the financial assistance of promotional funds from MAP, the California canned fruit industry markets canned fruit primarily in Canada, Japan, and Mexico.

Since 1997, the pear industry has worked on developing the ingredient market for California processed pear products. Efforts directed to the food manufacturing industry include the research and development of a variety of processed pear products that are affordable, versatile, and readily available for cereal, baking, or snack food applications.

REFERENCES

California Pear Advisory Board (CPAB). 2003. Annual report 2001–2002. CPAB Web site, http://www.calpear.com/brd/2001-02-ann_rpt.pdf.

U.S. Department of Agriculture, Foreign Agricultural Service (USDA/FAS). 2006. World pear situation. USDA/FAS Web site, http://www.fas.usda.gov/htp/Hort_Circular/2006/03-06/World%20Pear%20Sitation%202005-06.pdf.

4 Economic Considerations

KAREN M. KLONSKY, RACHEL B. ELKINS, AND STEVEN C. BLANK

To manage an orchard well you must make a great number of economic decisions over the long productive life of the orchard. You must first determine the orchard design, including the orchard location, water source, type of irrigation system, and the type of floor management, taking into consideration the topography, slope, and soil type at the site. Subsequent decisions require regular assessments of production practices such as pest and disease management, pruning, training, fertilization, and irrigation scheduling.

A *feasibility study* is a business analysis tool designed to determine whether a specific alternative is economically sound. Feasibility studies can take a variety of forms, depending on the question being addressed. The study may be partial or complete in focus, formal or informal in presentation. At one extreme, a complete feasibility study for the entire business would be very similar to a business plan, which reports or documents the overall performance of the business. In general, though, a feasibility study is narrower in scope and is a decision-making tool, as opposed to a reporting method. You should conduct a feasibility study any time you consider a change in your present farming operation. Your ultimate purpose in conducting a feasibility study is to avoid the costs associated with making a wrong decision.

We will use a feasibility study framework to present a brief discussion of the economic issues that pear growers deal with when establishing or replanting an orchard and also during production years. Use all or part of this discussion, as appropriate, when studying your orchard situation.

ASSESSING ECONOMIC FEASIBILITY

A feasibility study looks at the impact on the farm business of a particular change in the pear operation. You can divide a feasibility study for pears into four parts: market situation, resource constraints, production process, and economic analysis of alternatives. In the following sections each of these topics are introduced as they relate to the establishment or replanting of an orchard and to the management of an established orchard.

Market Situation

To make an informed decision related to pear production you need to know about historical and current sales and prices of pears so you will have a basis to project future sales patterns, competition, and market volatility. Information about marketing and distribution systems, numbers of producers and handlers, and yield and acreage trends gives you an understanding of the profit potential in the short run and the long run. In addition, competition from other major producers and trade agreements in effect have significant impacts on world supply and, consequently, on domestic planting and production decisions. Pears and apples are unique in that they can be stored for months, and certain varieties can be stored year-round. Pears imported from South America, Oregon, and Washington can still be on the shelf when California pears come into production. The absence of a marketing window decries the possibility of a high price for early fruit that is realized for other fresh fruits.

Like other fruits, pears can be sold into the fresh market or processed. Decreased demand from local canneries results in more pears in the fresh market, further depressing prices. Before marketing your crop in the fresh market, research potential handling and sales options. The first decision is whether to market directly (e.g., via direct-to-retail, farmers markets, roadside stands, or e-commerce), through a processor into the wholesale market, or both. Key criteria to be considered are the volume of fruit to be sold, the ability or desire to do your own packing and marketing, and access to a customer base.

If you choose to market through a processor, research available entities. Talk with other growers about returns, payment schedules, and services provided. Learn about their marketing strategies and customer bases. Do they have sufficient processing outlets to handle excess or lower-quality fruit? Is management accessible and willing to provide market updates and advice on delivering the best-quality fruit? It is up to the producer to ensure that his or her product is being handled and sold in the most desirable manner.

The primary processing outlet, and the one that generally pays the highest prices, is the cannery; other outlets include baby food, dry yard, juice, and concentrate. Selling to the fresh market as opposed to selling to a processor has additional costs for harvest, packing house charges, and pest control for russet control. For the fresh packout (the percentage of fruit that can be sold to the fresh market), growers often pick twice. The largest, best fruit is picked first, then the rest about 10 days after the start of the first pick. Obviously, picking twice increases labor and supervision. Also, the pickers go more slowly when picking for the fresh market because they do field sorting for defects, small fruit, and trash, further increasing labor and supervision costs. For selling directly to a processor, harvest is usually a once-over operation, starting somewhat later in the picking window to obtain maximum fruit size. Because of these and other considerations, the marketing plan impacts numerous other decisions concerning orchard management. See chapter 3, "Marketing," for further information about the market situation for pears.

The type of market information you need for a feasibility study varies depending on the decision you are considering. For example, in order to decide whether to establish a pear orchard, you need an in-depth understanding of the world and domestic pear markets, as well as a knowledge of competing fruits. This is a different set of information than you would need for selecting the site of a pear orchard.

Resource Constraints

A critical part of any management decision is assessing the resources needed for each alternative under consideration and the resources available to the business. Categories of resources to include in a pear feasibility study are land, water, labor, capital, and management.

The soil considerations for pear production are similar to those for most tree crops in that pears do best on deep, medium-textured, well-drained, fertile soil. Although these conditions are not ideal, pears will do reasonably well in heavy soils with high water tables that would not be tolerated by most other deciduous fruit trees. For these reasons, pears are found along the Sacramento River where the water table can be 2 feet (0.6 m) below the surface. Nonetheless, investment in prime ground may pay for itself with a higher net income in the future than would have been realized on poorer ground purchased at a lower price. A real estate purchase should not be based simply on the land's being priced at or below the going rate. If you cannot expect the orchard to generate adequate returns on your investment, what you are doing is speculating in real estate and not farming.

You may be able to lease an established orchard. Orchard leases can be developed in a wide variety of ways. A typical orchard lease covers several years. The cost of an orchard lease is typically based on a percentage of the gross income from the crop, but it can also be based on an annual cash rent. A development lease may suspend payment until the orchard comes into bearing, using some combination of cash and share rent or a graduated scale in the development years.

Leasing alternatives should not be compared solely on the basis of payments to the landlord. The landlord's contribution to operating and overhead costs can be a significant factor in your bottom line. The landlord may contribute all, part, or none of the operating inputs such as fertilizer, pesticides, and water. Furthermore, a lease can include the use of some equipment, including an irrigation system. The payment of property taxes, repairs to the irrigation system, and road maintenance are also negotiable points in an orchard lease.

Particularly during times when market prices are low or difficult to predict, you might want to expand production by managing an orchard for an absentee owner. These arrangements usually include a salary for services and may also include the use of equipment. The risk in the pear market is held by the owner and not by the manager.

Access to water is another important consideration that impacts the cost of land or an established orchard. Water can be either surface water or groundwater. Pumping water from a surface water source may mean lower-quality water than groundwater but it also costs less. These tradeoffs should be accounted for in making an orchard purchase, establishment, or lease decision.

When you make production decisions that affect the total required hours of labor, take the availability of labor into account and not simply the cost of labor. Also, the timing and skill level of the labor needed is critical. Some practices require different labor skills than others, so it is not enough to consider the costs of only the equipment and the labor. You must also consider the availability of trained labor or the cost of training your own employees. Custom hiring for specialized operations during the growing season or harvest reduces the farm's equipment and labor needs and also substitutes annual custom operator charges for long-term capital outlay for equipment and equipment storage facilities as well as for annual repairs and property taxes.

Because they are used over several years, equipment, buildings, and irrigation system costs are difficult to quantify on an annual basis. Calculation of the costs of equipment and facilities must take into account useful life, opportunity cost of ownership, taxes, insurance, housing for equipment, and repairs. If you use the equipment for more than one enterprise, you must divide the cost of ownership and allocate the costs to all enterprises that use the equipment.

The decision to expand, be it by purchasing an existing orchard, establishing a new orchard, or leasing an orchard, is constrained by the availability of all of the other resources available for the pear operation. Is adequate equipment, management, labor, and water available for the new orchard? What capital is needed for the purchase?

The capital for purchasing an orchard or land takes the form of credit and cash. By using cash, you reduce the amount of interest paid on borrowed capital. However, major cash outlays limit the availability of cash for supplies, wages, and equipment purchases. Leasing an existing orchard does not require a large capital outlay up front, but in a lease any appreciation in the value of the orchard is realized by the landlord and not by the lessee.

A feasibility study provides the basic information required for loan applications and will help you determine whether the orchard has the capacity to generate enough revenue to cover payments on the interest and principal associated with the loan after annual operating expenses have been covered. Whether you use cash, credit, or a combination of the two for purchasing land or establishing an orchard, calculate the expected return to capital based on realistic expectations of returns.

Careful analysis and projections of income and expenses are particularly important for pear producers for at least two reasons. First, pear orchards require a large capital outlay up front, whether for an established orchard or for planting a new orchard. Second, newly planted pear trees do not produce a marketable level of production until about the fifth year, even with quality management. Therefore, you must have enough capital available from outside sources to manage the orchard until it starts to generate an income.

Production Practices

Production practices for pears include pruning, thinning, fertilization, pest and disease management, irrigation, orchard floor management, frost protection, and harvest. An analysis of production practices includes identifying the resources required for each alternative under consideration and assigning costs to each resource. For each alternative under consideration, the feasibility study should include an accounting of resources required and impacts on crop yield, fruit size, and grade.

The period of time to be included in the analysis is also important. Using irrigation scheduling as an example, the cost of the water applied is not sufficient information to determine the amount of water to use. In a drought year, when growers are trying to conserve water, they may be tempted to cut the amount of water supplied to the trees. The long-term effect of water stress on the trees and the short-term effect on fruit size must be included in the decision-making process. Adequate irrigation in June increases fruit size and tonnage; large fruit are desirable for both fresh market and canning. It may be possible to cut off water in the fall to save on water use or skip an irrigation in the middle of the season. These approaches to cutting costs are not without risk but may be unavoidable in drought years.

When it is difficult or impossible to predict the effects on yield, fruit size, and grade corresponding to alternative management practices, break-even analysis is extremely useful. If you can make a reasonable estimate of all other factors, you can use break-even analysis to calculate the least acceptable difference in yield, fruit size, or grade that will justify the change in practices. For more information, see the section "Break-Even Analysis," below.

Economic Analysis

Several basic business analysis tools are invaluable components of feasibility studies for the orchard manager. These tools include whole farm budgets, enterprise budgets, partial budgeting, cash flow projections, and capital budgeting. The type of budget you choose for analysis depends on the decision you need to make. A whole farm budget is appropriate when the manager wants to know how a proposed change will affect the entire business, which may include income tax management. In particular, the timing and magnitude of resource requirements may impact the functioning of other enterprises on the same farm. An enterprise budget is appropriate when the manager is concerned with the potential profitability of the change and separate enterprises are not competing for resources. A partial budget is extremely useful when the impacts of the change can be clearly identified and unchanged costs can be ignored. Capital budgeting is useful for analyzing changes that have variable impacts over a long period of time. This may be important for orchard managers because pears tend to bear heavily in alternate years ("alternate bearing").

Examples of enterprise budgets for establishing and producing pears are shown in tables 4.1 and 4.2. Note that many of the costs used in these budgets are custom rates and could be different if growers used their own labor and equipment. These tables make no attempt to describe the mechanics of cash flow or capital budgeting;

Table 4.1 Sample costs per acre per year to establish a pear orchard, North Coast Region, California, 2003

	Cost per acre					
Year	**1st**	**2nd**	**3rd**	**4th**	**5th**	**6th**
Tons per acre	**0**	**0**	**0**	**0**	**4**	**7**
Planting						
Remove old orchard	375					
Land preparation: Rip 3×	250					
Land preparation: Disc 2×	17					
Land preparation: fumigate	1,800					
Land preparation: apply herbicide & disc	23					
Lay out orchard	210					
Auger tree holes	142					
Plant trees	96	3				
Trees: 182 per acre @ $5.80 ea., (2% in 2nd year)	1,056	23				
Tree guard & NPK fertilizer	227	4				
Head back trees	32	2				
TOTAL PLANTING COSTS	4,228	32	0	0	0	0
Cultural						
Train trees	32					
Pruning & training		95	284	568	757	757
Fertilizer: nitrogen	11	14	22	29	29	29
Weed control: disc 4×	33					
Weed control: mow middles 7×		61	61	61	61	61
Weed control: strip sprays	28	40	40	40	73	73
Pest control: dormant spray		23	74	74	74	74
Pest control: delay dormant spray					27	27
Pest control: gophers	8	8	8	8	8	8
Pest control: budbreak spray					30	30
Pest control: scab sprays					37	37
Pest control: blight			96	215	215	215
Pest control: blight and pear scab sprays					123	139
Pest control: cover sprays, codling moth					111	111
Pest control: psylla and mite sprays					235	235
Pest control: codling moth traps					2	2
Fruit sizing: hormone spray					31	31
Irrigate	153	153	176	176	149	149
Frost protection 18×					84	84
PCA fees			35	35	35	35
Leaf analysis				16	15	15
Pickup truck use	84	84	84	84	84	84
ATV use	71	71	71	71	71	71
TOTAL CULTURAL COSTS	420	549	951	1,377	2,251	2,267
Harvest						
Pick fruit					250	420
Haul to shed					32	55
Sort, pack, store, sell fruit					605	1,067
TOTAL HARVEST COSTS					887	1,542
Postharvest						
Irrigate					27	27
Pest control: postharvest oil					23	23
TOTAL POSTHARVEST COSTS					50	50
Assessments						
California Pear Advisory Board					38	68
California Pear Growers					5	5
TOTAL ASSESSMENT COSTS					43	73
Interest on operating capital @ 7.4%	303	14	24	29	53	61
TOTAL OPERATING COSTS/ACRE	4,951	595	975	1,406	3,284	3,993
Cash overhead						
Office expense	44	44	44	44	44	44
Sanitation fees	13	13	13	13	13	13
Liability insurance	7	7	7	7	7	7
Property taxes	108	105	107	107	109	111
Property insurance	15	13	14	14	16	17
Investment repairs	86	86	89	86	87	87
TOTAL CASH OVERHEAD COSTS	273	268	274	271	276	279
TOTAL CASH COSTS/ACRE	5,224	863	1,249	1,677	3,560	4,272
INCOME/ACRE FROM PRODUCTION					1,422	2,488
NET CASH COSTS/ACRE FOR THE YEAR	5,224	863	1,249	1,677	2,138	1,784
ACCUMULATED NET CASH COSTS/ACRE	5,244	6,087	7,336	9,013	11,151	12,935
Capital recovery (noncash overhead)						
Shop building	57	57	57	57	57	57
Worker housing	10	10	10	10	10	10
Fuel tank & pump	4	4	4	4	4	4
Shop tools	17	17	17	17	17	17
Sprinkler irrigation system	152	152	152	152	152	152
Deer fence: electric	32	32	32	32	32	32
Ladders: 16 each					4	4
Land @ $6,964/acre	536	536	536	536	536	536
Equipment	112	52	62	81	120	146
TOTAL CAPITAL RECOVERY COSTS	920	860	870	889	932	958
TOTAL COST/ACRE FOR THE YEAR	6,144	1,723	2,119	2,566	4,492	5,230
INCOME/ACRE FROM PRODUCTION					1,422	2,488
TOTAL NET COST/ACRE FOR THE YEAR	6,144	1,723	2,119	2,566	3,070	2,742
TOTAL ACCUMULATED NET COST/ACRE	6,144	7,867	9,986	12,552	15,622	18,364

for more information on these topics, refer to any standard textbook on farm management.

In general, the economic analysis portion of a feasibility study involves calculating the profitability of the proposed enterprise. However, cash flow and risks associated with the profits must be considered as well. Ignoring any of these three factors (profitability, cash flow, or risk) can lead to potentially harmful decisions. Also, if you are using borrowed funds, your lender will require that all three of these factors be evaluated.

Profitability

You can use estimated revenue and costs to project the profitability of an orchard. Estimated profit is the difference between expected revenue and expected costs. To project your expected revenue, simply multiply the expected price for the crop by the expected yield.

To determine the profitability of the operation, prepare either a whole farm budget or an enterprise budget for pears. Estimate the net (after tax) income for each year of the orchard's projected life. If you are deciding whether to expand the pear acreage by planting a new orchard or to renew an established orchard, the productive life of the new orchard could be over 50 years. The task of preparing a projected budget often goes neglected, yet it is critical to determining a proposed orchard's true economic viability.

A computer spreadsheet facilitates the work of showing how income levels vary widely from year to year during the life of an orchard. In particular, losses can be expected during the first several years because there are high development costs but the trees do not yet produce a marketable crop. As the trees mature, annual yields and sales receipts increase, on average, only to decline again at the end of the orchard's life. If you do not prepare a budget for each year, you will not know when profits will actually be earned or at what point the orchard's *economic* life will end (the point at which profits decline below the minimum level needed for economic viability).

One simple and effective way to analyze orchard performance over time is to compare the industry's average production figures to the individual orchard's production over the same time period. If the orchard under study is falling further and further below average it may be time for to change management practices or to replace the orchard.

The economic analysis of potential changes in cultural practices during production years can usually be accomplished using partial budgeting techniques. In partial budgeting, you compare the differences in revenue and costs associated with two or more alternative production practices and ignore the factors that are identical for the alternatives under consideration. This greatly simplifies the analysis and avoids the need to develop a complete enterprise budget.

During periods when market prices are low, lowering production costs may be your best strategy for increasing profit. Pear production requires an extensive pest management program (see part 4, "Pest Management in Pears"). You should stay abreast of the latest developments in pest control techniques that lead to a reduction in pesticide applications and costs. In tables 4.1 and 4.2 the pest management program is based on solely using insecticides to control codling moth. However, many growers use a combination of pheromone mating disruption and insecticides. Varying codling moth control strategies translate into cost changes.

Cash flows

For many growers, the completion of an enterprise budget has been the goal of a feasibility study, but this ignores the most important item in the study: the *cash flow summary*. A cash flow summary describes the timing of the inflows and outflows of cash from the farm business. The orchard may be generally profitable over a period of time, but the timing of the outflows may precede the inflows. This is particularly true during the years of orchard establishment. In such a situation, the enterprise requires some other source of capital. The provision for adequate working capital is perhaps the one item most critical for the successful operation of a business. The cash flow summary shows what the enterprise's cash needs will be and what sources of cash are available to meet those needs.

To be more specific, you will need to know how much capital will be required for month-to-month operation of the business and when this capital will be required. Monthly cash needs include expenditures on wages, inventories, utilities, and other inputs. Sources of cash include sales receipts, loans, and off-farm income.

The monthly cash flow summary is also critical to help determine the size of loans, the length of loans, probable pay-back periods, and the amount of interest and principal that can be paid back periodically. Developing a cash flow statement is especially important for pear producers because the business is usually not done on a cash basis. Even when credit periods are available to producers from input suppliers, inputs must be paid for long before income from the sale of harvested pears is received by growers. This means that the cash inflows lag behind cash outflows, resulting in the need for working capital (which may have to be borrowed).

The high-risk nature of pear production requires that cash flow planning be conducted for multiple years, as well as for one month to the next. You should prepare

Table 4.2 Sample monthly cash costs to produce pears, North Coast Region, California, 2003

	JAN	FEB	MAR	APR	MAY	JUN	JUL	AUG	SEP	OCT	NOV	DEC	TOTAL
Cultural													
Prune & train trees	819												819
Pest control: dormant	32												32
Pest control: delayed dormant		19											19
Weed control: strip spray 3×		54		11			10						75
Pest control: gophers 3×			8										8
Weed control: mow middles			9	9	9	17	17						61
Pest control: scab			73										73
Frost Protection: irrigate				42	42								84
Pest control: blight				106	84								190
Pest control: blight and scab				75	91								166
Pest control: blight and cover					80								80
Pest control: cover spray						88	61						149
Irrigate						48	48						96
Fertilize: nitrogen						27							27
Pest control: psylla & mites				101		101	101						303
Apply hormone								31					31
PCA fees	4	4	4	4	4	4	4	4	4				36
Leaf analysis							19						19
Pickup truck use	7	7	7	7	7	7	7	7	7	7	7	7	84
ATV use	6	6	6	6	6	6	6	6	6	6	6	6	72
TOTAL CULTURAL COSTS													2,424
Harvest													
Harvest fruit: 1st pick								353					353
Harvest fruit: 2nd pick								716					716
Haul to packinghouse								158					158
TOTAL HARVEST COSTS								1,227					1,227
Packing													
Sort, pack, sell fruit								3,395					3,395
TOTAL PACKING COSTS								3,395					3,395
Assessment													
California Pear Advisory Board								196					196
California Pear Growers Assn.								23					23
TOTAL ASSESSMENT COSTS								219					219
Postharvest													
Irrigate								53	27				80
Fertilize: nitrogen									27				27
Pest control: psylla									19				19
TOTAL POSTHARVEST COSTS								53	73				126
Interest on operating capital	5	6	6	8	10	12	14	43					104
TOTAL OPERATING COSTS/ACRE	873	96	113	369	333	310	287	4,985	90	13	13	13	7,495
TOTAL OPERATING COSTS/TON	44	5	6	18	17	16	14	249	5	1	1	1	377
Overhead													
Office expense	4	4	4	4	4	4	4	4	4	4	4	4	48
Liability insurance	7												7
Sanitation fee	1	1	1	1	1	1	1	1	1				9
Property taxes	84						84						168
Property insurance	28						28						56
Investment repairs	6	6	6	6	6	6	6	6	6	6	6	6	72
TOTAL CASH OVERHEAD	130	11	11	11	11	11	123	11	11	10	10	10	360
TOTAL CASH COSTS/ACRE	1,003	107	124	380	344	321	410	4,996	101	23	23	23	7,855
TOTAL CASH COSTS/TON	50	5	6	19	17	16	20	250	5	1	1	1	393

a plan covering at least 2 years that shifts cash received during high-revenue years into reserves for use during low-revenue years; 3 years may be more appropriate to account for two growing seasons and pool closings (when all the fruit from that season is sold). This is necessary because by the nature of the business, there is the risk that funds will not be available to cover costs, a problem that can cost producers significant amounts of money if they have to borrow cash during low-revenue years. Thus, real cost savings are available to producers who managing their money with a long-term perspective so as to minimize the amount of borrowing, even in periods of low interest rates.

Interplanting young trees into an older orchard that is either suffering from tree loss or widely spaced is a commonly used approach to increase revenue without interrupting cash flow by completely replacing an orchard. While this technique does result in a more even cash flow, it also creates the difficult management situation of simultaneously caring for young and old trees that have different water, fertility, and pest control requirements. Harvest may also prove difficult. In the long term the profitability of an interplanted orchard may be lower than it would have been if the entire orchard were replanted. Often, there is a tradeoff between profitability and cash flow that you should consider when making management decisions.

Break-even analysis

The uncertainty about profits combined with the extremely long time periods between cash outflows and inflows in agricultural production creates financial risk, a factor too often overlooked in feasibility studies. We recommended that you include two types of risk analysis in a feasibility study: *break-even risk* and *historical risk.*

You should estimate the break-even level of production when you need to know at what point an alternative course of action becomes profitable. The importance of a break-even estimate is in knowing the minimum level of production that must be attained to generate profits and how sensitive that level is to price and cost factors. For example, if the average revenue expected from expanding an existing orchard is above the break-even revenue required to justify such a change, the proposed expansion will probably be profitable. This knowledge may be enough to convince you to go ahead with the expansion even though you do not know the exact long-range profit potential.

To illustrate break-even analysis, we can take a look at its use in considering the change from a flood irrigation system to a sprinkler irrigation system. Suppose you have very little information concerning the yield response to sprinkler irrigation versus flood irrigation, but you do know the associated costs for both irrigation systems. Based on this, you can calculate the break-even yield difference—the difference in average annual yield that will make the two options equally profitable. The net profits for each of two alternatives can be written mathematically:

$$\text{Net profit}_f = (\text{Yield}_f \times \text{Price}) - \text{Irrigation cost}_f \quad (1)$$

$$\text{Net profit} = (\text{Yield}_s \times \text{Price}) - \text{Irrigation cost}_s \quad (2)$$

where

Net profit_f = Net profit per acre for a flood-irrigated orchard

Net profit_s = Net profit per acre for a sprinkler-irrigated orchard

Yield_f = Yield per acre for a flood-irrigated orchard

Yield_s = Yield per acre for a sprinkler-irrigated orchard

Price = Average price received per ton of pears

Irrigation cost_f = Irrigation cost per acre for a flood-irrigated orchard

Irrigation cost_s = Irrigation cost per acre for a sprinkler-irrigated orchard.

Setting equation (1) equal to equation (2) and collecting terms to solve for the break-even yield difference leads to:

$$(\text{Yield}_s - \text{Yield}_f) = (\text{Irrigation cost}_s - \text{Irrigation cost}_f)/\text{Price} \quad (3)$$

(3) By substituting the known values in the right-hand side of equation (3), you arrive at a specific number of tons per acre as the break-even yield difference. If the actual yield difference is greater than the break-even yield difference, sprinkler irrigation will be more profitable than flood irrigation. If the actual yield difference is less than the break-even yield difference, flood irrigation will be more profitable than sprinkler irrigation.

Tax considerations can easily be included in the analysis by considering the tax implications of long-term investments and changes in profit. In this case the revenue and costs should be calculated as after-tax.

Quite often factors enter into the decision-making process that are difficult or impossible to quantify in economic terms. In the irrigation example, the greater amount of water required for flood irrigation than for sprinkler irrigation may mean that at certain times there may not be enough water available for flood irrigation when there would be enough for sprinkler irrigation. Also, with sprinkler irrigation it is possible to irrigate more frequently with shorter applications, which is good for fruit sizing at the end of the season. Sprinkler irrigation has the additional benefit of frost protection. The sprinkler system can reduce tillage, lower irrigation labor costs, and make fertilizer application more efficient. It may also improve tree health. Further, it interferes much less with the timing of pesticide applications, which may lead to improved pest control and ultimately higher yield and quality. Aside from the cost of tillage and irrigation labor, none of these advantages or disadvantages are easy to measure.

Even if you cannot estimate the actual yield difference with much accuracy, you should be able to assign a probability to the break-even value. In this example, the manager may be 90 percent certain that sprinkler irrigation will result in more than the break-even yield. Then, without having information about the actual magnitude of the yield differences, the manager can decide to change to sprinkler irrigation with a 90 percent certainty that the change will be profitable.

Risk analysis

Historical risk analysis is useful when you are selecting between alternative investments. It involves evaluating the expected volatility of net income levels (profitability) over the life of an investment. The greater the variation in net income, the riskier the venture. By using past or expected annual income data, you can estimate how widely net income (profit) varies from year to year.

Whereas the basic feasibility decision is derived from observable quantities, risk analysis is based on estimates of future events and the probability that those events will occur. This means that the final decision is affected significantly by your skill in estimating the level of risk for each potential risk factor, and also by your willingness to accept such risks.

It is important to consider the whole-farm effects of any proposed action in the feasibility study. In other words, you need to know whether undertaking a project, such as changing the irrigation system, will affect existing operations and, if so, what the effect will be. Diversification across enterprises or expanding an existing enterprise to gain cost efficiencies may raise or lower your total level of income risk. The key is that you consider whole-farm factors in terms of their income and risk effects as part of a feasibility study.

THE FINAL CONSIDERATION

A basic economic feasibility analysis involves studying the factors that directly influence the farm enterprise, including market potential, resource constraints, and the production process. If you analyze all of these factors adequately and determine that they are favorable for economic operation, the venture should be profitable. However, the final profit-determining factor is *management*.

An orchard must have a competent manager who will follow through on planning, organizing, directing, staffing, and controlling in order to ensure a profitable undertaking. This means that decision makers must also assess their own management skills before reaching a final conclusion. A feasibility study helps a business avoid the costs associated with a wrong decision and indicates whether or not a proposed plan could succeed. But whether or not the plan *does* succeed is up to the manager who puts the study to work.

5 Varieties

CHUCK A. INGELS, DAVID J. BURKHART, AND RACHEL B. ELKINS

WORLD ORIGINS

Pears are believed to have originated in Central Asia. From this original form they evolved into two general groups of species—oriental species, native to eastern and northeastern Asia, and occidental species, native to Europe and northwestern Asia.

There are over two dozen species of pears, all of which are native to the Old World of the Northern Hemisphere. Two primary species are cultivated in North America today: *Pyrus communis* and *Pyrus pyrifolia*. *Pyrus communis* is native to southern Europe and Asia and constitutes the vast majority of U.S. pear production. *Pyrus pyrifolia* (syn. *Pyrus serotina*), the Asian pear, is indigenous to China and is also known as the Japanese, Chinese, or sand pear. A third species, *P. ussuriensis,* the Siberian or Ussuri pear, produces unpalatable fruit but has been crossed with *P. communis* to breed hardier *P. communis* varieties and better-tasting *P. ussuriensis* varieties. Some Asian pear cultivars are complex hybrids between *P. ussuriensis* and *P. pyrifolia*.

Early ancestors of *P. communis* were cultivated by the Greeks and Romans for medicinal as well as dietary purposes. Development of modern varieties began in the late 1700s in Belgium, which possessed ideal growing conditions for pear trees: cool, moist, fairly heavy soils and temperate, cool climate. The first noted breeder was Nicholas Hardenpont, a Belgian priest who is credited with producing from seedling trees the first of the varieties with soft, melting flesh that gave the best pears the nickname "butter fruit." He inspired other amateur breeders such as the pharmacist and physician Jean Baptiste Von Mons, whose "Nursery of the Fidelity" featured more than 80,000 pear seedlings. By 1867, over 900 varieties of *P. communis* had been recorded, all developed by amateur, mainly Belgian, breeders. Although its climate was less hospitable to pear trees, England was also the site of amateur breeding interest.

Asian pears are native to Mongolia and are widely cultivated in China and Japan. In the West, they have been crossed with European pears to produce varieties that are tolerant of heat and drought, thrive in sandy soils, and exhibit fire blight resistance. Some of these hybrids, e.g. Kieffer and LeConte, have been the mainstays of the small pear industry in the southern United States. Other species of Asian pears have been, or are being, utilized as rootstocks. For example, *P. pyrifolia* for a time replaced the domestic French seedling rootstocks of *P. communis* due to its fire blight resistance. However, it was susceptible to pear decline, caused the physiological disorder "black end," and lacked vigor, so its use ceased. *Pyrus calleryana* and *P. betulaefolia* are currently used in specific areas where increased vigor and blight resistance are desired.

CONSIDERATIONS IN SELECTING A VARIETY

Market demand and utilization are key factors in variety selection. Well-established markets have been developed for fresh-pack winter pears in the Pacific Northwest and fresh and cannery Bartlett pears in California and the Pacific Northwest. Probably over 95 percent of the pears grown in California are Bartlett, with small amounts of Bosc, Starkrimson, and a few other varieties.

Another factor is location and adaptability to the site. Winter pears such as Bosc and Anjou are grown primarily in Oregon and Washington because they are better adapted to these cooler regions, although Bosc also performs well in California's hot climate. Bartlett is better adapted to hotter regions. In the Pacific Northwest, Bartlett is used more as a pollinizer for the winter pears than as a primary variety.

Another factor, particularly in the Pacific Northwest, is timing of maturity. For example, growers usually have a gap in harvest following cherries, so they may plant Bartletts to utilize labor prior to the harvest of winter pears or early apples. Other factors include how the variety fits into the total orchard operation, the availability of storage facilities and qualified technical help, and the expertise of the grower, such as in growing processed versus fresh fruit.

Although fire blight susceptibility can be a factor in choosing a variety, it is often secondary in importance because control measures are usually successful. For example, fire blight conditions are usually worse in California's warm climate, but Bartlett, which is highly susceptible, is grown almost exclusively there. Nonetheless, fire blight susceptibility could someday be of greater importance. Of the varieties described in this chapter, all have low resistance to fire blight except Seckel and the HW Series varieties, which have moderate to high resistance. Bartlett and Red Clapp Favorite produce substantial secondary bloom (rat-tails), so their period of susceptibility is greatly extended. Forelle is extremely susceptible to fire blight.

EUROPEAN PEAR VARIETIES

Information on available bloom and harvest dates, as well as production and use statistics, for the pear varieties discussed in this chapter can be found in tables 5.1 and 5.2. Photographs of varieties can be found in the color plates beginning on page 199.

Anjou (Beurré d'Anjou)

Beurre d'Anjou (color plate 5.1) is commonly called d'Anjou or simply Anjou (in the United States it is usually pronounced to rhyme with "banjo") in the trade. Anjou originated in France and was brought into England in the early nineteenth century. It was introduced into the United States in about 1844 by Marshall Wilder of the American Pomological Society.

Anjou is the leading variety of winter pear and is second in production only to Bartlett. In the United States, it is grown almost exclusively in Washington and Oregon. Argentina is also a major producer. The Anjou pear is large and has a stocky neck. It is firm, juicy, and has a rich flavor. The fruit keeps well until spring with appropriate storage. The entire commercial crop is marketed as fresh fruit.

The Anjou tree is large, vigorous, and productive. It blooms prolifically each year and is pollenized by several cultivars, although Bartlett is primarily used. It is slow to come into bearing, but it bears consistently as a mature tree. Precocity can be influenced by rootstock selection. For example, the rootstock *P. betulaefolia* imparts delayed fruiting, whereas *P. communis*, specifically some Old Home × Farmingdale selections, has been shown to induce earlier bearing. Limb manipulation and other training techniques can also reduce the time from planting to production. The Anjou tree is moderately susceptible to fire blight, although the risk is reduced by the relatively early bloom period and lack of secondary, or rat-tail, bloom.

The Anjou pear ripens in early to mid-September. In relatively cool production areas, it can be left on the tree longer than most pear varieties. In warmer climates the fruit can have a high incidence of mealy breakdown of the flesh.

Table 5.1. Average full bloom dates for selected varieties in California and the Pacific Northwest

	Walnut Grove, CA (1992–2001)	Kelseyville, CA (1983–1988)	SOREC, Central Point, OR (1990–2004)	MCAREC, Hood River, OR (1990–2004)
Anjou: Columbia Red	—	—	—	Apr 15
Anjou: Green	—	—	Mar 30	Apr 12
Bartlett	Mar 20	Apr 1	Apr 4	Apr 18
Bosc: Standard	Mar 22	Apr 3	Apr 9	Apr 19
BPM	Mar 15	—	—	—
Comice	—	Apr 3	Apr 8	Apr 21
Starkrimson	—	—	Apr 4	—
Steckel	Mar 20	—	—	—

Sources: C. Baker, R. Elkins, M. MacCormack, G. Mielke, and P. VanBuskirk personal communications.

Note: SOREC = Southern Oregon Research and Extension Center; MCAREC = Mid-Columbia Agricultural Research and Extension Center.

Red Anjou

There are two commercially available red sports of Anjou (color plate 5.2). The first one, Gebhard Red Anjou, was found in the Gebhard orchard near Medford, Oregon, in the early 1950s. The Columbia Red Anjou was later discovered in the top of a green Anjou tree in the orchard of Gene Euwer near Parkdale, Hood River County, Oregon. Columbia Red Anjou matures at about the same time as green Anjou, but Gebhard Red Anjou matures nearly a week later.

Although the two strains appear to be almost identical in appearance, the Columbia Red strain appears to ripen more consistently and has less tendency to manifest physiological disorders. Both strains have maintained consistently good overall red color, which is unusual for many red pears. The red strains of Bartlett, for example, attain quite acceptable color in the preharvest period, but they tend to lose some of this color as the fruit ripens. Part of this loss of color may be due to warm temperatures during the preharvest period. Red Anjou does not lose color in this way. An occasional branch in a Red Anjou orchard may revert back to producing green fruit; these branches can be removed.

Bartlett

The Bartlett variety (color plate 5.3) was developed in the seventeenth century in Berkshire, England, by a schoolmaster named John Stair. Stair sold some of his pear tree cuttings to a horticulturist named Williams, who further developed the variety and renamed it after himself. After pear seedlings crossed the Atlantic with the early colonists, the Williams pear was propagated by Enoch Bartlett of Dorchester, Massachusetts. Bartlett, who was unaware of the pear's former name, distributed it under his own name. The Williams pear has been known as Bartlett in the United States ever since, but it is still referred to as Williams or Williams Bon Cretien in other parts of the world. Bartlett pear trees eventually came west in the covered wagons of people heading for the California Gold Rush.

Bartlett makes up about 75 percent of U.S. and Canadian pear production. California is the leading producer of Bartlett pears, and the great majority of California's pear acreage is planted to Bartlett. However, Bartlett production has declined in the past 10 years. Until recently, about one-third of the crop was sold fresh, and nearly two-thirds was canned; but an incresing amount is sold fresh. A smaller portion is dried or processed into baby food, juice, or puree. Bartlett is harvested from early July in the Sacramento Delta to late August or early September in the North Coast.

Bartlett is the standard against which all pears are compared. It is large, with the classic pear shape, and is yellow with blushes of brownish red. The flesh has a smooth texture and is juicy with a rich flavor. It is the only variety in the western U.S. to be used for fresh eating, canning, drying, and preserves. The large production in North America is due in part to its multiple uses and to its keeping quality, which can be up to 3.5 months after harvest in regular cold storage, and up to 6 months in controlled atmosphere storage. It ripens off the tree to develop the famous musky flavor.

The Bartlett tree is medium in size, compact, upright, and is well suited for dense planting. It is vigorous, precocious, very productive, and long-lived; many existing orchards have trees over 100 years old that continue to produce well. It is also adapted to a wide geographical range and a wide diversity of soil and climatic conditions. Bartlett is generally self-fruitful in California, but it requires a pollinizer in the Pacific Northwest in order to maintain consistent annual production, optimal fruit shape, and good fruit size. The major drawback is its susceptibility to fire blight, including both blossom and shoot blight, and it is prone to rattail bloom. Bartlett trees require 800 to 1,000 hours of winter chilling, depending on the rootstock used, for best performance.

Red Bartlett

The first commercialized red sport of Bartlett was discovered as a limb on a green Bartlett tree near Zillah, Washington, in 1938. It was named Max Red Bartlett (color plate 5.4) when patented in 1947. The Red Bartlett fruit has the same internal characteristics of green Bartlett, but the tree lacks vigor and often produces smaller fruit. Sensation Red Bartlett, often referred to as Red Sensation Bartlett, was subsequently discovered in Australia and soon found its way to the United States. It appears to have more consistent and brighter red color than Max Red Bartlett. Rosired Bartlett is the third major red strain in the market, but it has failed to gain widespread acceptance due to its susceptibility to russetting.

The red Bartlett strains bloom and harvest 2 to 5 days after Bartlett. Rosired Bartlett ripens a few days later than the other red strains, and it lacks the characteristic stripes of Max Red Bartlett and Sensation Red Bartlett. Red Bartlett fruit grows as fast as green Bartlett, but when left on the tree too long, red color and storage life diminish drastically. Conversely, when harvested prematurely, color and storage life are good but the size is too small.

Bosc (Beurré Bosc)

American and British authors have generally attributed the origin of Bosc (color plate 5.5) to Belgium, on the assumption that it is the variety introduced by Van Mons as Calabase Bosc in 1807. However, the descriptions of Calabase Bosc given by European authors in no way resemble the characteristics of the pear now recognized as Bosc or Beurré Bosc in England and America. The present-day Bosc pear of American and English pomology is more likely the variety Beurré Bosc. This variety originated in France and was named in honor of the eminent French horticulturalist Louis Bosc. Calabase Bosc, on the other hand, was named by Van Mons after the French naturalist, M. Bosc, who was no relation to Louis Bosc.

The Bosc fruit is large, with a long, tapering neck. The yellow skin is almost completely covered with an attractive brown russet, which varies in intensity and

Table 5.2. California Bartlett pear production

	Processed		Fresh	
Year	(tons)	(metric tons)	(tons)	(metric tons)
1996	205,100	186,000	77,000	69,800
1997	208,600	189,200	68,000	61,700
1998	216,400	196,300	53,000	48,100
1999	220,000	199,500	84,000	76,200
2000	182,000	165,000	81,000	73,500
2001	170,000	154,200	67,000	60,800
2002	160,000	154,100	69,000	62,600
2003	151,000	136,900	62,000	56,200

Source: California Pear Advisory Board 2003.

character depending on climatic and cultural conditions. Bosc flesh is yellowish white and very juicy and buttery but does not melt. The rich, aromatic flavor rates among the best in dessert quality when properly grown.

The tree, particularly when young, produces vigorous, upright, leggy growth with little branching. This characteristic makes it difficult to train during the formative years, but the tree can be well shaped and stately in maturity. The tree blooms after Anjou and Bartlett, along with Comice, and it is a good pollinizer for late-blooming varieties. It is fairly precocious and usually produces a heavy crop. It matures shortly after Bartlett in California and up to 3 to 4 weeks later in the Pacific Northwest. Bosc is quite susceptible to fire blight due to its vigorous, elongated growth habit. Unlike Bartlett, however, it does not have rat-tail bloom, so the period of susceptibility is much shorter.

Several better-russetting strains of Bosc have become available, including Golden Russet, Bronze Beauty, and Noble Russet. These strains are all similar to Bosc in fruit size and shape. Many growers have replaced or topworked standard Bosc orchards to russetted strains due to their higher market value.

Bronze Beauty

Bronze Beauty (color plate 5.6) was discovered by Bob Wimmers in his orchard west of Hood River, Oregon. It begins to russet earlier in the growing season than standard Bosc and has a solid, heavy russet that completely covers the fruit. When ripe, the fruit takes on a rich bronze color. Harvest is generally 5 to 7 days after standard Bosc.

Golden Russet

Golden Russet Bosc (color plate 5.7) is a sport of Bosc that was found in the Fukui orchard near Parkdale, Hood River County, Oregon. It matures at the same time as standard Bosc. It has natural russetting on a golden yellow background with dense, yellowish white flesh. It is the most widely planted of the russetted Bosc cultivars.

Noble Russet

Noble Russet (color plate 5.8) was discovered by Noble Law in the Okanogan region in north-central Washington. This very smooth, thoroughly russetted Bosc also becomes russetted early in the season.

Butirra Precoce Morettini

This cultivar, usually referred to as BPM, is an Italian variety with Bartlett parentage. It precedes Bartlett harvest by about a week. The fruit has a bright, blushed, yellow skin when ripened, with a melting texture and tart flavor. The tree crops well and regularly with cross-pollination. BPM is grown only where there is little threat of frost due to its early bloom period.

Comice (Doyenne du Comice)

Originating in France around 1850, this well-known dessert pear is a specialty of growers located in the Rogue River Valley near Medford, Oregon. Known as the "Queen of Pears," it is called Comice in the trade (color plate 5.9). It matures in late August to early September, just before Anjou.

The Comice pear is perhaps the smoothest-textured and juiciest of all varieties. It develops its best flavor after a month of storage. Often called the Christmas Pear, Comice often appears in gift boxes and baskets during the winter holidays. The fruit is large, with a rosy-blushed yellow-green skin. However, it is one of the most easily bruised pear varieties.

The tree is medium-sized, vigorous, and upright. It has historically been known to be a light bearer. Precocity can be improved with certain rootstocks. For example, many growers in the Medford District use quince or *P. communis* rootstocks but not *P. betulaefolia*. The Comice tree blooms in the Bosc season after Bartlett, and the two varieties are good pollinizers for each other.

Forelle

The origin of Forelle (color plate 5.10) is uncertain. Some claimed that it originated in northern Saxony at the beginning of the eighteenth century. It was brought to Massachusetts from England in 1823. The cultivar became common in the eastern United States during the first half of the eighteenth century. Forelle is currently grown mainly in Washington and Oregon and is available from October through late winter.

The Forelle pear is small, attractive, and of good quality. It is bright in color with trout-like speckling (*Forelle* means "trout" in German). The skin is smooth, yellow, overlaid with red, and covered with grayish russet dots. The white flesh is juicy, melting, and richly flavored.

The tree is large, vigorous, and spreading. Plantings are limited due to its extreme susceptibility to fire blight. Forelle blooms prior to Bartlett and is a self-unfruitful variety that requires a pollinizer. Harvest season is normally in mid to late September. There is mixed opinion on the need for thinning Forelle, but since larger fruit usually bring more money, some growers remove small fruit.

Packham's Triumph

A discussion of pear varieties would not be complete without some reference to Packham's Triumph (color plate 5.11), the main winter pear grown in the Southern Hemisphere. Small plantings of this pear exist in the Pacific Northwest. The Packham's Triumph pear came out of a breeding program in New South Wales, Australia, around the turn of the twentieth century.

The fruit is usually large and is shaped like Bartlett. Northern Hemisphere growers sometimes mistake the Packham's Triumph tree for a Bartlett when the fruit are observed during the growing season. Whereas the fruit has a relatively smooth, russet-free surface when grown in the Southern Hemisphere, it often has a bumpy, uneven surface in the Northern Hemisphere. Fruit quality is excellent because of its melting texture and juiciness.

The tree is vigorous and upright and is susceptible to fire blight. The blossoms are less attractive to bees than other pear varieties, so in some areas fruit set has been relatively low in many years. Production has been consistently high in other areas. It blooms with Bartlett but it matures about 30 days after Bartlett.

Red Clapp Favorite

Red Clapp Favorite (color plate 5.12), also known as Starkrimson, Kalle, Crimson Red, Super Red, and Red Crimson, is one of the best-colored early-season red pears available for the pear grower today. It has an attractive red fruit with good eating quality, melting texture, and sweet, juicy flavor. It blooms with or slightly before Bartlett but is harvested about a week earlier. Because of its consistently dark maroon color and competitive advantage for early-season shelf space, this pear has contributed to making the early-season sales of Red Bartlett difficult. Key disadvantages are that the fruit ripen fairly quickly, so bruising is a major problem that requires extra care during harvesting and handling, and it has a very short shelf life.

Red Clapp Favorite trees are large, upright, and vigorous, and they are slow to come into bearing. It often bears slightly less heavily than Bartlett. It produces even more rat-tail bloom than Bartlett and is thus very susceptible to fire blight.

Seckel

Seckel (color plate 5.13) is one of the few varieties of pears that originated in the United States. In the late eighteenth century, a cattle dealer near Philadelphia had a single tree on land that was later owned by a Mr. Seckel, who gave the pear his name and introduced it.

The small, yellowish brown fruit with pale russet and red cheek has smooth, very sweet, and very juicy flesh. The flesh, along with the spicy, aromatic flavor of the skin, makes it an excellent dessert pear. Seckel is well suited for pickling, spicing, and canning whole. It is one of the few winter pears that can be eaten off the tree and does not require ripening or chilling. The small fruit size, the limited market, and the need for thinning make this variety less attractive for many pear growers.

The Seckel tree is productive and slow growing, and it reaches only 15 to 20 feet (4.6 to 6.1 m) in height. It comes into bearing late but then bears heavily and reliably, and it is adapted to a wide variety of climatic conditions. Seckel is more resistant to fire blight than most other European types. It is self-fertile but sometimes sets seedless fruit, which may be small and possibly more susceptible to preharvest drop. Cross-pollination usually eliminates seedless fruit but may increase fruit set and the need for additional thinning.

RECENT PEAR INTRODUCTIONS TO THE WEST COAST

New apple, peach, and cherry varieties that have originated from a breeding program, a foreign country, or a grower's orchard as chance seedlings or mutations are quite common in the tree fruit industry today. It is a different story with pears. With the exception of the red and the better-russetted pear selections, growers are still growing the same pears now that were grown 200 years ago. Several theories have been advanced as to why this is the case:

- Pear breeding programs have been reduced or eliminated in many countries around the world.
- Pears appear to mutate less frequently than do apples.
- Pears normally ripen off the tree and are thus difficult to evaluate.
- Growers and packers are reluctant to try new pear varieties. The long life of pear trees compounds this.

Several new pears have been identified as possible new commercial selections. They include Taylor's Gold from New Zealand, Concorde from the United Kingdom, Abate Fetel from Italy, and Rosemarie from South Africa. Banjo and Cinnamon are mutations or sports from the Hood River District. The HW Series and the USDA releases from Beltsville, Maryland, come from breeding programs. A pear breeding program under the direction of Dr. Richard Bell at the USDA's Appalachian Fruit Research Station at Kearneysville, West Virginia, produced several fire blight–resistant pears. Whether any of these recent introductions will be accepted by pear growers and consumers is unknown.

Abate Fetel

Abate Fetel (color plate 5.14), also known as Abbé Fetel, is a partially russetted Italian winter pear that is a favorite in Europe. The fruit is large, elongated, and yellow, often with a slight russet and possibly a blush on its cheek. The pear has excellent flavor and keeps well into spring under regular storage conditions. Its harvest season is between Bartlett and Anjou. The fruit can be difficult to ripen.

Abate Fetel is seldom planted in the western United States, but it appears to be best adapted to warmer climates, such as in California and southern Oregon. The fruit buds do not appear to be winter-hardy in cold climates.

Blake's Pride

Blake's Pride (color plate 5.15) combines a high degree of fire blight resistance with aromatic, juicy fruit and consistent yield. It was created by a cross in 1965 and it has parentage that includes Bartlett, Seckel, and Comice. The fruit is shaped like a Bartlett and the skin is golden to light yellow, with some light tan russetting. The flavor and aroma are more similar to Comice than to Bartlett. It ripens 10 to 14 days after Bartlett in the Pacific Northwest but closer to Bartlett in California.

The tree is moderate in vigor on Bartlett seedling rootstock and is upright-spreading in habit. Blake's Pride requires pollination with Bartlett, Bosc, Harrow Delight, and Packham's Triumph.

Cinnamon

Cinnamon (color plate 5.16), a limb mutation in a Bartlett tree, was discovered by Nancy and Don Biehn in their orchard south of Hood River, Oregon, in 1979. First thought to be another russetted Bartlett, it was later determined to be a winter pear that was harvested in the Anjou season. The fruit is covered with a netlike russet similar to the new, improved Bosc strains. The fruit flavor has been described as having a sweet, spicy, cinnamon taste. Adequate fruit for testing has been limited, though it appears that Cinnamon may have a relatively short storage life.

Tree habit appears moderately vigorous, with spreading, upright growth. It has been pollenized by either Bartlett or Bosc, which grow near the original tree. It appears to be winter-hardy.

Concorde

Concorde (color plate 5.17) is a cross of Comice and Conference (color plate 5.18) from the East Malling Research Station in England. It has a shape similar to Conference, the leading winter pear of Europe, and the green skin color of Comice. It may have a slight russet, depending on the weather or the site on which it is grown. As the fruit matures the skin color changes to yellow, which may not indicate ripeness, according to the East Malling researchers.

Excellent flavor, high potential yield, and medium fruit size all make Concorde attractive to the pear industry. Concorde has been described as a grower-friendly tree because of its precocity and its compact growing habit. The tree has dark green leaves that appear to be tolerant of mites and psylla. Bartlett, Comice, and Bosc appear to be excellent pollinizers for Concorde. Concorde is picked in the Anjou season and stores into the spring months.

Although there have been several plantings in the Pacific Northwest, there has been some resistance by packers because of its partial russetting. Pacific Northwest packers, unlike their European counterparts, seem to want either a fully russetted pear or one that is perfectly clean and free of russet. Preliminary observations indicate that best fruit quality is achieved under cooler growing conditions.

Harrow Sweet

New pear varieties known as the HW Series are being developed at the Harrow Research Centre in Ontario, Canada, under the direction of Dr. David Hunter. Harvest Queen and Harrow Delight, two high-quality pears that ripen prior to Bartlett, were released for the home garden and fresh market in Canada. Because of their uneven ripening characteristics and short storage life the two pears have not been planted commercially in the western United States.

Harrow Sweet, perhaps the most promising of the HW series, is a later-maturing pear that picks approximately 3 weeks after Bartlett. The medium to large fruit are slightly smaller than BPM on unthinned trees. The fruit is pyriform shaped, with a shallow medium basin and an open calyx. The flesh is creamy white, very sweet, juicy, and has excellent flavor. The Harrow Sweet tree blooms with Bartlett. In tests at the Harrow Research Centre, the Harrow Sweet appears to be more precocious than Bartlett. It is resistant to fire blight. Testing has been limited in the western U.S. pear-growing areas.

Potomac

Potomac (color plate 5.19) originated from a cross of Moonglow and Anjou made in 1961 and tested with other seedlings at the USDA Beltsville Agricultural Research Center. Moonglow was created by crosses originating with Bartlett and Seckel. The fruit shape, size, and texture of Potomac are similar to Anjou, and the fruit have occasional slight bumpiness. Potomac is harvested about 2 weeks after Bartlett.

The tree is moderately vigorous and spreading.

Full bloom occurs with Anjou, about 2 to 4 days before Bartlett. Potomac requires cross-pollination with Bartlett or Anjou. Preliminary yield tests showed that Potomac was less precocious than Bartlett, but its yields rapidly surpassed Bartlett after year 6, due in part to loss of limbs to fire blight in Bartlett trees.

Rosemarie

Rosemarie (color plate 5.20) was the first pear cultivar to be bred in South Africa, and it is now being grown on a commercial scale in that country, although acreage has declined due to lack of red color retention. A cross between Forelle and the strain of Red Bartlett called Bon Rouge, Rosemarie is also being used for canning and drying. The fruit is similar to Bartlett in shape and has a bright orange to red blush over a yellow ground color. The fruit is medium in size, slightly smaller than Bartlett, and often requires thinning to attain optimal size.

Rosemarie grows vigorously on good soils. In South Africa it is planted on the BP 1 rootstock, which produces a tree about 70 percent of the size of Bartlett seedling rootstock. This rootstock is very susceptible to pear decline, so it would not be used in the United States (pear decline is spread by pear psylla, which is not yet present in South Africa). Most South African growers train the Rosemarie to an open center in order to maintain the blushed fruit. The tree blooms slightly earlier than Bartlett and is pollinated by pears that bloom in the same period. One-year-old shoots of the Rosemarie tree are reddish brown with distinctive white lenticels. Vegetative buds are much larger than either those on Bartlett or Forelle and are pointed and dark in color. The harvest season is shortly after Bartlett.

Taylor's Gold

One comparatively new pear from New Zealand that has made a small entry into U.S. markets is the russetted Comice called Taylor's Gold (color plate 5.21). First discovered by Michael King-Turner in his Riwaka pear orchard near Nelson, New Zealand, in 1985, this Comice mutation was immediately recognized for its sweet, delicately perfumed flavor. The fruit finish is similar to Golden Russet Bosc, with its yellow-gold undercolor and fine netlike russet. The russeting and the firmer skin appears to resist the marking that is so common with the regular Comice. However, recent trials have shown that it does not russet adequately in arid districts such as southern Oregon and California. Only in higher-rainfall areas of Oregon and Washington does it adequately russet.

According to New Zealand growers, Taylor's Gold russets well in cooler climates similar to the pear-producing districts of the Pacific Northwest. The tree is very vigorous, upright, and like Comice, it tends to set lights crops. New Zealand growers initially used Bosc as a pollinizer for Taylor's Gold, but they now find that using earlier-blooming varieties like Bartlett or Red Clapp Favorite set better crops. The fruit is picked after Gala apples in New Zealand, which would be the early Comice season in California.

Source of Germplasm

The National Clonal Germplasm Repository at Corvallis, Oregon, is one of the locations of the USDA/ARS National Plant Germplasm System (NPGS). The repository is a genebank that preserves genetic resources of selected temperate crops, including pears. Individuals can obtain specific varieties, rootstocks, and strains through the repository. For more information, see the NPGS Web site, http://www.ars-grin.gov/npgs.

REFERENCES

Bell, R. L., T. van der Zwet, R. C. Blake, C. K. Chandler, and J. C. Scheerens. 1996. 'Potomac' pear. HortScience 31(5): 884–886.

California Pear Advisory Board. 2003. Annual report. CPAB Web site, http://www.calpear.com/brd_annreporttables.cfm.

Flaherty, D. C., and S. E. Harvey. 1988. Fruits and berries of the Pacific Northwest. Edmonds, WA: Alaska NW Publishing Co.

Hedrick, U. P. 1921. The pears of New York. Albany: J. B. Lyon.

Lombard, P. 1986. Principal green and new pear varieties for USA. In M. Willett and R. Stebbins, eds., Pear production in the Pacific Northwest. Proceedings of the 1986 Pacific Northwest tree fruit short course. Pullman: Washington State University Office of Conferences and Institutes. 45–67.

Mid-Columbia Agricultural Research and Extension Center. 2005a. Bartlett packout report 1993 to 2002. MCAREC Web site, http://oregonstate.edu/dept/mcarec/prodstats.php.

———. 2005b. Bloom dates: Pears. MCAREC Web site, http://oregonstate.edu/dept/mcarec/stats/bloom_dates/bloom_dates_pears.html.

Pear Bureau Northwest. 1998. USA Pears. Pear Bureau Northwest Web site, http://www.usapears.com.

van der Zwet, T., and N. F. Childers. 1982. The pear: Cultivars to marketing. Gainesville, FL: Horticultural Publications.

Williams, M. W., H. M. Couey, H. Moffitt, and D. L. Coyier. 1978. Pear production. Washington, DC: U.S. Department of Agriculture, Agriculture Handbook 526.

6 Propagation and Rootstock Selection

WILBUR O. REIL, JOHN IRELAND, AND RACHEL B. ELKINS

PROPAGATION

Successful pear propagation requires the selection and preparation of an appropriate nursery site. In general, the primary work of propagation consists of two steps: propagating rootstocks from seedlings, hardwood cuttings, or micropropagation; and joining scions of pear varieties chosen for their bearing qualities to the rootstocks by budding or grafting.

Nursery people propagate complete trees through budding or topworking, both of which are vegetative means of propagation. These methods are used for three reasons. First, neither varieties nor rootstocks grow true from seed in pears. Therefore, vegetative propagation is the only way to maintain the specific qualities of a particular variety. Second, pear cuttings do not root easily, although in a few cases it can be done. Third, budding or grafting imparts the advantages of the rootstock to the variety to which it is joined.

Specific rootstocks offer qualities related to anchorage, size of tree, disease or nematode resistance, yield, nutrition uptake, and tolerance of adverse soil conditions. The rootstocks most commonly used with pear trees are Winter Nelis seedlings, Bartlett seedlings, Old Home × Farmingdale rooted cuttings, Quince rooted cuttings, and *Pyrus betulaefolia* seedlings. Choosing the best plant material in any given case means selecting the rootstock best suited for the orchard site and the most productive variety that is compatible with the rootstock and local conditions.

Budding

Most pear trees are grown in the nursery for 2 years before digging and transplanting to the orchard site. Two budding techniques are in common use: fall budding and spring budding. Trees are not budded from mid-May to August because there is insufficient growing weather to produce a desirable tree by fall. T-budding (also called shield budding because the bud and the adjoining stem resemble a shield when cut) is the preferred technique used in all three techniques. For T-budding to be successful the bark of the rootstock must be slipping (the bark is easily removed from the subtending wood).

Fall budding

Nurseries generally insert buds into pear rootstocks in August. The union heals during the fall, but the remaining shoot (above the bud) keeps the bud from growing. In spring, just prior to growth, a worker cuts off the shoot and the bud begins to develop. When the tree is sold at the end of the season, the rootstock is 2 years old and the scion is 1 year old.

The process of T-budding begins when workers collect budsticks and remove leaves from the current season's growth. They then cut the bud from the budstick (fig. 6.1). A worker cuts a T-shaped slit into the bark of the rootstock (fig. 6.2). The slit should be a few inches above the soil line. The worker slides a bud into the cut (fig. 6.3) and wraps the union with a budding rubber (fig. 6.4) to hold the bud tightly in place.

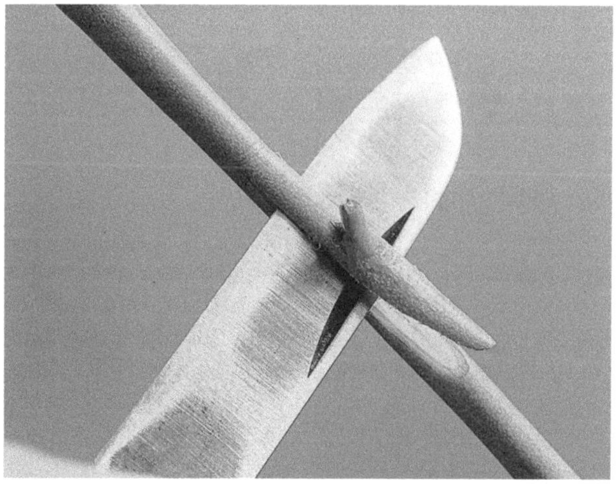

Figure 6.1. A T-bud is cut from a scion. *Photo:* J. K. Clark.

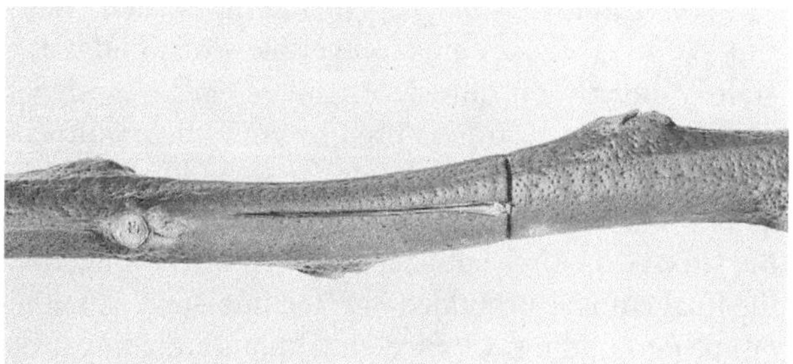

Figure 6.2. A T-cut made in a rootstock. *Photo:* J. K. Clark.

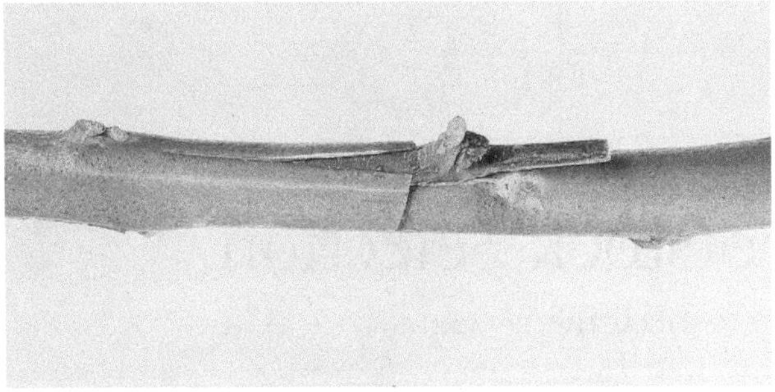

Figure 6.3. A bud inserted into a rootstock. *Photo:* J. K. Clark.

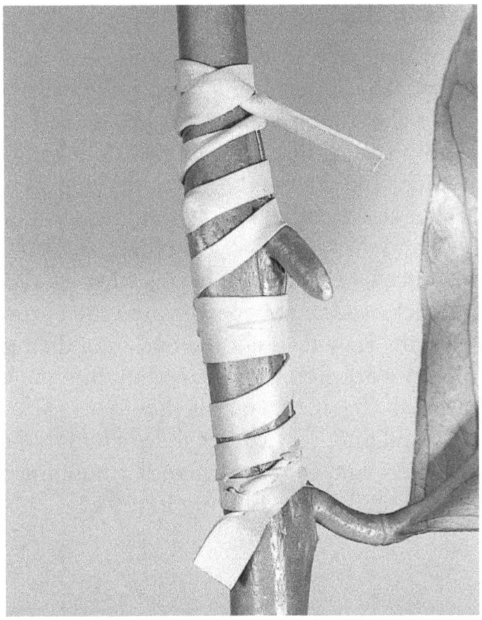

Figure 6.4. A budding rubber holds the bud in place. *Photo:* J. K. Clark.

Spring budding

The process of spring budding begins in winter, when workers collect dormant wood. The wood is stored at 36° to 38°F (2° to 3°C) until needed, in mid-April to mid-May. T-budding is usually used for spring budding. All shoots are removed from the rootstock 20 days after the bud is inserted. Spring budding is sometimes used on quince cuttings, especially if an interstem is required. It is also used if insufficient trees were fall-budded the preceding year.

Several cultivars of pear, including Bartlett, Bosc, and Winter Nelis, are not compatible when budded or grafted directly on quince. An interstem or sandwich of Old Home or Hardy is used to grow these cultivars on quince. The interstem cultivar is budded into the quince cutting using spring or June budding during the first year. After sufficient growth of the interstem the final cultivar is budded into the interstem using fall or spring budding. Comice and Anjou are compatible with quince.

Topworking

Older trees in an existing orchard must sometimes be topworked to change to a scion variety that is more suitable or to add pollinator varieties. Topworking involves several steps: collecting and, possibly, storing the scion wood; using one of several methods to graft the scion to the rootstock; and performing appropriate follow-up procedures.

In older orchards topworking an existing tree offers advantages over planting a new tree. Grafted trees are easier to manage and care for than newly planted ones. In addition, topworked trees can compete effectively with established trees. The disadvantage of topworking is that the grafting process can introduce viruses from an infected rootstock or scion. Younger trees are easier to graft than older trees. On older trees cuts into large limbs sometimes fail to heal adequately.

Wood collection and storage

Scion wood to be grafted is collected during winter, when buds are dormant. It may be grafted immediately or stored. To store the wood, place it in plastic bags or bins with plastic liners and refrigerate it at temperatures slightly above freezing. The wood must not dry out during storage, so it must be checked occasionally to make sure there is adequate moisture present in the plastic. The temperature must be cold enough to keep buds from growing.

Grafting

Nursery personnel can choose from a number of grafting methods: whip grafting, wedge grafting (also know as saw kerf grafting), cleft grafting, and bark grafting. Consult *Propagation of Temperate-Zone Fruit Plants* (Hartmann and Beutel 1981) or *Plant Propagation: Principles and Practices* (Hartmann, Kester, and Davies 1990) to learn the specific techniques for each method.

Growers usually use whip grafting on young trees. Wedge grafting and cleft grafting are used during the dormant season, because they do not require that bark be slipping. Bark grafting is done when the bark slips in spring, usually after bloom.

Follow-up procedure

After the scions are grafted workers may have to rewax the union, remove suckers, provide support to protect the tree from wind, or prune branches in summer to begin training the limbs. Many orchardists paint exposed limbs and trunks with white interior latex paint to protect them from sunburn and invasion from flatheaded borers. Borers often try to penetrate the bark at the base of the scion where it may take some time for the graft union to heal over completely. In summer, when the grafted limbs are well established, the nurse limbs are removed. At the end of the first season's growth, workers select and prune the desired scaffolds and can remove or bend horizontal surplus scions. Bending the surplus scaffolds can decrease some of the excess vigor that is usually present in older grafted pear trees. It will also increase early fruit production.

Removal and Storage of Nursery Trees

Before being dug from the soil, nursery trees must be dormant. To ensure timely dormancy in the fall, nurseries decrease watering in late summer and early fall. Trees should not be removed until most of the leaves have fallen. In November many nurseries defoliate the trees by spraying them with zinc sulfate, Nursery trees are usually ready for removal in December: trees must be removed by late January, when root growth starts.

Removal involves using a curved cutting bar to undercut the root system. The bar is pulled through the soil at a depth of 14 to 16 inches (36 to 40 cm). An extension on the back of the cutting bar lifts the roots and soil—this loosens the soil from the roots. The tree is dropped back into the hole and then pulled from the ground by hand. Each tree is graded according to uniformity and quality. Trees are then sorted by size. Size is determined by the diameter, or caliper, of the stem 2 inches (5 cm) above the bud union. Workers bundle the trees into groups of 5 large trees or 10 small ones. The trees are heeled in using sand or wood shavings until needed. Roots must be kept damp by maintaining the moisture of the sand or shavings.

Pear trees to be planted after late February usually require cold storage while dormant. The temperature of the storage unit should be 32° to 36°F (0° to 2°C). Trees stored at these temperatures will remain dormant well into summer. The roots of stored trees need humidity; they should be wet down if necessary. Some nurseries store trees in bins with damp wood shavings around the roots. The shavings should be fresh and free of pathogens and resins. Redwood or cedar shavings should not be used. Other nurseries pile bareroot trees on pallets and periodically mist the roots to prevent drying.

While trees are in cold storage, nurseries make sure not to expose them to ethylene and do not store them with fruits that produce ethylene, such as apples and pears. Extended exposure to low rates of ethylene can retard tree growth. For the same reason, trees are not stored in environments that contain fumes produced by gas or propane-powered vehicles.

Orchard Planting

The orchard planting site needs to be properly prepared before planting. If the ground is too wet on the day appointed for planting, workers must wait for suitable conditions. Trees are planted as early as possible, however; the earlier they are planted, the larger they will be by the end of the season.

In early spring, if weather is cool and moist, trees taken from cold storage or delivered directly from the nursery can be kept bareroot for several days. To prevent overheating and desiccation, trees are kept out of direct sun. Roots must also be kept from freezing. A temperature just slightly below 32°F (0°C) can kill rootlets.

In late spring or early summer, the planting crew should remove from cold storage only those trees that can be planted that day. Trees are watered the same day they are planted using only enough water to settle the soil around roots and exclude air pockets. Routine irrigation is not begun until the trees are growing. For more information, see chapter 7, "Orchard Establishment."

TYPES OF ROOTSTOCK PROPAGATION

There are two types of pear rootstocks: seedling and clonal. Seedling rootstocks are very easy and inexpensive to propagate, and they generate large quantities of rootings in a moderate period of time. Fruit is harvested from the desired rootstock tree when the fruit is mature and seed is extracted, stratified, and planted in the nursery row. One disadvantage of seedlings is that each individual plant is genetically different from all the rest. This variability can cause differences in performance between trees. The root systems of seedling trees are very symmetrical, have fey twists and kinks, and tend to grow downward.

Clonal rootstocks are derived from a single parent tree. They are genetically the same, so variability among trees is minimal. Propagation of hardwood cuttings requires collecting 1-year-old shoots in the fall, conditioning the cuttings, and then planting them in the nursery row. Generally this requires that a substantial number of parent trees be maintained to provide wood for cuttings. Also, as trees mature they lose juvenility; cuttings generally become harder to root from these mature trees. The root systems of clonal rootstocks may be asymmetrical and tend to grow laterally as opposed to downward.

Stooling, or layering, is a more expensive means of clonal propagation. It is useful when cuttings are not successful. The root system of layered trees is usually symmetrical.

Micropropagation, or tissue culture, is a clonal technique that may be successful when cuttings and layering fail. The juvenility of plant source material can be maintained, which should greatly improve rootability. Micropropagation could allow nurseries to have a few parent trees for rootstock material while being able to clonally generate large populations of rootstocks in a short period of time. Stand establishment of nursery rootstocks could be greatly improved. Micropropagation, however, is relatively costly. Also, the roots of micropropagated trees may have twists or kinks and often tend to grow laterally as opposed to downward.

CUTTING-BASED PROPAGATION

Pear is a difficult-to-root woody species, and successful propagation of such plants by rooted cuttings requires careful consideration and control of genotypic, seasonal, and environmental parameters, as well as application of auxin. Depending on the genotype, rooting percentages of 30 to 90 percent are possible. Three methods of conventional propagation have been used to root pear and quince trees: stooling, hardwood cuttings, and softwood cuttings.

Stooling, or Layering

Stooling, or layering, is a method of rooting shoots in a bed. It is the method used to root quince rootstock because quince roots easily, but it has generally not been successful in rooting pear. To establish a stooling bed, stock trees, preferably of clonal material on their own roots, should be pruned low to the ground during dormancy. The following spring, vigorous shoots will grow. Stake these shoots down horizontally in May or June to serve as a source of shoots for stooling. Girdle the branches by tying the stems with copper wire near the base of the shoot about ¼ to ½ inch (6.5 to 12.5 mm) below a node. An application of from 2,000 to 5,000 ppm of IBA (auxin) above the girdle may be beneficial. Applying auxin to stooled shoots is difficult, however, and no single method can be recommended. After girdling the branches, bury the entire stool bed with wood shavings at least 3 to 4 inches (7.5 to 10 cm) above the girdle point. Keep the bed moist with overhead sprinkler irrigation. In the winter, when the trees are dormant, harvest cuttings and line them out in the nursery row.

Hardwood Cuttings

Hardwood cuttings are collected after the plants become dormant and lose their leaves, preferably soon after leaf drop. Cuttings harvested through January continue to root, but the percentage declines with time. Make hardwood cuttings from current-year shoots. Trim cuttings to 10 to 12 inches (25.5 to 30.5 cm) long. Make the bottom cut ¼ to ½ inch (6.5 to 12.5 mm) below a node. Stem diameters from ¼ to 1 inch (6.5 to 25.5 mm) will root, but it is preferable to have material with a diameter of ⅜ to ½ inch (9.5 to 12.5 mm). Subterminal cuttings are more successful than terminal cuttings. Apply auxin either as a quick dip of 4,000 to 8,000 ppm IBA or as a 24-hour soak of 200 to 300 ppm IBA. After the auxin application, stick cuttings in individual rooting plugs and place them in a humid environment. Mist is not needed. Maintain basal heat of 75° to 80°F (24° to 27°C) for 30 to 50 days to induce root formation. After the hardwood cuttings root, plant the plug and cutting directly in the nursery row.

Softwood Cuttings

Small, leafy softwood cuttings collected in early summer with leaves attached have given good rooting of some hard-to-root pear rootstocks, especially when treated with high rates of IBA. After treatment, the cuttings are placed in porous media with good drainage in a greenhouse environment. They are maintained in intermittent mist to wet the leaves until rooting starts. The cuttings are then gradually hardened from 100 percent humidity conditions before being transferred to the nursery in fall or early winter. One disadvantage of using softwood cuttings is that they must be collected in early summer, which does not allow the collected cuttings to grow adequately before winter. The cuttings are usually quite small the following spring, so they are often grown for 1 year in the nursery and either fall-budded or spring-budded and lifted from the nursery 2 years after planting.

TISSUE-CULTURE-BASED PROPAGATION

In the future, tissue culture (micropropagation) methods will become increasingly common in propagating pears. These methods are currently employed mostly for research and for limited commercial propagation. As the methods are improved and scaled up, however, they will likely become widely used in commercial production.

The principal advantage of tissue culture over conventional propagation is its ability to rapidly produce large numbers of shoots from a very small amount of initial material. This is particularly valuable when developing new cultivars or rootstocks. For routine propagation, efficient application of these methods could reduce propagation costs, but the ability of micropropagation to compete economically with conventional propagation remains to be proven. A disadvantage of tissue culture is the high cost of the labor and equipment needed to care for material and maintain sterility.

In addition to rapid multiplication, tissue culture should offer several new options for growers. Continued improvements in the efficiency of rooting micropropagated shoots should facilitate the development of clonal rootstocks and self-rooted cultivars. The ability to provide disease-free stock material may prove beneficial in the future.

Micropropagation continuously forces the production and elongation of cultured shoot buds. Like conventional grafting, micropropagation multiplies shoot pieces from a single tree to make many genetically identical trees. Unlike grafting, micropropagation is conducted under aseptic (sterile) conditions in the laboratory, where material can be grown continuously. Even a very small amount of material can rapidly produce a large number of shoots through micropropagation.

A microculture begins with seedling tops, suckers, or young shoots from a mature tree that are cut into small segments, each containing an axillary bud. These segments, or explants, are then surface-sterilized, usually by a brief soak in a diluted bleach solution. The soaking removes surface bacteria and fungi that would kill the plants when cultured. Under sterile laboratory conditions, the stem pieces are then placed on a nutrient-enriched sugar-agar medium. Recent research suggests that combining the agar-solidified medium and a liquid medium into a double-phase culture produces a higher number of axillary buds and improves plantlet performance and survival. Plant hormones are incorporated into this medium. Exposing the buds to long-day-length light stimulates the axillary buds to develop and elongate, forming new shoots. By repeatedly cutting and reculturing emerging shoots, many genetically identical individual shoots can be rapidly produced from a single bud.

Micropropagated shoots can be rooted to produce trees for the field. For the production of clonal rootstocks or self-rooted trees, shoots are rooted by applying a high-concentration IBA treatment at the base. A period of a few days of darkness is also generally used, followed by exposure to short-day-length light.

In preparing micropropagated plants for the field, proper acclimatization is critical. Tissue-cultured shoots are grown in small containers at a relative humidity near 100 percent. The shoots do not regulate moisture loss well; therefore, they are extremely sensitive to changes in humidity. To avoid desiccation of new plants, the ambient humidity must be reduced very gradually, allowing time for new leaves to grow and for plant parts to develop a waxy cuticle. High humidity must be maintained until the plants are well established. Plants can be acclimated in the greenhouse by using a humidity chamber to slowly reduce the humidity and increase the amount of "dry" time.

ROOTSTOCK SELECTION

Rootstocks provide anchorage and absorb nutrients and water for the tree. The rootstock can also impart characteristics to the grafted cultivar (the scion), including better control of diseases, insects, and nematodes; tolerance to adverse soil and water conditions; size or vigor of the tree; precocity; and yield. Fruit size may also be indirectly controlled by the rootstock. Pear rootstocks can cause problems in the grafted cultivar, such as black end (hard end), cork spot, pear decline, and possibly the susceptibility to fire blight.

CRITERIA FOR SELECTION OF ROOTSTOCKS

The primary criteria for rootstock selection by nurseries are grower demand and ease of propagation. Nurseries raise trees to sell only if a high percentage of successfully propagations can be achieved each year. Rootstocks must be adaptable to many scion varieties and areas of production, and they must be compatible with budding or grafting.

For proper selection of a rootstock the characteristics of the orchard soil must be well understood. Also helpful is the history of the site and personal experiences in farming the site. The first step is to identify problems at the site and rate the seriousness of each problem. The most significant problems will narrow the choice of rootstocks.

The next consideration is the anticipated scion cultivar, tree spacing, and tree training system. Each of these decisions affect the choice of rootstock. For example, a vigorous cultivar grown in close spacing would benefit from a less-vigorous rootstock with good precocity; wide tree spacing benefits from a vigorous rootstock to fill the space quickly; a close spacing benefits from a less-vigorous rootstock with good precocity; and a low-production cultivar can benefit from a precocious rootstock. A further additional consideration is cultivar compatibility with the rootstock: for example, Bartlett is incompatible with quince rootstock.

Tree size can be modified through pruning. Extremely vigorous trees in close plantings can be pruned to the tight spacing, but the vigorous vegetative growth would cause excessive shading, leading to poor development of fruit buds and poor fruit set (Fig. 6.5). Similarly, a low-vigor tree planted in wide spacing will not maximize the production potential of the orchard (Fig. 6.6).

Once a decision is made on the desired rootstock and cultivar, contact the nursery as quickly as possible, because not all rootstock-cultivar combinations will be available on short notice.

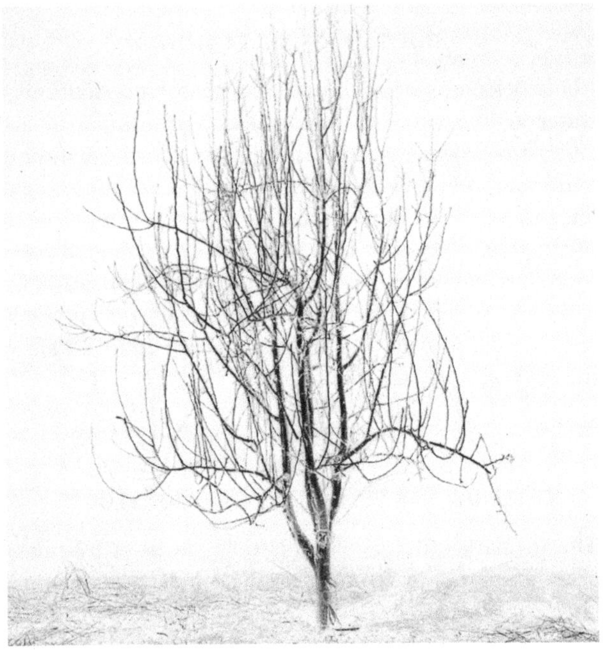

Figure 6.5. Wide-spaced pear trees generally require high vigor to fill the allotted space. *Photo:* T. Borland.

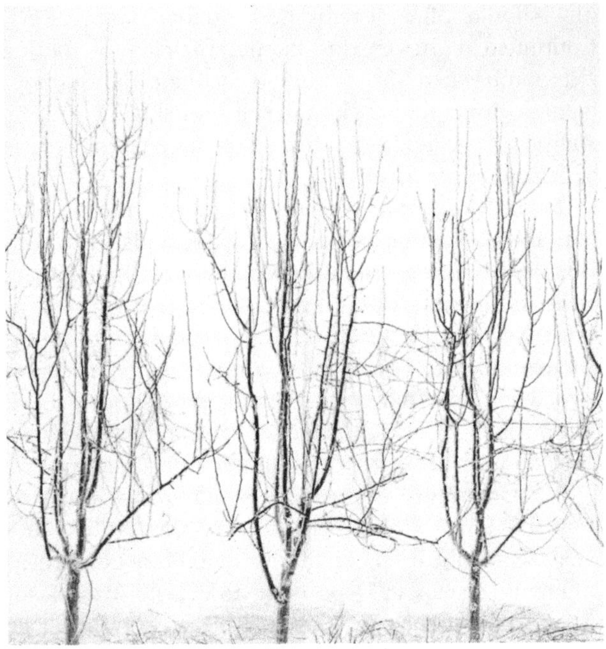

Figure 6.6. Lower-vigor pear trees can be planted closer together. They fill the allotted space quickly and can be reasonably short. *Photo:* T. Borland .

After planting, growers want trees that begin to bear fruit and attain full size as quickly as possible and then produce heavy crops without the tree getting larger. Given high land costs and interest rates, early production, or precocity, is extremely important.

ROOTSTOCKS OF THE PAST

Several pear rootstocks have been used in the past. The first pear trees grown in California were probably seedlings introduced from Europe. In time, many of these early seedlings were grafted to better varieties as these new varieties were introduced. The use of European-type rootstocks continued into this century: Old French, Domestic French, and Kirschensaller were used in the early 1900s, but the yield from them varied considerably. Many trees grown on these rootstocks produced many root suckers and churnbottom bases that were much larger than the graft. Sources of seed also varied, producing inconsistent nursery trees. Many trees in Europe used as sources of seed were destroyed during World War I.

Since then, local nurseries have established local seed sources. In the 1920s and 1930s many pear trees in California were propagated on *Pyrus pyrifolia (P. serotina)*, *P. ussuriensis*, and certain Asian rootstocks. These proved to be easy to propagate, fast-growing, and productive. When pear decline reached California in the 1960s, orchards of Bartlett pears grown on *P. pyrifolia* rootstock were devastated.

Old Home clonal rootstock was planted during the first half of the twentieth century. It has the advantage of being resistant to fire blight, and it was promoted and planted because of this resistance. In rootstock trials it has been slow in coming into production (low precocity) and has also given low yields. Currently it is sometimes used as an interstock between quince rootstocks and Bartlett scions.

ROOTSTOCKS FOR PEARS IN CALIFORNIA

Winter Nelis seedling is the most widely planted rootstock in California. Because the seed is collected from Winter Nelis trees grown in Bartlett orchards, the seed are identified as Winter Nelis × Bartlett. Trees grown on Winter Nelis rootstock are usually very uniform in size and productivity, though they are not as precocious as trees grown on some clonal rootstocks. The trees tolerate a wide range of soil conditions, although they do poorest in heavy clay soils (fig. 6.7); they are also susceptible to fire blight. Bartlett trees grafted on Winter Nelis rootstock withstand moderate or heavy psylla feeding without severe symptoms of pear decline; they are very susceptible to crown gall. The tree vigor is high. Winter Nelis is the standard against which the tree vigor of other rootstocks is measured.

Bartlett seedlings are widely used in the Pacific

Northwest for rootstocks, and some are also used in California. The seed usually comes from cull fruit from packing houses or seed removed in processing at canneries. Even though the fruit is Bartlett, the flowers must be cross-pollinated by a different cultivar within the orchard, such as Winter Nelis, Anjou, Comice, or Bosc. Because pollen from each of the pollinizer trees produce a different progeny, the performance of the seedlings can be variable. The seedlings from Bartlett × Winter Nelis (the reciprocal of Winter Nelis seedlings) should grow and produce the same tree as Winter Nelis seedlings. However, Bartlett seedlings are often quite variable in vigor and perform less consistently as rootstocks than do Winter Nelis seedlings. This suggests that the pollen for the Bartlett source came from cultivars other than Winter Nelis or from several different cultivars.

Pyrus betulaefolia seedlings are the most vigorous rootstocks currently used for pears. They form trees that are 20 to 25 percent more vigorous than trees grown on Winter Nelis, but the trees are generally slower coming into full bearing. Seedling uniformity is good. This rootstock performs well under adverse soil conditions and tolerates wet clay soils better than other rootstocks. Most seed sources currently used (Reimer strain) are fairly resistant to fire blight. Bartlett trees grown on *P. betulaefolia* rootstock are more resistant to pear psylla feeding and pear decline than Bartletts grown on any other current rootstock. A few black end pears have been seen when trees were grown under severe stress, but this has not been seen under normal growing conditions. *Pyrus betulaefolia* rootstock is recommended for Asian pear plantings, and, due to its vigor, it is good as a replant in old orchards growing on heavy clay soils. Vigorous cultivars that tend to set small crops, such as Comice, set even smaller crops when grown on *P. betulaefolia*. One drawback to *P. betulaefolia* is its susceptibility to iron chlorosis (though it is less susceptible than quince rootstocks), which can be a problem in heavy clay soils.

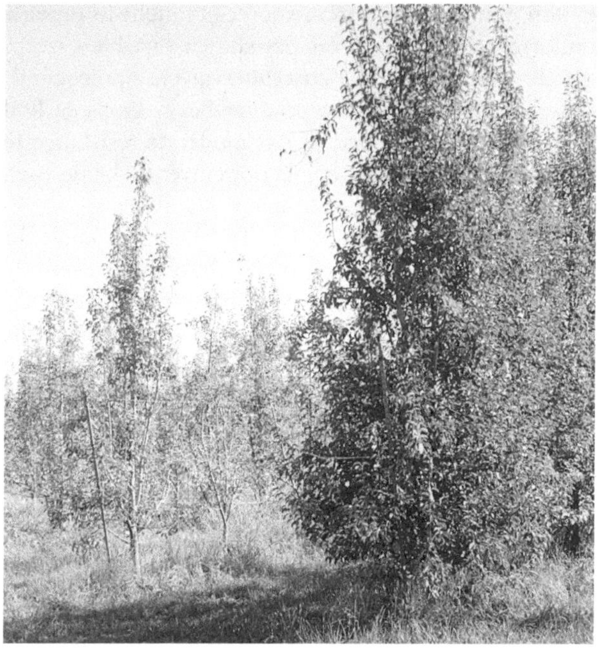

Figure 6.7. Seven-year-old Bartlett pear trees grown on two different rootstocks planted on clay soil. Small trees at left are on Winter Nelis rootstock, while vigorous trees at right are on *Pyrus betulaefolia* rootstock. *Photo:* W. Reil.

Pyrus calleryana seedlings are usually quite vigorous, forming trees 10 to 20 percent larger than Winter Nelis when not subjected to pear psylla pressure. Pear psylla feeding and pear decline can cause variable growth and performance, with some trees exhibiting severe shock while others exhibiting very mild symptoms. Some researchers have suggested that pure *P. calleryana* is resistant to pear decline, while hybrids exhibit pear decline susceptibility. This is probably part of the expla-

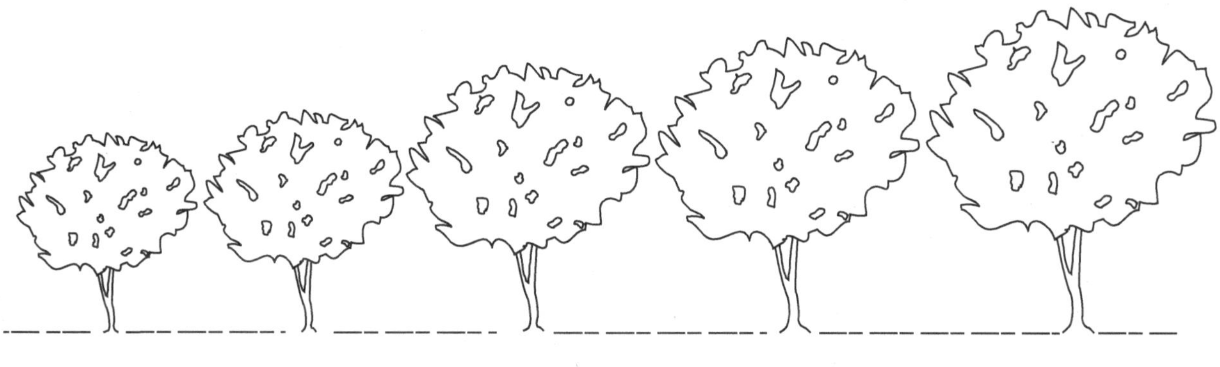

Figure 6.8. Comparative size of Bartlett pear trees grown on various rootstocks. *Source:* Adapted from Westwood 1993.

nation, but even seed from sources thought to provide uniform, true-to-type trees has shown variable susceptibility to pear decline. *Pyrus calleryana* is a precocious, fast-growing tree that can produce heavy crops early if pear psylla is controlled. It has moderate resistance to fire blight. *Pyrus calleryana* is not currently being used as a rootstock for new plantings.

Old Home × Farmingdale (OH×F) clonal rootstocks are crosses made in Oregon that have been carefully evaluated over the last 40 years. They provide a range of vigor from standard to about 60 to 70 percent of the vigor of Winter Nelis rootstock. They are resistant to fire blight and range from tolerant to resistant to pear psylla feeding and pear decline. The evaluation of these rootstocks has been time-consuming. As the parent trees matured, many of the selections taken from it became harder to root. Several clones that first showed promise in evaluation trials have been eliminated by nurseries because of poor rootability. Precocity and productivity have been comparable to or better than Winter Nelis rootstock in California trials. Recent evaluations of Bartlett grown on some Old Home × Farmingdale clones have shown that some of these rootstocks produce greater yields than Winter Nelis rootstocks in mature, vigorous blocks. If some of the Old Home × Farmingdale rootstocks used for size control continue to perform well, they might have considerable promise in higher-density, smaller-height orchards. Rootstock numbers 97, 217, and 267 have produced Bartlett trees that are about 80 to 90 percent as tall as Bartlett grown on Winter Nelis rootstock. Slightly less vigorous or more precocious rootstocks are numbers 40, 51, 69, 87, 220, 230, 333, 513, and 516. Clones 87, 69, and 40 are being extensively tested and appear to provide vigor control of approximately 60 to 70 percent of the vigor of Winter Nelis, with early, high yield.

Figure 6.8 shows comparative size differences that can be expected from Bartlett pear trees grown on some currently planted rootstocks. Comparative size differences can also be influenced by soil texture, nutrition, and management.

Bartlett clonal rootstock or Bartlett cuttings have been tried in several trials since 1968. This tree has no bud union. It has performed very well in all trials, being one of the most productive. It is a fast-growing, precocious tree. When it starts to bear heavy crops, growth slows, forming a tree about 80 percent of the size of Bartlett on Winter Nelis rootstock. It is susceptible to fire blight but tolerant to pear psylla feeding and pear decline. It does well on most soils except on heavy, wet clays. While this tree would be highly recommended, nurseries do not grow it because of problems with propagation. Even with techniques such as bottom heat to callus cuttings and using cuttings from juvenile trees, stands have sometimes been poor. Perhaps if Bartlett plantings increase, this rootstock could warrant development of micropropagation techniques for better rootability.

Quince clonal rootstocks have been used for many years. Bartlett and Bosc are not compatible when grafted directly on quince but require an interstock of Comice, Old Home, Hardy, or some other compatible interstem. Quince is susceptible to fire blight. Pear psylla feeding and pear decline have caused moderate to severe stunting of Bartlett grown on quince rootstock. The trees are also subjected to chlorosis on high-pH or calcareous soils. Quince rootstock forms a tree that is 50 to 60 percent the size of trees grown on Winter Nelis rootstock. Anchorage is fair. The trees are very precocious but need to be planted close together for maximum production. Fruit size is large. Three selections of quince are currently being propagated: Provence Quince BA 29, EM Quince A, and EM Quince C.

Provence Quince BA 29 is the most widely planted selection in California. It forms a tree about 60 percent the size of trees grown on Winter Nelis. Bartlett trees grown with an interstem on Provence Quince have not shown as much damage from pear decline as Bartlett trees grown with an interstem on EM Quince A or EM Quince C. Bartlett fruit grown on Provence Quince rootstock have been smoother in appearance and are more elongated in shape than when grown on other quince selections, although all fruit from quince-rooted Bartlett trees are shorter in length and wider than those rooted on Winter Nelis.

EM Quince A and C are the most widely planted pear rootstocks in Europe. EM Quince A forms trees about 50 percent the size of trees grown on Winter Nelis. Pear psylla feeding and pear decline have caused severe decline of vigor and reddening of foliage in these trees. EM Quince C has not been tested in trials in California, but it would be expected to perform about the same as EM Quince A.

Quince is usually the desired rootstock for Comice pears. Comice is directly compatible with quince and forms medium-sized trees. Comice is more precocious on quince than on other rootstocks and produces good to excellent crops with large fruit. Not enough trials have been conducted to compare Comice performance when grafted on the various quince rootstocks. It is imperative, however, that both the rootstock and Comice scion wood be virus-free. Viruses in Comice can delay fruiting and stunt tree growth.

Table 6.1 summarizes the principal attributes of selected pear rootstocks.

Table 6.1. Comparative attributes of selected pear rootstocks

	***Pyrus betulaefolia* seedling**	**Winter Nelis seedling**	**Bartlett seedling**	**OH×F 333**	**OH×F 97**	**OH×F 87**	**OH×F 69**	**OH×F 40**	**Quince BA 29**
Soil type adaptability	tolerates very heavy soils	tolerates heavy soils	tolerates heavy soils	tolerates heavy soils	tolerates heavy soils	tolerates heavy soils	tolerates heavy soils	tolerates heavy soils	tolerates very heavy soils
Chlorosis	moderately susceptible	tolerant	tolerant	tolerant	tolerant	tolerant	tolerant	tolerant	very susceptible
Wet soil	very tolerant	tolerant	tolerant	tolerant	tolerant	tolerant	tolerant	tolerant	very tolerant
Anchorage	excellent	excellent	excellent	good	good	good	good	good	fair
Vigor	very high	high	variable	high	high	moderate	moderate	moderate	low
Precocity*	moderately low	moderately low	moderately low	moderately low	moderate	moderate	moderate	moderate	high
Yield efficiency†	moderately low	moderate	moderate	moderate	moderate	moderately high	moderately high	moderately high	high
Cold hardiness	−15°F (−26.1°C)	−20°F (−28.9°C)	−20°F (−28.9°C)	−20°F (−28.9°C)	−20°F (−28.9°C)	−20°F (−28.9°C)	−20°F (−28.9°C)	−20°F (−28.9°C)	−5°F (−20.5°C)
Pear decline tolerance	tolerant	tolerant	tolerant	tolerant	tolerant	tolerant	tolerant	tolerant	moderately tolerant
Fire blight resistance‡	good	moderate	moderate	good	good	good	good	good	low
Crown gall resistance	good	low	moderate	good	good	good	good	good	good
Root lesion nematode resistance	high	low	low	low	low	low	low	low	high
Oak root fungus resistance	moderate	moderate	moderate	moderate	moderate	moderate	moderate	moderate	moderate
Compatibility	excellent with all cultivars	excellent with all cultivars	excellent with all cultivars	excellent with all cultivars	excellent with all cultivars	excellent with all cultivars	excellent with all cultivars	excellent with all cultivars	poor with most cultivars
Propagation	seed	seed	seed	cuttings	cuttings, tissue culture	cuttings, tissue culture	cuttings, tissue culture	cuttings, tissue culture	layering

Notes:

*Precocity can be varied with pruning and training techniques.

†Yield efficiency is a rating of total fruit by weight per unit of land area occupied by a tree.

‡Fire blight resistance refers to resistance by the rootstock only and does not refer to resistance imparted to the cultivar by the rootstock.

ROOTSTOCKS OF THE FUTURE

Pears are grown on a wide range of soil types. Generally the heavy clay soils require a rootstock that provides vigor and that may overcome adverse conditions such as poor drainage, high water table, calcareous soil, stratified soils, and very shallow soil depth. The current higher-vigor rootstocks such as Winter Nelis and *Pyrus betulaefolia* will probably continue to be adapted to these heavier soils.

On sandy loam, loam, or silt loam soils, most growers want a low-vigor, highly precocious rootstock. Lowering the tree height to about 6 to 10 feet (1.8 to 3 m) is a desirable goal to eliminate or reduce the height of ladders needed for harvest and pruning and to improve spray coverage. Rootstocks currently used do not have the desired characteristics, so growers continue to experiment with newer rootstocks. The U.S. National NC 140 research trials evaluate pear rootstocks. New plantings are currently being made to evaluate rootstocks originating in foreign countries as well as rootstocks developed in the United States. A few of these rootstocks may prove to be ideal but many will either be unsuccessful or will adapt to a very small niche location. It will take many years to prove rootstock adaptability to multiple conditions.

Many of the Old Home × Farmingdale clones have not been adequately tested. Old Home × Farmingdale numbers 333, 97, 87, 69, and 40 are commercially available, and plantings continue to be made using them, although 333 tends to produce small fruit. They are also included in the NC 140 trials.

Many new rootstocks are being introduced to the United States from breeding programs around the world. The majority of these programs had the specific objective of developing a rootstock that results in small trees with high yield efficiencies and large fruit size. These new rootstocks are being tested to learn their attributes for California growers. Generally, these new rootstocks are available only on a very limited basis through nurseries. The attributes of new selections mentioned in this chapter have not been confirmed by testing in the United States or in California. The rootstocks mentioned are some of the first to enter testing in the United States.

Additional rootstocks continue to be imported and planted in test plots. From all the new rootstocks, only a few will become useful to pear production in California. As testing progresses, information on new rootstocks will be available through University of California Cooperative Extension and through nurseries.

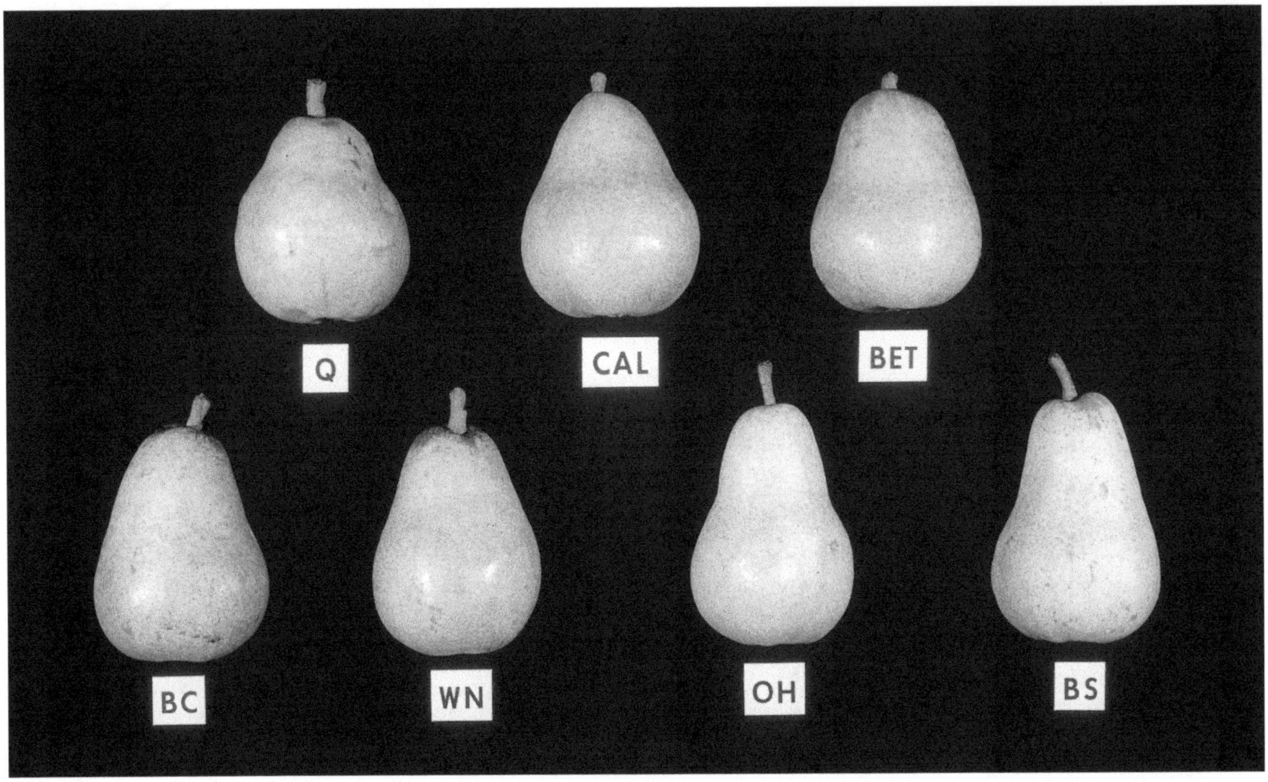

Figure 6.9. Bartlett pear fruit shape from one trial showing the shorter height in relation to diameter caused by fruit grown on Quince (Q) or *Pyrus calleryana* (CAL) compared to fruit grown on Winter Nelis (WN), Old Home (OH), or Bartlett seedling (BS). *Pyrus betulaefolia* (BET) and Bartlett cuttings (BC) are intermediate. *Photo:* J. K. Clark.

Horner Series are clonally propagated rootstocks developed in the Hood River region of Oregon. Some selections have shown size reduction, precocity, and good yields. The original crosses were Old Home × Farmingdale. Other attributes expected to be similar to Winter Nelis.

Pyrodwarf is a clonally propagated rootstock developed at Geisenheim, Germany. The tree reportedly will be 65 percent the size of a tree on Winter Nelis rootstock. There are no known incompatibilities. Low-vigor cultivars such as Concorde may be too weak on this rootstock. Precocity is rated higher than Quince rootstocks. Other attributes are expected to be similar to Winter Nelis.

Pyro 2-33 is another clonally propagated rootstock developed at Geisenheim, Germany. Cultivars grown on it should be standard-sized trees, and there are no known incompatibilities. Precocity is rated similar to Quince rootstocks. Other attributes are expected to be similar to Winter Nelis.

708-36 is a clonally propagated rootstock developed at Horticultural Research International, East Malling, England. Tree size may be 75 percent of trees grown on Winter Nelis with yield efficiency slightly higher than with Quince rootstock. Compatibilities and other attributes are expected to be similar to Winter Nelis.

QR 193-16 is a Quince rootstock developed at Horticultural Research International, East Malling, England. Tree size is similar to trees grown on other Quince rootstocks. Reportedly, the fruit are larger than with other quince rootstocks. Compatibilities and other attributes are expected to be similar to other quince rootstocks.

Fox Series are clonally propagated rootstocks originating at the University of Bologna, Italy. They were developed to be a size-reducing rootstock that is tolerant of lime soils. The size-reduction attributes have been questionable by Italian standards.

Pyrus heterofolia is a clonally propagated rootstock from INRA, Angers, France. It appears to be size-reducing with good yields. Little is known about this selection.

OH-11 (Pyriam) is a clonally propagated rootstock from INRA, Angers, France. It may reduce tree size to slightly larger than trees grown on quince, with good productivity. It supposedly performs well in sandy, droughty soils. Other attributes are expected to be similar to Winter Nelis.

Brossier Series are clonally propagated rootstocks from INRA, Angers, France. They may reduce tree size to smaller than trees grown on quince rootstocks, with good yields. These rootstocks have been difficult to propagate. Other attributes are unknown.

EFFECT OF ROOTSTOCK ON FRUIT SHAPE

In California trials, Bartlett fruit grown on Winter Nelis, Bartlett Seedlings, Old Home × Farmingdale, and Old Home rootstocks were the most elongated. Ratios of about 1.27:1 occurred when comparing the length from the stem to blossom end to the fruit diameter (fig. 6.9). Self-rooted Bartlett (Bartlett cuttings) and *Pyrus betulaefolia* rootstocks usually caused the fruit to be slightly shorter, with ratios about 1.23:1. Fruit from trees rooted on Provence quince usually had a ratio of about 1.2:1. The shortest Bartlett fruit came from trees on EM Quince A and *P. calleryana* rootstocks with ratios of about 1.15:1. Generally, the more elongated fruit is preferable.

EFFECT OF ROOTSTOCK ON YIELD

When trees are young, difference in yield is a function of both tree size and precocity. The fastest-growing trees often produce higher yields per tree because they have a larger bearing surface. In trials with trees planted the same distance apart, small trees can be less productive until they fill the planting space unless they are very precocious.

In rootstock trials evaluating mature Bartlett trees growing on the more vigorous orchard sites, the moderately vigorous rootstocks of Old Home × Farmingdale and self-rooted Bartlett yielded more than the very vigorous trees on *Pyrus betulaefolia* and *P. calleryana* rootstocks. Trees on Winter Nelis rootstock were intermediate in yield. Yields within these orchards appeared to be closely correlated to tree shade, with the less-vigorous trees having a less-dense canopy and producing larger yields. The very vigorous trees were the least productive, probably because the interior fruiting area of these trees was shaded out.

All the rootstocks that are currently being evaluated in mature orchard trials on vigorous orchard sites produce trees that fill the allotted space. The very vigorous trees produce long shoots in the top of the trees, creating a dense canopy with poor light penetration to the lower fruiting wood. These long shoots can be pruned off in the winter, but new vigorous shoots regrow the following year, continuing the heavy shade. The less-vigorous trees allow more light to filter through the canopy, promoting additional fruiting spurs.

Growers should carefully consider soil type and orchard growing conditions, and then select a rootstock, cultivar, and training system that fully utilize the orchard's potential. Vigorous rootstocks are needed on poor soils, in orchard replant sites, and for weak-growing cultivars. On excellent soils these vigorous rootstocks will give excessive growth even with weak cultivars, causing serious shading and low production unless properly managed. Less-vigorous rootstocks would improve light penetration and improve yield on these soils. The current trend of planting higher-density orchards and growing shorter trees will continue. Rootstocks that tend to reduce tree height and vigor and increase precocity will be increasingly planted on better soils.

REFERENCES

Batjer, L. P., H. A. Schomer, E. J. Newcomer, and D. L. Coyier. 1967. Commercial pear growing. Washington D.C.: United States Department of Agriculture, Agricultural Research Service, Agricultural Handbook No. 330.

Day, L. H. 1947. Apple, quince and pear rootstocks in California. Berkeley: University of California College of Agriculture, California Agricultural Experiment Station Bulletin 700.

Dolcet-Sanjuan, R., D. W. S. Mok, and M. C. Mok. 1990. Micropropagation of *Pyrus* and *Cydonia* and their responses to Fe-limiting conditions. Plant Cell, Tissue, and Organ Culture 21:191–199.

Griggs, W. H., J. A. Beutel, W. O. Reil, and B. T. Lwakiri. 1980. Combination of pear rootstocks recommended for new Bartlett plantings. California Agriculture 34(10): 21–24.

Hartmann, H. T., and J. A. Beutel. 1981. Propagation of temperate zone fruit plants. Oakland: University of California Division of Agriculture and Natural Resources Leaflet 21103.

Hartmann, H. T., D. E. Kester, and F. T. Davis Jr. 1990. Plant propagation and practices. 5th ed. Englewood Cliffs, NJ: Prentice-Hall.

Kappel, F., et al. 1993. Pear rootstock research in British Columbia. Acta Horticulturae 367:342–351.

LaRue, J. H., and R. S. Johnson, eds. 1989. Peaches, plums, and nectarines: Growing and handling for fresh market. Oakland: University of California Division of Agriculture and Natural Resources Publication 3331.

Lombard, P. B., and M. N. Westwood. 1987. Pear rootstocks. In R. C. Rom and R. F. Carlson, eds., Rootstocks for fruit crops. New York: Wiley Interscience. 145–183.

Lombard, P. B., M. N. Westwood, and R. L. Stebbins. 1984. Related genera and *Pyrus* species for pear rootstock to control size and yield. Acta Horticulturae 146:197–202.

Micke, W., ed. 1996. Almond production manual. Oakland: University of California Division of Agriculture and Natural Resources Publication 3364.

Mielke, E. A., and T. J. Facteau. 1993. Rootstock for pear production. Acta Horticulturae 475:135–141.

Ohlendorf, B. L. 1999. Integrated pest management for apples and pears. 2nd ed. Oakland: University of California Division of Agriculture and Natural Resources Publication 3340.

Ramos, D. E., ed. 1998. Walnut production manual. Oakland: University of California Division of Agriculture and Natural Resources Publication 3373.

Reil, W. O., and R. Elkins. 1996. Evaluation of established long-term rootstock trials for Bartlett pears. California Pear Advisory Board 1995 research report. Sacramento: CPAB.

Reil, W. O., and W. E. Howell. 1998. Yield, tree growth and fruit measurements of six selections of mature Bartlett pear trees growing on several rootstocks. Acta Horticulturae 475:157–168.

Reil, W. O., and E. Sutter. 1996. Propagation. In W. C. Micke, ed., Almond production manual. Oakland: University of California Division of Agriculture and Natural Resources Publication 3364. 64–69.

Reil, W. O., C. A. Leslie, H. I. Forde, and J. R. McKenna. 1998. Propagation. In D. E. Ramos, ed., Walnut production manual. Oakland: University of California Division of Agriculture and Natural Resources, Publication 3373. 71–83.

Viseur, J. 1987. Micropropagation of pear, *Pyrus communis* L., in a double-phase culture medium. Acta Horticulturae 212:117–124

Webster, A. D. 1998. A brief review of pear rootstock development. Acta Horticulturae 475:135–141.

Westwood, M. N. 1993. Rootstocks. In Temperate-zone pomology, physiology and culture. 3rd ed. Portland, OR: Timberline Press. 141–143.

7

Orchard Establishment

WILBUR O. REIL

Establishing a successful pear orchard requires many decisions. Generally speaking, the sooner the orchard begins bearing and the quicker it reaches full production, the more profitable the orchard will be. The high cost of land, planting, and tree care during establishment requires early returns to compensate for interest, return on investment, and establishment costs.

Growers must be aware of the many steps needed to establish a successful planting. These steps include site selection and preparation, irrigation system design and installation, rootstock and variety selection, tree ordering or propagating, site layout and planting, and proper care of the young trees.

If each step of orchard preparation is not done properly and in a timely fashion, planting will be delayed, which compromises tree size and growth the first year. This size decrease and poor growth delays the precocity of the orchard, ultimately costing the grower money.

ESTABLISHING THE ORCHARD

Site Selection

Pears are grown in only a few areas of California. While pear trees grow well throughout the state, profitable production has been limited to areas where a heavy, consistent fruit set occurs almost every year, fire blight can be managed economically, and trees can be protected from spring frost.

Pears require a certain amount of winter chilling to properly break dormancy. Approximately 1,000 hours below 45°F (7.2°C) is required to break the rest period. If there is much less chilling than this, the early spring bloom period can be long, and a great deal of late bloom can occur when fire blight pressure is high; fruit set is also usually less in years with extended bloom. Pears also require spring temperatures that are warm enough to prevent the bloom or small fruit from being killed or damaged by frost. While pears will tolerate colder temperatures in the spring better than many other fruits, temperatures of 25° to 28°F (–3.9° to –2.2°C) for any extended period can cause serious damage.

Bartlett and Bosc pears can set fruit either by pollination or parthenocarpically. In most California districts, Bartlett sets an excellent crop each year without pollination, and orchards are generally planted as solid blocks without pollenizer trees. In orchards planted with pollinizer trees, hives are placed in the orchard before bloom and kept in the orchard through bloom. For detailed information on parthenocarpy and pollination, see chapter 8, "Vegetative Growth and Fruit Development."

Although pears can tolerate a wide range of soil types, the highest yields usually occur on deep clay loam to sandy loam soils. Most pear rootstocks tolerate heavy clay soils, but production is usually much less than on better soil. Sandy soils require more frequent irrigation and fertilizer applications. Pears will grow on soils with high water tables, but they do best where the water table is a minimum of 5 feet (1.5 m) deep during the entire growing season. The trees can tolerate some flooding during the winter.

A soil pH of slightly to moderately acidic or neutral is ideal for pears. Alkaline soils, especially those with high lime or bicarbonate content, cause serious yellowing of trees and iron chlorosis, especially if certain rootstocks are used (see table 6.1 in chapter 6, "Propagation and Rootstock Selection").

Site Preparation

Any proposed site for planting should be evaluated for soil uniformity and potential problem areas. The USDA soil survey for the county is an excellent reference. This evaluation generally provides some idea of potential problem areas, as well as identifying clay or hardpan conditions that may exist. Use a backhoe to dig soil-sampling holes in different areas of the field to complement the initial soil survey. The holes reveal changes of soil texture as well as any layering, compaction, claypans, or hardpans.

Observing the soil profile allows you to develop a procedure to modify any structural differences in soil layers. Usually the ideal time to correct soil problems is the summer before planting. Changing soil structure, ripping, slip plowing, leveling, and drainage work must be done before planting and usually require dry soil conditions for best results.

If the orchard is going to be furrow- or flood-irrigated, the land must be level or have a uniform slope. Irrigation applications are much more uniform if the slope is uniform. Sprinkler- or micro-sprinkler-irrigated orchards generally require leveling only to remove low areas where runoff water might collect.

A ripper can be used in areas of hardpan. The ripper shank must be deep enough to go through the hardpan and shatter the rocklike material. The hardpan will not reform after shattering, so a single properly executed ripping should permanently improve the site.

Soil layering or claypans require more extensive soil modification. Generally, a slip plow that removes a band of soil 12 to 18 inches (30.5 to 45.5 cm) wide per pass and mixes this band with topsoil before it falls back into the created trench is required to achieve best results. The slip plow must penetrate through or below the compacted layer to accomplish the desired effect.

Root lesion, root knot, and dagger nematodes can be a problem on pear trees. If the site is old orchard ground, especially old pear orchard ground, soil and root samples should be analyzed for nematodes. If nematodes are present in the soil, fumigation before planting is recommended.

Weeds, especially perennial weeds such as johnsongrass (*Sorghum halepense*), bermudagrass (*Cynodon dactylon*), or field bindweed (*Convulvulus arvensis*), are much more easily controlled or eradicated before planting trees. Herbicide applications and tillage can help control weeds (see chapter 13, "Orchard Floor Management").

Irrigation

Pear trees require an adequate supply of irrigation water during the spring and summer. Because pear fruit grow at a uniform rate throughout the growing season, moisture stress during any part of the growing season reduces the final size of the fruit (see chapter 14, "Water Relations"). Pears usually require 10 to 20 percent more water than the reference evapotranspiration (ET) for the area during the late spring and summer. If irrigation will be used for frost protection, this extra use must be considered during system design.

Site Layout

Pears have been planted with many different spacings between trees. Older orchards tend to have wider spacings, with trees planted 18 to 24 feet (5.5 to 7.3 m) apart. These spacings allowed orchardists to cultivate between trees as well as in the tree rows. At the wider spacings the trees must grow vigorously to eventually fill the space. Tighter or higher-density orchards generally have several advantages over the wider spacing. They start bearing economical crops sooner, produce more, and usually have smaller trees that are easier to manage and require less ladder work. Given high labor costs, less ladder work usually means less labor needed and substantial cost savings. The disadvantages of higher-density orchards are the increased costs of trees, planting, and orchard establishment. However, these costs are usually recaptured very early in the life of the orchard and can be made up for when the orchard begins to produce.

The density of a pear orchard should be governed by three main factors:

- The soil characteristics and the potential vigor the soil will impart to the trees. Weaker or shallower soils require higher-density plantings.
- The vigor of the rootstock and cultivar. Low-vigor rootstocks or cultivars need higher-density planting.
- The grower's management skills and techniques. Higher-density plantings require more intensive management.

The soil characteristics impact the growth and ultimate size of the tree. The depth of soil, due either to restrictions or water table, limits root development. While pears are quite tolerant of wet soils, roots will not grow in saturated soil. If the water table or root-restricting layer is closer to the surface than 5 to 6 feet (1.5 to 1.8 m), tree size will generally be suppressed. As the soil depth decreases, the size of trees decreases until trees may be too small for economical growth. Pear trees grow on a wide range of soil textures but do best on uniform-textured loams or sandy loams. On clay loam or clay soils, water management requires more skill to maintain adequate moisture without saturating the soil. Sandy soils usually require frequent irrigations to maintain the desired moisture.

Rootstock vigor can greatly impact the ultimate size of the tree (see chapter 6, "Propagation and Rootstock Selection"). The rootstock should be selected based on the soil, desired tree size, and the cultivar being grown. Cultivar vigor must be considered along with rootstock vigor. A weak-growing cultivar on a low-vigor site must be managed much more carefully than a vigorous cultivar on the best soils.

Water restriction, reduced nitrogen fertilizer, cover crops, and summer pruning can all limit the vigor and growth of trees. Light pruning, limb tying, and positioning shoots on wires can increase early fruit bud formation. Also, the amount of crop set can impact the growth and size of the tree.

Currently, most pear orchards are medium-density,

with row spacing approximately 14 to 15 feet (4.3 to 4.6 m). It becomes increasingly more difficult to manage cultural operations such as cultivating, spraying, and brush removal at row spacings tighter than 14 feet (4.3 m).

The number of trees in the tree row depends on the three factors listed above, as well as grower economics and a desire for early return. Trees on the most vigorous sites and conditions may be planted 15 feet (4.6 m) apart, and trees on weaker sites may be planted as close as 6 feet (1.8 m) apart. With trellis systems, trees usually tend to be spaced 6 to 10 feet (1.8 to 3 m) apart. Most newer orchards are planted in a hedgerow configuration, with trees directly opposite each other. Since the rows are separated from each other by an open space without tree limbs, there is no need to offset trees in one row in relation to trees in adjoining rows. Longer rows should be periodically broken by cross-drives to allow for turning and ladder movement.

ORCHARD PLANTING

The orchard planting site must be properly prepared before planting. If the ground is too wet on the day appointed for planting, workers should wait for suitable conditions. Trees should be planted as early in the growing season as possible because the earlier they are planted, the larger they will be by the end of the season.

In early spring, if weather is cool and moist, trees taken from cold storage or delivered directly from the nursery can be kept as bare root plants for several days. To prevent overheating and desiccation, trees should be kept out of direct sunlight. Roots must also be kept from freezing. A temperature just slightly below 32°F (0°C) can kill rootlets.

In late spring or early summer, the planting crew should remove from cold storage only those trees that can be planted that day. Care must be taken to prevent the tree roots from drying. Keep the tree roots out of direct sunlight and prevent the wind or air from blowing on roots when transporting the trees to the orchard site. Keep the roots covered with moist sacks and sprinkle them frequently, as needed.

If the trees must be heeled in or stored at the planting site, they can be temporarily planted in a trench or other area in the field. Make sure the area has received the same care and fumigation as the site being planted so that the trees are not contaminated by this temporary storage.

The field should be laid out and staked before planting time. Remove or kill all vegetation and weeds before planting the trees. Remove excessive vegetation or organic matter from the planting hole, as it may produce toxic gases and can also deplete the amount of nitrogen available to the tree. On heavy soils, planting on berms allows for better drainage and increases the temperature and air volume around the root system.

Holes for planting trees can either be dug by a shovel or augered. Do not allow the sides of the hole to become glazed. In wet ground, the hole can be dug with a shovel without glazing several days sooner than can be done by a mechanical auger. Excessively long, broken, or crossed roots should be trimmed at planting. Root tips that have callused should not be cut. In sites being replanted or in soils suspected of having crown gall bacteria, spray the roots with *Agrobacterium radiobacter* before planting.

Plant the tree so that after settling the tree will be no deeper than it was in the nursery. Anticipate that the tree will settle as much as 3 inches (7.5 cm). Do not allow dirt to cover the graft union. Once the tree has been planted, do not pull it up at a later date, as this will damage the feeder roots.

Place the roots in the hole by hand so that the roots are straight and pointed downward. Make sure the roots are not curved around the hole, and make sure the root ends are not curved upward. Backfill dirt into the hole, placing friable loose dirt around the roots. Press the soil firmly around the roots, making sure that no air pockets exist around the crown. Fertilizer or other soil amendments should not be added to the planting hole in most situations.

If trees are planted in the winter with ideal moisture, no additional watering is necessary until after growth starts in the spring. If the orchard is planted late or if the soil is dry or cloddy, the trees should be watered enough at planting to settle and moisten the soil, but not enough to saturate the planting holes. No additional watering is necessary until after the trees start growing.

Trees should be pruned or headed at planting time to a height of about 24 to 28 inches (0.6 to 0.7 m). All side branches below that height should either be removed or stubbed to approximately 2 inches (5 cm) long if they are located in desirable scaffold positions. On 2-year-old trees, leave at least three stubbed branches because there are not enough buds on the trunk for adequate scaffold selection and positioning.

Paint the trunk with white latex paint diluted 50 percent with water to help prevent sunburn. Be sure to paint to the soil line or just below to prevent unexposed bark. Ideally, the painting should be done before the last two or three shovels of dirt are placed around the tree so that the tree remains painted even after the soil has settled. A tree protector can also be used to prevent rodent damage and herbicide contact with the trunk.

Immediately after planting, a safe preemergence herbicide can be strip-sprayed on soil in the tree rows to avoid the need for a contact herbicide or hand hoeing (see chapter 13, "Orchard Floor Management").

Caring for Young Trees

After the initial irrigation to settle young trees at planting, trees should not need to be irrigated again for some time. If trees are planted in late winter or early spring, rainfall should be sufficient to supply their needs until continued warm, dry weather prevails in late spring. By then, trees have enough foliage to increase their water usage and to require frequent irrigations. Climate also dictates amount and frequency. The amount of irrigation should increase as summer continues.

Irrigating

Sandy soils require more frequent irrigations than do loam or clay soils. As the tree grows and more water is transpired on hot summer days, irrigation frequency and volume should be increased. Furrows should be placed far enough away from trees so that the crown is not wet frequently or for long periods of time, but furrows should also be placed close enough to the trees so that water will easily penetrate to the root area. Frequent checking with a shovel or soil tube will indicate whether the tree has had the proper amount of water.

Drip emitters, foggers, or micro-sprinklers, if used, should be placed near enough to trees to allow irrigation water to penetrate to the root system. Care should be taken not to allow emitters to puddle water at the tree base. For trees that are 2 or 3 years old, drip emitters and micro-sprinklers should be moved so that the tree crown remains dry during irrigation. When it becomes necessary to irrigate more often in hot weather, frequent light irrigations are most efficient in supplying water needs of young trees. Roots of young trees are confined to a small area near the base of the tree. A young tree may need less than 1 gallon (3.8 l) of water per day in spring, but up to 5 gallons (19 l) of water per day as growth increases and during hot weather. For more information on tree water needs and irrigation, see chapter 14, "Water Relations."

Fertilizing

Young trees usually have enough nutrients stored in their roots when planted in the orchard site to grow early foliage. It is seldom necessary to apply fertilizer (especially nitrogen) to newly planted trees until after growth has begun in spring or early summer. Sufficient nitrogen should then be made available to allow the tree to grow as rapidly as possible in its first growing season. Some soils may have enough residual nitrogen remaining from a previous crop to supply the tree for the entire growing season. Sandy soils may have very little available nitrogen, and young trees planted in them will become nitrogen deficient before the end of spring. Some irrigation water contains enough nitrogen to supply some or all of the nitrogen the tree needs.

Light, frequent applications of nitrogen will be less likely to burn the roots than heavier rates applied at a single time. Two to four applications of ½ to 1 ounce (14 to 28 g) of actual nitrogen per tree during the growing season will stimulate good growth. For more information, see chapter 18, "Nutrition."

Pest Control

Weeds compete for moisture and nutrients and can greatly reduce tree growth. Weeds can be eliminated by careful cultivation, hoeing, or application of herbicides. Preemergence herbicides can eliminate most, if not all, weeds in spring and early summer. Other controls may be needed as the season progresses after irrigations leach preemergence herbicides out of the soil. Special care should be taken to control perennial weeds, which are difficult to eliminate once they become well established. Care should also be taken when applying postemergent herbicides to avoid spraying the trunk or any green growth on young trees to prevent phytotoxicity. Follow all label instructions when applying herbicides. For more information, see chapter 13, "Orchard Floor Management."

Vertebrate pests such as pocket gophers, meadow mice, deer, rabbits, and ground squirrels can damage young trees. Pocket gophers and meadow mice can be especially troublesome, girdling the young tree at or below the crown. Good weed control around the trees can reduce the potential for damage by removing cover for vertebrate pests. Trapping or using poison baits can control pocket gophers, meadow mice, and ground squirrels, while fencing and tree protectors are used for deer and rabbits. If baits are used, follow all label instructions during their use.

Shaping Young Trees

Light summer pruning sometimes helps shape trees during the first year's growth. When the trees are small, pinch back vigorously growing shoots positioned too low on the trunk to later form into main side branches. Pinch back or spread upright-growing limbs when they are about 8 to 12 inches (20.5 to 30.5 cm) long. This helps promote side branching and tree spreading early in the life of the tree. If the pears are grown on a wire trellis do not bend or tie scaffolds at an angle more than 30° to 45° from the vertical or growth will be retarded (see chapter 10, "Training Young Trees").

In most cases it is probably best not to intercrop young trees because of the risk of reduced tree growth due to competition from the other crop. It is also difficult to manage the irrigation, spraying, cultivation, and harvest of separate crops.

REFERENCES

Beutel, J., and R. Elkins, eds. 1990. Pear growing in the 1990s. Proceedings of a short course held November 27–29, 1990, Kelseyville, CA. Davis: University of California, Davis, Department of Pomology.

Hartmann, H. T., D. E. Kester, and F. T. Davis Jr. 1990. Plant propagation: Principles and practices. 5th ed. Englewood Cliffs, NJ: Prentice Hall.

LaRue, J. H. 1989a. Establishing the orchard. In J. H. LaRue and R. S. Johnson, eds., Peaches, plums, and nectarines: Growing and handling for fresh market. Oakland: University of California Division of Agriculture and Natural Resources Publication 3331. 21–30.

———. 1989b. Propagation and nursery practices. In J. H. LaRue and R. S. Johnson, eds., Peaches, plums, and nectarines: Growing and handling for fresh market. Oakland: University of California Division of Agriculture and Natural Resources Publication 3331. 12–15.

Ohlendorf, B. L. 1999. Integrated pest management for apples and pears. 2nd ed. Oakland: University of California Division of Agriculture and Natural Resources Publication 3340.

Reil, W., and E. Sutter. 1996. Propagation. In W. C. Micke, ed., Almond production manual. Oakland: University of California Division of Agriculture and Natural Resources Publication 3364. 64–69.

Willett, M., and R. L. Stebbins, eds. 1986. Pear production in the Pacific Northwest. Proceedings of the 1986 Pacific Northwest tree fruit short course. Pullman: Washington State University.

8

Vegetative Growth and Fruit Development

RACHEL B. ELKINS, BAS VAN DEN ENDE, AND JAMES BEUTEL

VEGETATIVE GROWTH PATTERN

Most European pear varieties grown in California normally begin to flower prior to leafing out. If leafing out occurs prior to bloom in years when winter chilling has been low (less than 800 hours), it may lead to poor flowering and fruit set. Trees that have suffered water stress or other types of stress may especially suffer during years when leafing out precedes bloom. There must be adequate stored nitrogen and carbohydrates in the spurs and other storage organs to ensure strong flowering.

Roots function at a very low rate when the scion is dormant, but as dormancy is broken root activity increases. Spring root activity begins prior to flowering and initial vegetative growth. This flush of new root growth in the early spring occurs before competition for nutrients by vegetative and fruit growth inhibits root development. Conditions that debilitate early-season root development, particularly saturated soil conditions and related soilborne pathogens, should be avoided if possible.

Vegetative growth of the scion is the strongest sink for carbohydrates early in the season. Rapid shoot growth continues through early summer given ample supplies of water. After this "grand period of growth," carbohydrates are used by rapidly growing fruit and to replenish storage organs (woody parts and roots) as vegetative growth slows down. The end of rapid shoot extension occurs in mid to late summer and is signaled by terminal bud set. It is generally accepted that about 12 to 18 inches (30.5 to 45.5 cm) of shoot extension during the season is adequate to provide enough carbohydrates for vegetative, fruit, and root growth, as well as storage for the following season. It is common for pear shoot length in California, however, to reach 24 to 60 inches (0.6 to 1.5 m) in vigorous trees, which causes shading. Eliminating spring nitrogen applications and pruning for a balanced crop should be used in the case of excessive shoot growth.

FRUIT BUD INITIATION

The developmental pattern of the pear fruit is basically the same worldwide, being only slightly modified by variety, rootstock, climate, and cultural practices. Pear fruit development begins well before fruit are harvested. Environmental conditions and cultural practices during the year before harvest can impact ultimate fruit size and yield.

As with most deciduous fruit trees, pear flower buds are initiated the season preceding bloom. Initiation about occurs 60 days after full bloom, or mid-May to early June in California. Flower buds of all cultivars form terminally on spurs, or short (less than 4 inches, or 10 cm, long) shoots, which develop on 2- and 3-year-old wood (fig. 8.1A). The majority of Bartlett flower buds, however, form at the terminal ends of

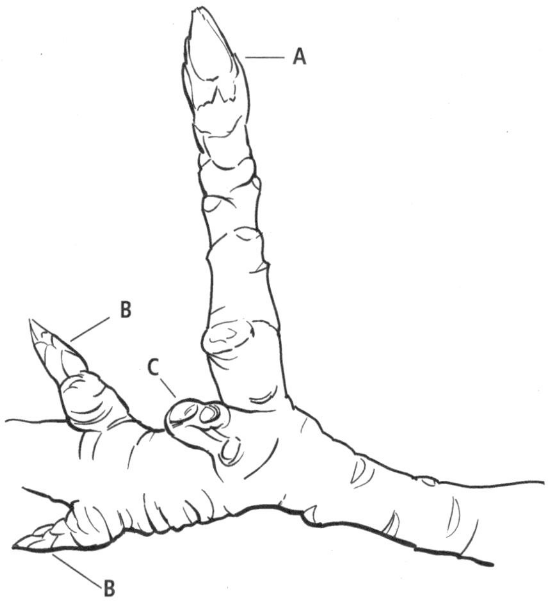

Figure 8.1A. Fruiting habit of pear trees. Note the terminal fruit bud at A and two lateral leaf buds at B. Fruit was produced at C during seasons preceding that in which buds A and B were formed. Growth of B buds will give rise to characteristic branching of older pear spurs. *Source:* Davis and Tufts 1941, p. 46.

1-year-old shoots (fig. 8.1B). Differentiation between vegetative and fruit buds occurs sometime in early summer, and fruit buds can be detected under a microscope by the last week of June or first week of July. They are visible in the orchard by August.

The main factor controlling whether a bud will have fruit or be vegetative is the exposure of fruiting wood to sunlight. Other factors may include previous crop load, nutritional or water stress, and insect or mite damage. Pear flower bud development is sensitive to light; once fruiting wood becomes shaded, it is difficult to regain productivity, especially in tall, vigorous trees. As pear trees grow and age, it is common to see nearly half the crop in the upper one-third of the tree, a sign of nonproductive lower fruiting wood. Fruit that is formed on shaded limbs is often small.

Because pears are so sensitive to shading, they must be managed to promote sunlight penetration into the lower canopy. This is achieved by

- limiting tree height to 80% of row width; this is especially important for closely spaced hedgerows and orchards where trees are planted in rows with east to west orientation
- pruning to keep fruiting limbs exposed to sun
- avoiding nitrogen rates exceeding 100 units per acre, especially in the spring
- avoiding excessive irrigation in the spring

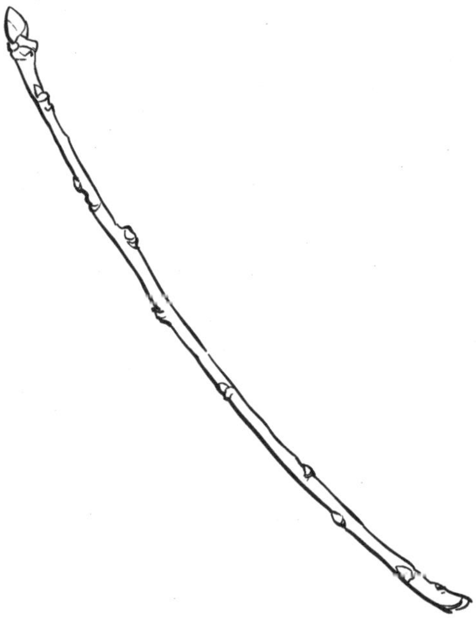

Figure 8.1B. Fruit bud on terminal of 1-year-old lateral shoot greater than 4 inches (10 cm) long. These shoots seldom spur up and should be replaced regularly after fruiting to allow new laterals to develop. *Source:* van den Ende 1996.

Fruit bud formation may also be diminished by very poor vigor due to excessive water stress, insect and mite damage, poor nutrition, and overcropping. These stresses can impede photosynthesis and subsequent carbohydrate manufacture.

Once fruit buds form, they are susceptible to damage by herbicides, excessive or late stop-drop (NAA) sprays, and caustic chemicals such as lime sulfur and spray adjuvants. Dormant and summer oil sprays have also been reported to impact bud survival and vigor, though often this is difficult to verify and more likely due to excessive rates, inappropriate timing, or impurities in the oil.

DORMANCY TO FLOWERING

Flower buds enter the resting stage sometime in fall prior to leaf fall. This rest period is probably triggered by short days and cool nights that stimulate physiological changes within the tree. At some point in the fall, winter dormancy sets in; this is the stage in which rest cannot be broken via "forcing" bloom by exposing buds to warm temperatures.

Winter dormancy is a critical phase for healthy flowering. Using one method of calculation, a minimum of 1,000 chilling hours at or below 45°F (7.2°C) is generally thought to be required for most European pear cultivars grown in the western United States, including Bartlett, to break dormancy and bloom normally, although normal crops have been observed with only 800 hours. This chilling requirement currently restricts European pear production to regions with relatively cold winters. Inadequate chilling is seldom a problem in the main Bartlett-producing regions of California. Winter fog compensates to some extent for below-normal chilling in the Sacramento and San Joaquin Valley growing areas, while winter temperatures in mountain districts regularly fall below 45°F. Consequences of less-than-adequate chilling include reduced vegetative and flower bud break, fewer viable flowers, a delayed and extended bloom period, and reduced fruit set.

Pear growers have learned through the years to time dormant oil pest control sprays to help promote uniform bloom and partially offset the effect of low-chill winters. Oil applied in early winter, after trees have received about 500 chilling hours (mid-January to early February), has been observed to promote uniform bloom. Applications in December or early January are used to promote early bloom. Applications after the first week of February can delay bloom 1 week or more. The latter tactic is employed where frost is a concern and other frost protection methods are inadequate; at this time, lighter oils are used to avoid damage to buds.

BLOOM AND INITIAL FRUIT SET

Pear bloom normally occurs from mid-March through mid-April in California. The pear flower has five sepals, five petals, and several stamens. The sepal, petal, and stamen bases are fused and define a nectar-producing area at the center of the flower. There are five inferior ovaries embedded in the hypanthium, or accessory tissue. The hypanthium, which is sometimes incorrectly referred to as the ovary, is the source of the edible tissue of the ripe pear. Extending from each ovary is a style that terminates in a stigma. Each of the five ovaries contains two ovules. The ovules are potential seeds; if pollination and fertilization occurs, they will develop into seeds (fig. 8.2). Bud stages from dormancy through fruit set are shown in figure 8.3.

Air temperature at bloom has a major impact on fruit set. The majority of Bartlett pears in California set fruit by vegetative parthenocarpy, that is, without pollination or fertilization. Bartlett pear is self-fertile in California, not because it pollinizes itself, but because fruit growth occurs without pollination.

Temperature during bloom is thought to be the primary stimulant of parthenocarpic fruit set. Parthenocarpic fruit set depends on temperatures reaching 60°F (15.6°C) for 72 hours or more for about a 10-day period during bloom, or more ideally, exceeding 60°F for 48 to 50 hours and 72°F (22.2°C) for 24 hours. It has been suggested that under the above conditions, gibberellins, plant hormones that are normally produced in seeds in most fruit, may also be produced in the fruitlet pericarp at high enough levels to induce and maintain fruit set. The pericarp is the fleshy tissue derived from the hypanthium that surrounds the seed of the pear fruit.

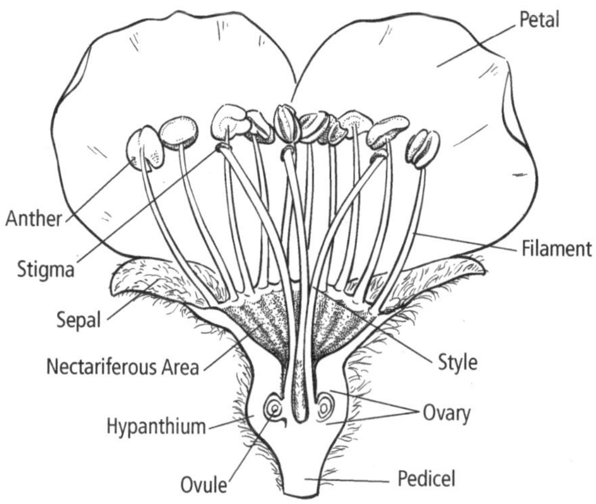

Figure 8.2. A Bartlett pear flower sectioned to show its component parts. *Source:* Redrawn from McGregor 1976.

Temperatures from 70° to 85°F (21.1 to 29.4°C) even for 1 day during flowering tend to result in heavy crops; years when temperatures are below those ideal for parthenocarpic fruit set generally result in lighter crops. Trees with parthenocarpic fruit set have less tendency toward alternate bearing than heavily cropped pollinized trees in most years. Fruit thinning is thus rarely required in California Bartlett orchards.

Although environmentally influenced, the tendency toward parthenocarpy is genetic. Among widely grown pear varieties, Bosc, Clapp's Favorite, and Bartlett have been reported to be variably parthenocarpic, while Passe Crassane is consistently parthenocarpic. Parthenocarpy in Bartlett is generally limited to California (although it is reportedly less pronounced in the upper Sacramento Valley) and southern Oregon; thus, this variety can be grown in solid blocks, unlike the northwestern and eastern United States, where it is interplanted with pollinizer varieties and bees are used to facilitate pollination.

In most years, yields from pollinized Bartlett orchards in California are no greater than yields from orchards depending on parthenocarpic fruit set. Bee pollination is sometimes used to enhance fruit set in years when conditions for parthenocarpic set are poor. If bees are used, about 2 hives per acre (5 per ha) should be placed in the orchard before flowering and kept through bloom. Bee activity in cold years can continue down to 55°F (12.8°C), 5°F (2.8°C) below the threshold for parthenocarpic set, with several hours above this temperature during bloom for adequate pollen transfer and fertilization of the fruit. Additionally, for pollinated fruit set, pears require a compatible pollinizer. This is because pears are self-sterile; one variety must have pollen from a different, compatible variety. Thus, if bees are to be effective, adequate provision for cross-pollination must be made, such as planting a pollinizer variety as every third tree in every third row. Solid rows of the pollinizer trees can be planted if the pollinizer has high value. All pear varieties seem to be cross-compatible with each other, providing there is good overlap in flowering. Bosc and Winter Nelis are commonly used for Bartlett. A few of the red sports, however, do not have pollen that will pollinate the green fruit pears of the same variety; for example, Max Red Bartlett will not pollinate Bartlett.

The effects of rain on crop set may be easily confused with the effects of temperature. Rain by itself is unlikely to reduce fruit set in parthenocarpic trees. However, in temperate California, rainy periods are usually accompanied by cold temperatures, and cold can inhibit fruit set. Rain does affect crop quality, especially by increasing fruit russetting.

Figure 8.3. Bloom and fruit set stages of pear. *Photos:* J. K. Clark.

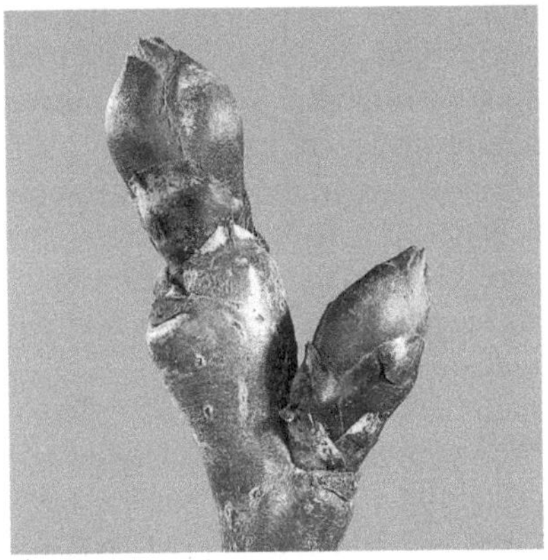
Dormant

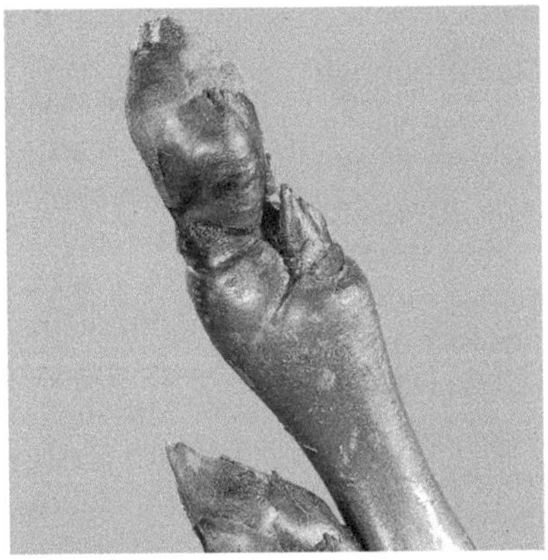
Delayed dormant

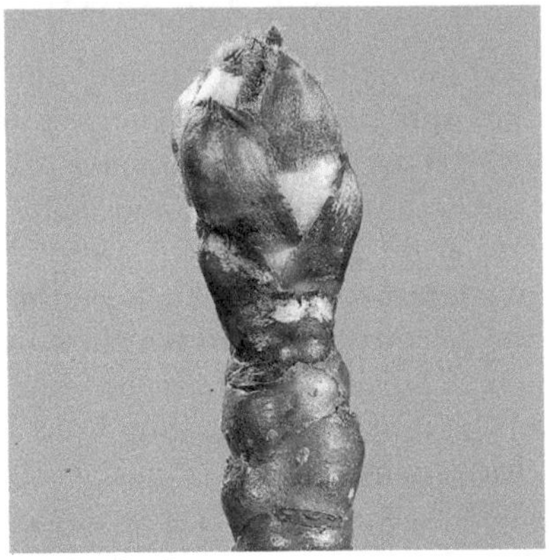
Swollen bud

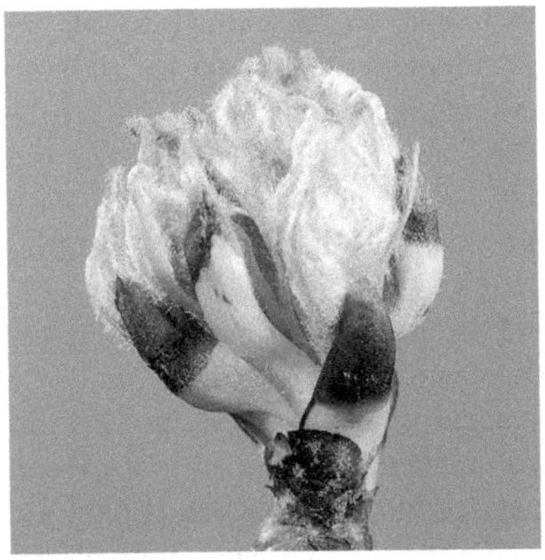
Cluster bud

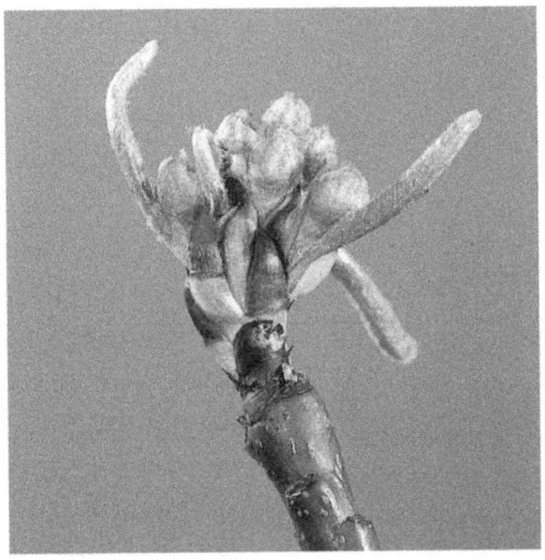
Tight cluster

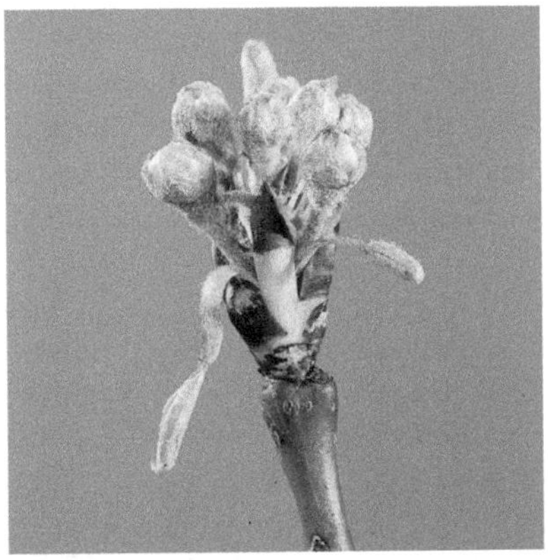
Finger bud

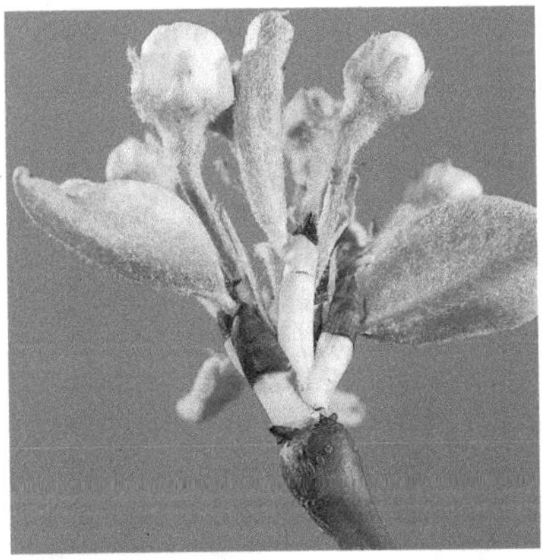
White bud

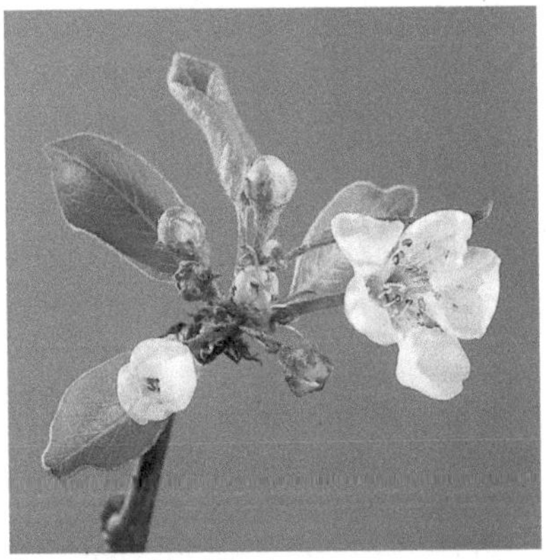
First open bloom

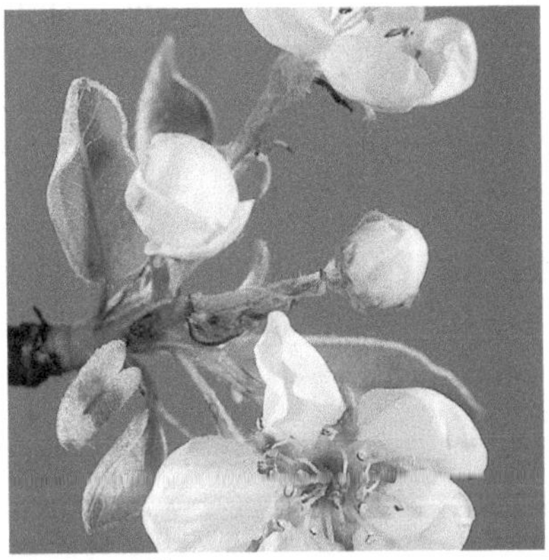
50% open flowers

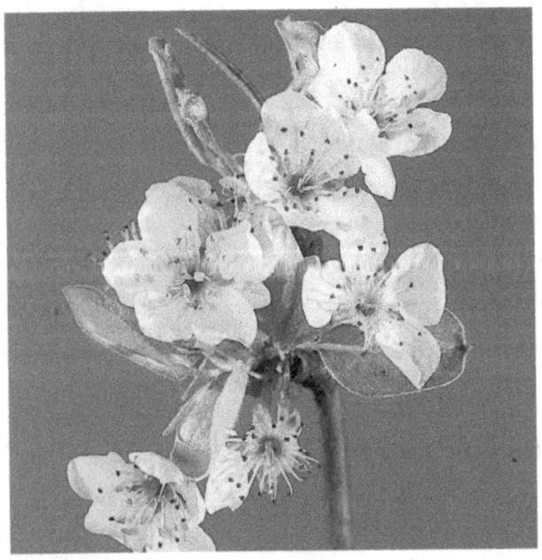
Full bloom

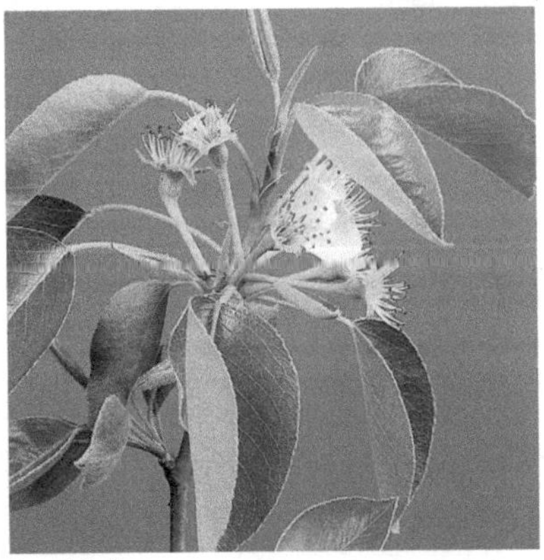
Petal full

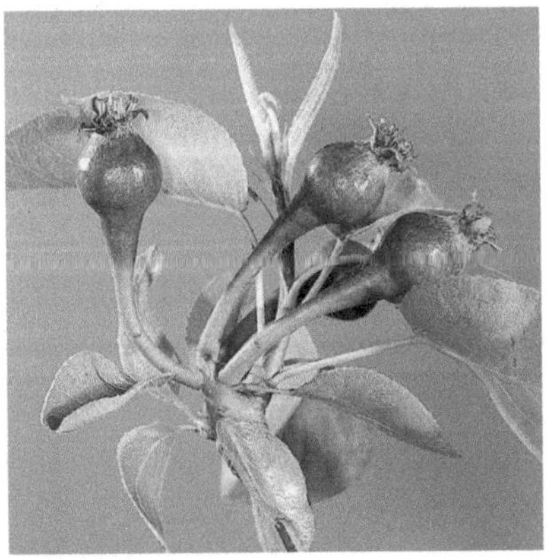
First jet

PATTERN OF FRUIT GROWTH

Cell division in the pericarp lasts for 7 to 9 weeks after full bloom. Once the period of cell division ceases, cell enlargement starts and continues up to the time of harvest. Ultimate fruit size is determined by the number of cells in the fruit flesh and the rate at which these cells enlarge.

The first flowers to set fruit tend to be larger than other flowers. All other conditions being ideal, larger flowers compete better for carbohydrates and nutrients and have more cells than smaller, later-setting flowers. Trees on the periphery of an orchard often set fruit earlier than trees inside the orchard because it is usually warmer on the outside edges during flowering. Clean orchard floor management may also hasten bloom and early set due to a warmer environment. One drawback to early set may be the increased likelihood of exposure to frost.

The length of time required for fruit to grow from bloom to harvest increases from south to north in California (table 8.1). A key predictor is spring temperatures from bloom through initial fruit set. The accumulated mean temperatures above 40°F (4.4°C) during the first 36 days after full bloom has been shown to be highly correlated with the number of days required to mature fruit. Warmer temperatures in the spring normally accelerate harvest, barring other weather effects on fruit development such as cool summer weather (fig. 8.4).

Cell enlargement continues through the season after the period of cell division. Harvest size and shape is predictable about 60 days after full bloom. Fruit diameter increases linearly 3 to 10 percent per week (about 1/8 inch, or 3.2 mm), doubling 2½ times from the time the fruit turns down from its own weight (full bloom plus 40 days) until harvest. Fruit weight and volume increase 20 to 25 percent per week, doubling every 20 to 30 days (figs. 8.5 and 8.6).

Harvesting the largest pears in the first pick (where multiple picking is employed) improves the size of later-harvested pears. Smaller fruit may gain 50 percent in weight during the 3 to 4 weeks from the start of first pick to the second pick.

While predictable, pear fruit volume is difficult to measure directly because fruit shape may vary

Table 8.1. Differences in length of pear fruit growth period by region

Region	Days from full bloom to harvest
Sacramento Valley, California *	114
North Coast, California †	121
Medford, Oregon‡	126

Notes:
*Average for 4 years of data, Sacramento and Solano Counties.
†Average for 33 years of data, Lake and Mendocino Counties.
‡Average for 12 years of data.

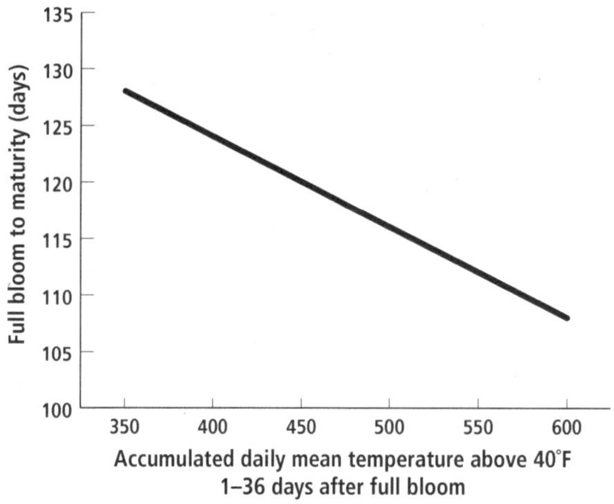

Figure 8.4. Regression of Bartlett maturity (days from full bloom to maturity) on the accumulated daily mean temperature above 40°F (4.4°C) during the 36 days after full bloom (r = –.882). *Source:* Lombard, Gordy, and Hanson 1971 (reprinted with permission).

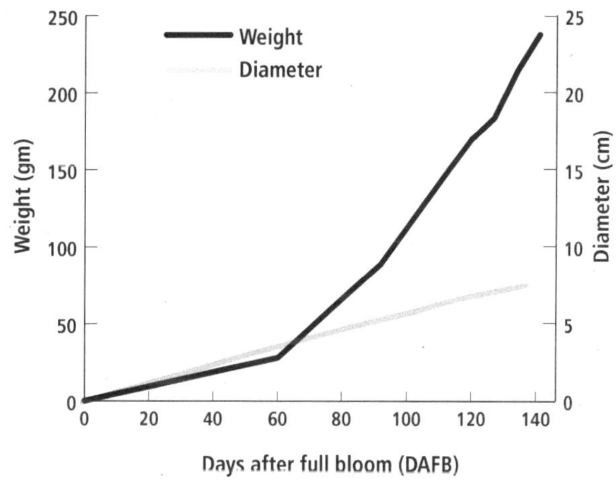

Figure 8.5. Pattern of change in fruit diameter versus fruit weight after full bloom.

Table 8.2. Standard growth curve for Bartlett pears

Days after full bloom	Fruit diameter (in)							
60	1.06	1.17	1.26	1.34	1.43	1.51	1.63	1.75
65	1.17	1.28	1.37	1.46	1.55	1.63	1.75	1.87
70	1.28	1.39	1.48	1.57	1.66	1.75	1.87	1.99
75	1.37	1.49	1.58	1.68	1.78	1.87	1.99	2.11
80	1.47	1.58	1.69	1.79	1.90	1.99	2.11	2.23
85	1.57	1.68	1.80	1.90	2.01	2.11	2.23	2.35
90	1.67	1.78	1.90	2.01	2.12	2.23	2.35	2.47
100	1.87	1.98	2.10	2.22	2.33	2.44	2.56	2.68
110	2.04	2.15	2.27	2.40	2.52	2.63	2.76	2.88
120	2.19	2.30	2.43	2.55	2.68	2.80	2.93	3.04
125	2.25	2.36	2.49	2.62	2.74	2.87	3.00	3.12

Source: Williams, Billingsley, and Batjer 1969, p. 596.

Note: To use this table to predict final harvest size, find the current number of days after full bloom in the column on the left, then find the present fruit diameter to the right in that same row. When you have found the size closest to the present fruit diameter, follow that column down to the expected harvest date (number of days after full bloom) to find the predicted harvest diameter.

Table 8.3. Effect of Bartlett pear size on yield

Fruit diameter (inches)	Fruit count per box	Yield: Percent*	Yield: Tons per acre	Yield: Boxes per acre†
1.95	193	100	7.2	317
2.10	180	111	8.0	352
2.20	165	122	8.8	387
2.30	150	135	9.7	427
2.40	135	149	10.7	471
2.50	120	169	12.2	537
2.60	110	187	13.5	594
2.70	100	206	14.8	651
2.80	90	233	16.8	739
2.95	80	264	19.0	836
3.10	70	301	21.7	955

Notes:
* Percent is based on a starting-point diameter of 1.95 inches.
†Assumes 45.5 boxes per ton based on the 44-pound standard Western carton.

Table 8.4. Estimating pear fruit volume

Pear diameter (d) (cm)	d^3 (cm)	Coefficient for calculating volume from diameter*	Estimated volume, $d^3 \times$ coefficient (cm^3)
1	1	0.830	0.83
2	8	0.743	5.94
3	27	0.696	18.8
4	64	0.664	42.5
5	125	0.641	80.1
6	216	0.622	134.3
7	343	0.607	208.1

Source: Adapted from Westwood 1993, p. 278.
Note: *Coefficients are calculated from the equation 0.83×2.839, which incorporates the volume of the neck and a constant specific to Bartlett.

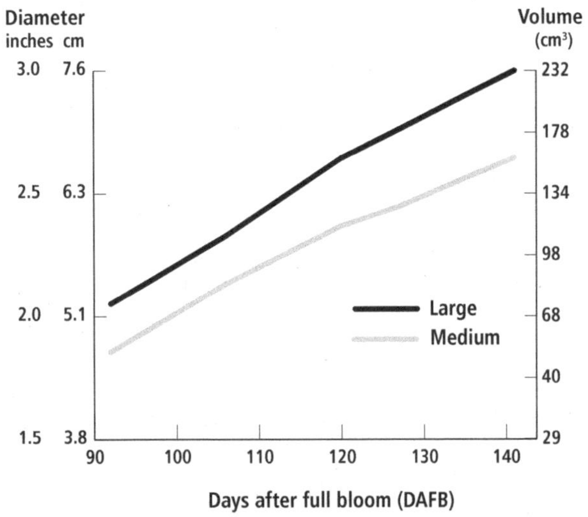

Figure 8.6. Relative gain in pear diameter and volume Lake County, California, 1969.

due to climatic differences among districts and years. Influences include geographical latitude, temperatures 1 to 6 weeks after bloom, and rootstock. The neck (narrow section of the fruit extending from the stem end) of parthenocarpic fruit generally is more elongated and "pearlike" than pollinized fruit, and necks tend to be longer on terminals than on lateral spurs (fig. 8.7). Two methods of determining volume are provided below. The fruit length (from stem to blossom end) to width (diameter) ratio may be 1:1 to 1:5 (fig. 8.8).

Despite variation, pear size and shape fit a fairly standard pattern in any particular orchard, which then may be used to predict final fruit size. This predictability is often used by packinghouse and cannery field personnel to schedule fruit thinning and harvest (tables 8.2 and 8.3).

Cell turgor drives fruit growth, and most growth occurs at night. Measuring daily fruit growth each morning between 6:00 A.M. and 8:00 A.M. gives the maximum fruit size each day.

Methods to Determine Pear Fruit Volume

One way to accurately determine pear fruit volume is to dip the fruit in a container of water and measure the amount of water displaced by the pear. However, fruit weight and volume can be calculated mathematically from the diameter, using the formula given below, since the specific gravity of pear is approximately 1 (i.e., 1 gram = 1 cubic centimeter):

$$\text{volume in grams} = cm^3 = (0.5236\, d^3 \text{ or } \tfrac{4}{3}\pi r^3) + 15 \text{ to } 20\ cm^3$$

where:

d = diameter of fruit in centimeters

15 to 20 cm^3 = volume of fruit neck

$\pi = 3.14$

r = radius (or ½ diameter) in cm

Fruit volume may also be calculated using published coefficients in the following formula, in which the coefficient is obtained from table 8.4.

$$\text{volume in grams} = (d^3 \times \text{coefficient})$$

Here is an example using the above formulas. Assume that the diameter of the fruit (d) = 6.0 cm (about 2⅜ inches).

$$\text{volume in grams} = [0.5236 \times (6.0 \times 6.0 \times 6.0)] + 15 \text{ to } 20$$
$$= 113 + 15 \text{ to } 20\ cm^3$$
$$= 128 \text{ to } 133$$

$$\text{volume in grams} = (6.0 \times 6.0 \times 6.0) \times 0.622 \text{ (from table 8.4)}$$
$$= 216 \times 0.622$$
$$= 134$$
$$= 216\ cm^3 \text{x}\ 0.622 = 134\ cm^3$$
$$= 134 \text{ grams}$$

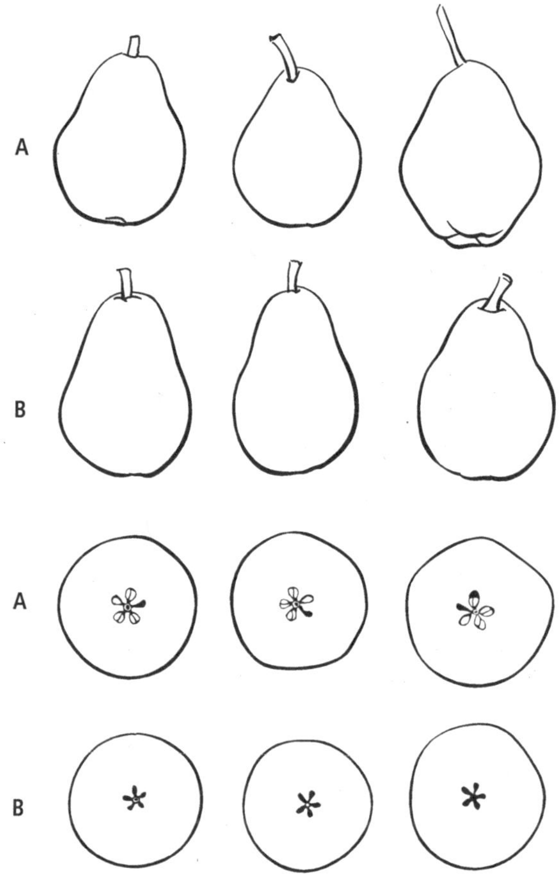

Figure 8.7. Factors affection fruit growth. Top section: A. Bartlett pears from naturally pollinated blossoms in an orchard interplanted with a pollinizing variety. B. Bartlett pears from naturally pollinated blossoms in an orchard planted solidly to Bartlett. Bottom section: Cross-sections of the fruit above showing seed content of cross-pollinated fruit and seedless fruit developed in solid blocks of Bartlett. Fruit was picked two days before first commercial picking. *Source:* Griggs and Iwakiri 1954.

FACTORS AFFECTING FRUIT GROWTH

It is thought that a minimum of 20 to 40 leaves per fruit is needed to produce good fruit growth and quality. Well-managed central leader and V-shaped canopies often size fruit better than large, multileader trees because they encourage better light distribution within the tree canopy. Only about one-third of the leaves on many large trees get adequate sunlight. Spurs on 2- to 3-year-old wood produce larger fruit than spurs on older wood. Yields are higher and fruit are larger on sun-exposed spurs than on spurs in the shaded part of the tree. Fruit on terminal shoots longer than 4 inches (10 cm) are generally smaller than fruit on shorter spurs.

A favorable orchard environment enhances fruit growth rate. Key factors are:

- uniform, well-drained soil
- favorable conditions for early fruit set
- adequate water status
- balanced nutrition
- good penetration of sunlight to all parts of the tree
- balanced crop load

Adverse conditions during the growing season that can reduce fruit growth include:

- temperatures over 85°F (29.4°C) or below 55°F (12.8°C)
- low humidity in the orchard and wind, which limit water movement to the fruit
- significant loss of leaf efficiency due to pests such as mites and psylla
- root damage due to wet or dry soils or soilborne pests and diseases
- heavy crop load

Severe tree water stress may reduce ultimate fruit size and total yield. It is especially important to maintain adequate water status in the 6 to 8 weeks prior to harvest, when fruit sizes rapidly.

Pear fruit size can be substantially reduced by potassium deficiency, and it is slightly reduced by low nitrogen and by deficiencies of the micronutrients zinc, iron, and boron. Nutritional imbalances can also delay maturity and reduce storage life (see chapter 18, "Nutrition").

The timing of harvest affects final fruit size. Harvesting too early reduces final tonnage because fruit gain weight rapidly during the final weeks before harvest (see fig. 8.5 and table 8.3). Delaying harvest too long may allow for inadequate time or labor to harvest the entire crop before fruit soften, resulting in reduced

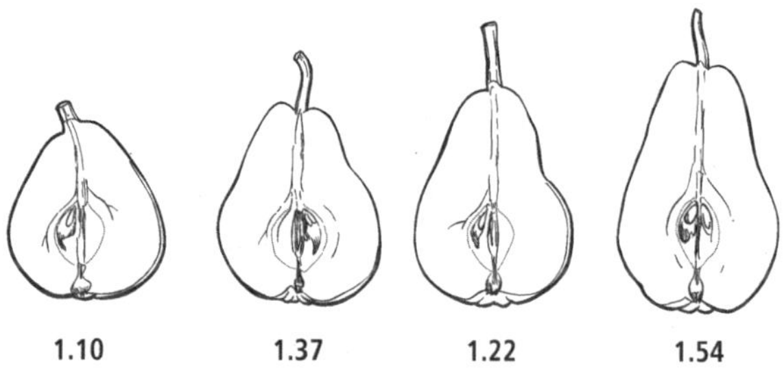

Figure 8.8. Longitudinal sections of Bartlett pears showing length-to-diameter ratios of fruit of various shapes. *Source:* After Davis and Tufts 1941.

fresh shipping quality or even unacceptable firmness (less than 15 pounds (6.8 kg) force; see chapter 22, "Postharvest Biology and Technology").

FINAL DEVELOPMENT STAGE

Cell walls break down gradually as the fruit reaches maturity, resulting in a lower cell wall pressure, or firmness. Cells in the pear stem near the spur also break down concurrently, creating an abscission zone. This causes the pear to eventually separate from the spur and fall to the ground. Fruit drop tends to be worse on older trees rather than young trees bearing their first crops. 1-naphtaleneacetic acid (NAA) is normally applied 7 to 10 days before harvest at 21 to 22 pounds (9.5 to 10.0 kg) firmness to keep the cells in the pear stem growing and maintaining their attachment to the spur. It may take 3 to 4 days before the hormone becomes effective if temperatures are unusually low (see chapter 12, "Plant Bioregulators").

A serious problem caused by low temperatures before harvest is premature ripening of the fruit. This can occur when daily maximum temperatures drop below 75°F (23.9°C) and night temperatures drop to 40° to 50°F (4.4° to 10.0°C). If cool temperatures persist for 3 to 4 days during the month before harvest, Bartlett fruit pressure can suddenly drop, causing them to become too soft for storage and shipping to the fresh market or cannery. Fruit color changes rapidly from green to yellow. Prior to overall color change, a more subtle sign of premature ripening is "pink end," a characteristic pink color at the blossom (calyx) end. Premature ripening occurs mainly in the mountain districts of California. It is a more serious and common problem in the high-elevation orchards of Hood River, Oregon. The only practical correction when these cold periods occur is rapid harvest and precooling.

REFERENCES

Beutel, J., and R. Elkins, eds. 1990. Pear growing in the 1990s. Proceedings of a short course held November 27–29, 1990, Kelseyville, CA. Davis: University of California, Davis, Department of Pomology.

Dadant and Sons. 1975. The hive and the honey bee. Hamilton, IL: Dadant and Sons.

Davis, L. D., and W. D. Tufts. 1941. Pear growing in California. Berkeley: California Agricultural Extension Service Circular 122.

Faust, M. 1989. Physiology of temperate zone fruit trees. New York: Wiley.

Griggs, W. H., and B. T. Iwakiri. 1954. Pollination and parthenocarpy in the production of Bartlett pears in California. Hilgardia 22(19): 643–678.

Lombard, P. B., C. B. Gordy, and B. Hansen. 1971. Relation of post-bloom temperatures to 'Bartlett' pear maturation. Journal of the American Society for Horticultural Science 96(6): 799–801.

McGregor, S. E. 1976. Insect pollination of cultivated crop plants. Washington, DC: USDA/ARS Handbook 496.

Meheriuk, M., et al. 1994. Postharvest disorders of apples and pears. Ottawa: Agriculture Canada Publication 1737/E.

Mellenthin, W. M., and R. L. Stebbins. 1980. Control of premature ripening of Bartlett pears. Corvallis: Oregon State University Bulletin F5241.

Ohlendorf, B. L. 1999. Integrated pest management for apples and pears. 2nd ed. Oakland: University of California Division of Agriculture and Natural Resources Publication 3340.

Ravetto, D. J. 1975. Premature pear ripening; an unsolved problem. Fruit Grower 95(6): 18B–30C.

Ryugo, K. 1988. Fruit culture: Its science and art. New York: Wiley.

Stebbins, R. L. 1992. Pruning mature pear trees in commercial orchards. Oregon, Washington, and Idaho Extension Services Publication PNW 405.

Stebbins, R. L., W. M. Mellenthin, and P. B. Lombard. 1981. Pollination and commercial varieties of pear in Oregon. Corvallis: Oregon State University Cooperative Extension Circular 986.

van den Ende, B. 1996. Fruiting habits of Bartlett and Packham. Northern Victoria Fruitgrowers Association Technical Bulletin 5:6.

Wang, C. Y., and W. M. Mellenthin. 1972. Induction period and threshold temperatures for premature ripening in 'Bartlett' pears. Journal of the American Society for Horticultural Science 97(4): 557–560.

Westwood, M. N. 1993. Temperate zone pomology. Physiology and culture. 3rd ed. Portland, OR: Timber Press.

Williams, M. W., H. D. Billingsley, and L. P. Batjer. 1969. Early season harvest size prediction of 'Bartlett' pears. Journal of the American Society for Horticultural Science 94(6): 596–598.

Zoller, B. N.d. Some characteristics of the growing season as they have affected fruit quantity and maturity, Lake County, 1974–1987. Unpublished manuscript.

9

Canopy and Light Management

THEODORE M. DEJONG

A pear tree can be viewed as a massive network of solar energy collectors. The individual solar cells are located in chloroplasts, microscopic structures within green cells of the leaves. Each leaf contains thousands of these solar cells, and the tree, in turn, has thousands of leaves. In this analogy the woody framework of the tree can be viewed as providing the structure by which the tree is capable of exposing maximum numbers of solar cells to the sun's rays. In addition to providing the structural framework for optimal light exposure, the wood and bark provide a vascular system for transporting water and nutrients to the leaves and chemical energy from the solar cells (chloroplasts) in the leaves to other parts of the plants.

The efficiency of the pear tree as a solar energy collector network depends on the capture of light energy, its conversion into chemical energy (photosynthesis), and the subsequent transport, storage, and use of that chemical energy. The pear orchard manager must maximize the efficiency of these processes to produce the pear crop. This requires a basic understanding of these processes and the factors that influence them.

PHOTOSYNTHESIS

Photosynthesis is the process by which energy from the sun is trapped in green pigments (chlorophyll molecules), converted into chemical energy, and used to convert carbon dioxide and water into a simple carbohydrate that eventually becomes sugar. Oxygen is also given off in the process (fig. 9.1). The reverse process, whereby all plants and animals recover energy from these simple carbohydrates, is called respiration. It is important to note that all the energy and structural materials used for tree growth, maintenance, and fruit production are derived from the process of photosynthesis.

Because photosynthesis uses light energy, the amount of photosynthesis that can occur in a leaf depends on the amount of light that is intercepted by the leaf. However, in most tree crops an individual leaf exposed to full sunlight can use only approximately one-third of that amount of sunlight for photosynthesis. Therefore, the photosynthetic apparatus of most leaves is light-saturated at approximately one-third full sunlight (i.e., the slope of the line becomes level at about 30 percent full sunlight; see fig. 9.2). Only leaves on the outer surface of the tree canopy will be exposed to full sunlight, and they are exposed only for a part of the day as the sun moves across the sky. Thus, most leaves on a mature tree function most of the day at the lower portion of the light response curve. Each leaf located in different zones of the tree canopy has its own ever-changing light environment. Light becomes more limiting to photosynthesis from the outer edge to the center of the foliar canopy. If tree canopies are allowed to become too dense, the inner parts become so shaded that leaves can no longer carry on an appreciable amount of photosynthesis. Those interior leaves will contribute insufficient energy to the adjacent spurs, and eventually the spurs will die.

LIGHT AND RESOURCE PARTITIONING

Because leaf photosynthesis depends on light exposure and because tissue growth and maintenance depends on photosynthesis, light has a large influence on how

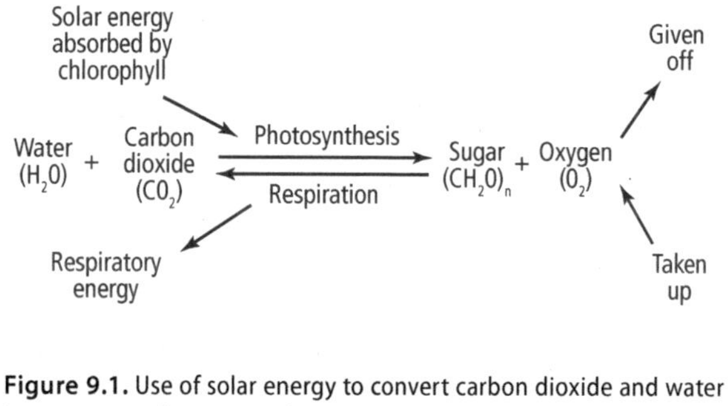

Figure 9.1. Use of solar energy to convert carbon dioxide and water into energy-rich sugars and oxygen is called photosynthesis. In the reverse reaction, respiration, sugar is used by living cells to release chemical energy, yielding carbon dioxide and water.

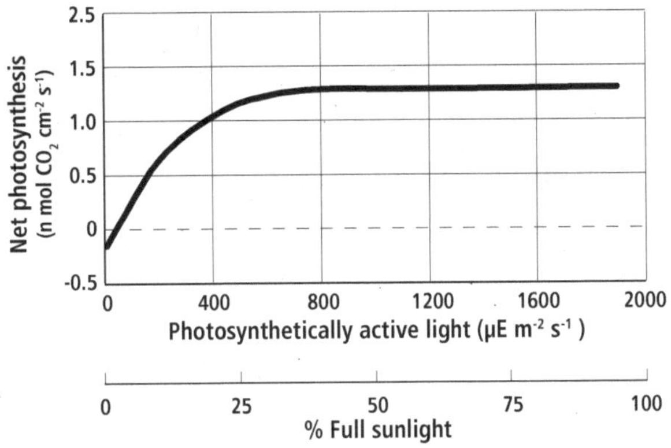

Figure 9.2. Photosynthetic response of pear leaves to differing levels of irradiance. The photosynthetic response of single leaves is saturated at approximately 30 percent of full sunlight.

resources are partitioned within tree canopies. Leaves that are exposed to the most light are generally thicker and have more nitrogen per unit of leaf area than leaves growing in more shaded areas of the canopy. Research with peach (fig. 9.3) has shown that leaf nitrogen per unit of leaf area is directly related to leaf light exposure at all levels of tree nitrogen status. Research with apple, pear, prune, and walnut indicates that this is a general phenomenon among fruit tree species (table 9.1). These relationships indicate that not only do leaves in exposed positions of the tree carry on more photosynthesis, they also have more nitrogen on a leaf-area basis. There is other evidence that exposed leaves may also have higher amounts of other nutrients. The relationships among light exposure, photosynthesis, and nutrient partitioning mean that exposed spurs or shoots generally have greater access to carbohydrates and nutrients than leaves in shady positions. This has direct effects on spur health and productivity. Research in apples has shown that spur flowering in spring is directly related to spur light exposure the previous season (fig. 9.4). Other research has also shown that flower bud size and fruit set are affected by light exposure the previous season. The data in figure 9.4 indicate that the light threshold for flower bud development in apples is approximately 30 to 40 percent of full sunlight. This minimum light threshold probably varies between apple and pear, but there is little doubt that the same general relationships exist in these two related species.

CANOPY AND LIGHT MANAGEMENT

Because of the relationships between light and leaf photosynthesis, leaf quality, spur health, flowering, and fruit quality it is essential to properly manage light within tree canopies. The objective in light management is not only to maximize light interception in the orchard but to have the light distributed as evenly as possible throughout the tree canopy. This is primarily attained by developing tree pruning, training, and planting systems that are compatible with the growth

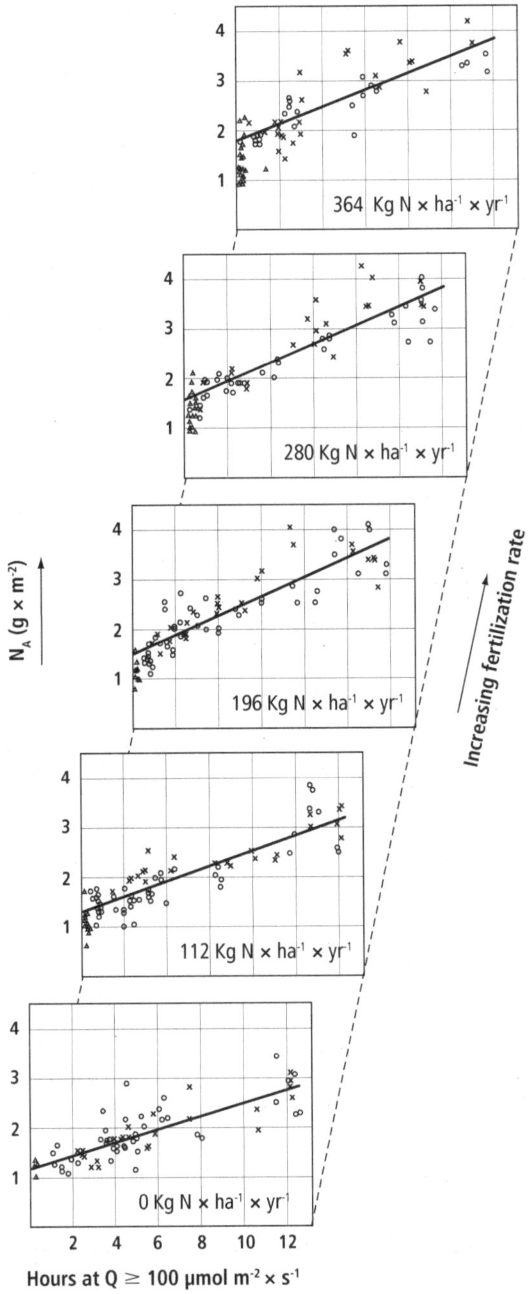

Figure 9.3. Relationship between nitrogen per unit leaf area (N) and leaf light exposure (measured as the number of hours in a day that the leaf was exposed to Q ≥ 100 μmol $m^{-2}s^{-1}$) for peach trees in each of five N fertilization treatments. *Source:* DeJong, Day, and Johnson 1989, p. 91.

habit and the vigor of the tree. There are many different types of these systems, and there is probably no single "best" system for any crop. The system that the grower chooses must be adaptable to the grower's management style, labor availability, equipment, and so on, and it must also achieve adequate light distribution characteristics to provide good yields of quality fruit.

However, there are few generalizations that can be made about pruning, training, and planting systems. One is that the greater the density in which the trees are planted, the higher the management intensity required to optimize light distribution in those canopies. Although high-density plantings may increase total orchard light interception, there is also a tendency toward internal shading in high-density orchards as trees mature. This is particularly true in vigorous orchards where crop loads or rootstocks are not sufficient to slow down the growth. Therefore, to perform efficiently, high-density orchards often require a summer pruning program and more precise winter pruning.

A second generalization is that as trees age, there is a natural tendency for the top of the canopy to overgrow and shade the bottom of the canopy, and the pruning-training-planting system should prevent or deal with this problem throughout the life of the orchard. Many "new" systems look good for the first 5 years in the orchard, but after that they can turn into a disaster because they were not designed to deal with this problem. To prevent this, a grower must have a basic understanding of tree growth and be able to visualize what the trees will look like in 10 to 20 years after planting and assess whether that visualization is practical.

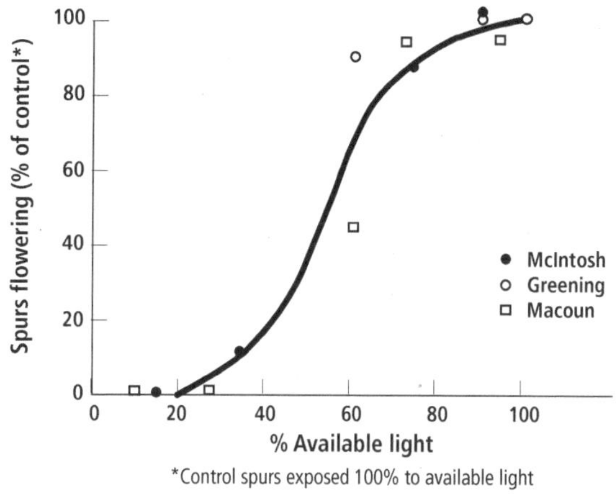

Figure 9.4. Effect of light on spurs at different positions within the tree on their flowering the following spring. All marked spurs were deflowered in the spring of the year in which light values were measured. *Source:* Cain 1971.

Table 9.1. Correlation coefficients (the larger the number, the stronger the correlation) for the relationship between canopy light levels and spur and fruit characteristics of Oregon Spur Delicious apple

	Correlation coefficients (r)	
Spurs and fruit traits	**Vegetative spurs**	**Fruiting spurs**
Spur characteristics		
Specific leaf weight	0.650*	0.733*
Leaf dry weight per spur	0.555*	0.542*
Average leaf dry weight	0.533*	0.600*
Leaf area per spur	0.334*	0.046
Leaf number per spur	0.319	0.254
Bud diameter	0.248	0.477*
Leaf area	0.168	−0.201
Spur length	0.102	0.328*
Fruit characteristics		
Percent small fruit ≤ 160 g	—	−0.568*
Soluble solids	—	0.522*
Percent large fruit ≥ 204 g	—	0.510*
Nitrogen content	—	−0.506*
Agtron blue	—	0.488*
Starch	—	0.426*
Agtron green'	—	0.426*
Fruit weight	—	0.351*

Source: Barritt et al. 1987, p. 404.
Note: Asterisks (*) indicate statistically significant correlations, and minus signs (–) indicate negative correlations.

A third generalization is that there are no short-cuts to good pruning and light management. Mechanical pruning without follow-up hand pruning is not a long-term option for maintaining high yields of good-quality fruit. Although mechanical pruning may have applications in some situations, the exclusive use of mechanical pruning generally leads to increased shading because of the spur-bearing habit of pear and the growth stimulated from indiscriminate pruning cuts made by mechanical saws. Generally, the optimal tree canopy is not a uniform, solid wall created by a hedging machine but an irregular canopy into which shafts of light can penetrate as the sun moves across the sky. Most traditional hand-pruned canopy forms are designed to maintain irregular fruiting surfaces to optimize light distribution, and these are practically impossible to achieve with mechanical pruning.

REFERENCES

Barritt, B. H., C. R. Rom, K. R. Guelich, S. R. Drake, and M. A. Dilley. 1987. Canopy position and light effects on spur, leaf, and fruit characteristics of 'Delicious' Apple. HortScience 22:402–405.

Cain, J. C. 1971. Effects of mechanical pruning of

apple hedgerows with a slotting saw on light penetration and fruiting. Journal of the American Society for Horticultural Science 96:664–667.

DeJong, T. M., K. R. Day, and R. S. Johnson. 1989. Partitioning of leaf nitrogen with respect to within canopy light exposure and nitrogen availability in peach *(Prunus persica)*. Trees 3:89–95.

Grossman, Y. L., R. B. Elkins, and T. M. DeJong. 1997. Summer pruning does not appear to increase Bartlett pear crop yield on reduced-height trees in Northern California. Acta Horticulturae 451:543–549.

Mika, A. 1986. Physiological response of fruit trees to pruning. Horticultural Reviews 8:339–378.

Rom, C. R., and B. H. Barritt. 1987. Management of apple fruiting spurs for fruit quality and profitability. Good Fruit Grower 38(21): 14–23, 58–59.

Strydom, D. K. 1993. Training and pruning of pear trees in countries characterized by long, hot summers and where dwarfing rootstocks cannot be used or are not available. Acta Horticulturae 367:130–137.

PART 2

PEAR ORCHARD AND TREE MANAGEMENT

10

Training Young Trees

RACHEL B. ELKINS, BAS VAN DEN ENDE, ROBERT STEBBINS, AND WARREN C. MICKE

Pear trees, like all fruit trees grown for commercial production, must be trained to attain a vigorous, structurally sound tree with good branch angles that is capable of producing and supporting optimal amounts of high-quality fruit. Well-trained trees can also be managed efficiently in terms of labor, pest control, and other orchard operations. The concepts and methods of training and pruning pear trees have evolved as orchardists have adapted higher-density plantings in order to hasten bearing and economic return.

Many ideas and techniques used in pears have been adapted from apples. Pear trees, however, are more difficult to train than apple trees for several reasons. First, pear trees show very strong apical dominance. This refers to the inhibition of lower buds as auxins move down the stem from the upper buds, causing growth that is strong in the top and weak in the bottom (figs. 10.1A and 10.1B). Pear trees tend to form few side branches, and the side branches that do grow often have narrow crotch angles that make the trees very upright. Blind wood, that is, lacking buds, and tip bearing make it difficult to develop fruiting wood. Second, pear trees are highly sensitive to shade and require specialized and much more detailed canopy management than do apple trees (see chapter 9, "Canopy and Light Management"). Pear leaves need about 20 percent more light than apple trees to produce the necessary photosynthates for growth and fruit production, and they are thus more negatively affected by poor light distribution within the trees than apples. Poor light distribution effects spur vigor, flower bud development and quality, fruit set, and fruit size (see chapter 8, "Vegetative Growth and Fruit Development"). For these reasons, pear trees must be trained early to maximize canopy development. Only a canopy structure capable of intercepting at least 70 percent of available light can produce high yields of good-quality fruit. Obtaining high early yields depends to a large extent on how soon this threshold is reached.

As with other U.S. pear production areas, true dwarfing rootstocks that enable apple growers to manage small trees using a wide variety of training systems such as spindles and vertical axes have been tried unsuccessfully on pears in California. Dwarfing rootstocks, mainly strains of quince, are being successfully used on pears in some European and Southern Hemisphere countries. They are often used in conjunction with plant growth regulators that control vegetative growth. However, quince has failed to produce acceptable yields for most varieties in the western United States (see chapter 6, "Propagation and Rootstock Selection"). Seedling rootstocks such as *Pyrus calleryana* have been successfully used with the Tatura trellis training system in Australia.

This chapter addresses traditional and newer training systems, as well as other techniques used to manage pear vegetative growth and promote fruit development.

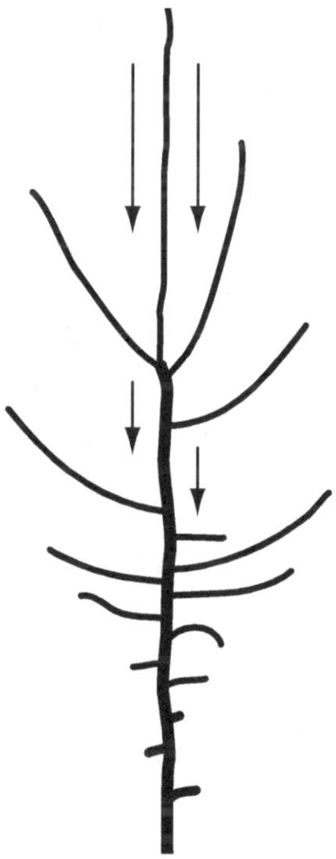

Figure 10.1A. Effect of apical dominance of the terminal bud(s) on the growth of lower buds and shoots. *Source:* After Stebbins 1992, p. 3.

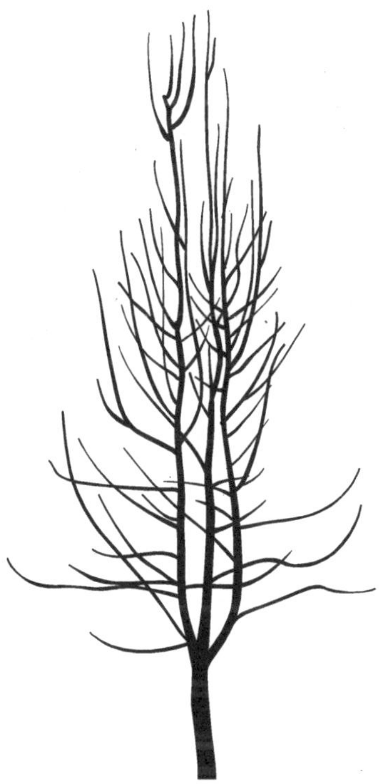

Figure 10.1B. A 3-year-old pear tree showing the strong upright growth of upper shoots and sparser, flatter growth of lower shoots. *Source:* After Ross 1974, p. 8.4.

TRAINING AND PRUNING TERMINOLOGY

Parts of the Tree

Knowledgeable pruners should know the various parts of a pear tree and understand how these parts respond to being cut at different times of the year (fig. 10.2):

- The leader, or trunk, is the primary woody structure.
- Scaffolds, or major limbs, arise from the leader(s), and together these form the permanent structure of the tree. Lower scaffolds should be stronger and longer than upper ones.
- Branches are small limbs older than 1 year that have formed spurs, shoots, and laterals.
- Fruiting units are pieces of 1-, 2-, and 3-year-old wood that grow from scaffolds and branches, and they are continually renewed by pruning.
- Spurs are a compressed leafy shoot with either a fruit or leaf bud that grow on 2-year-old wood. New spurs can originate from older spurs after several years.

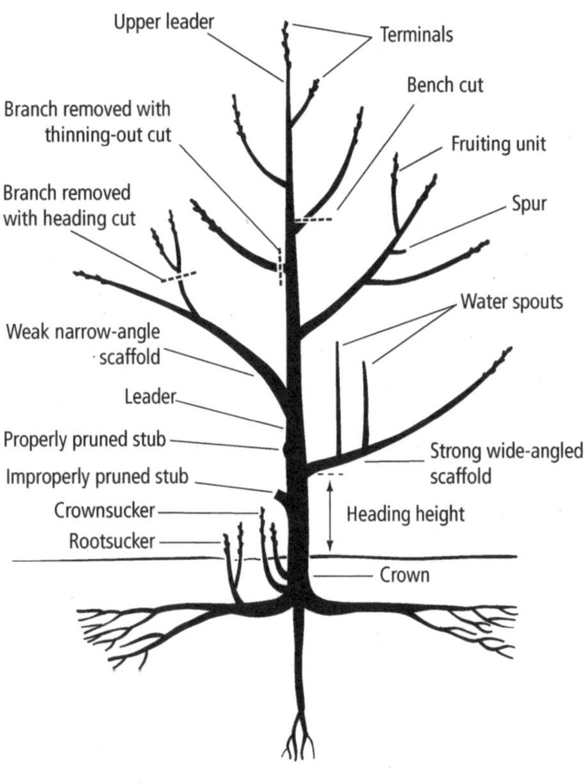

Figure 10.2. Parts of the pear tree and types of pruning cuts. *Source:* After Ohlendorf 1999, p. 34.

- Shoots are long lengths of new growth that are 1 year old or younger.
- Laterals, or pencils, are short shoots with a fruit bud at the tip.
- Terminals are shoots that develop from the tip of a branch.
- Feathers are side shoots that are less than 1 year old that grow on the leader while the tree is growing in the nursery (fig. 10.3A).

Types of Pruning Cuts

Certain basic terms apply to types of pruning cuts regardless of the training system being used.

Heading

Heading is cutting a shoot or branch part way off rather than at a branching juncture (see fig. 10.2). The effect of heading differs according to when it is done. Heading in the dormant season usually stimulates two or three buds immediately below the cut to form vigorous upright shoots with narrow angles (fig. 10.4A). Severe heading, in which more than half of the shoot is removed, may force more buds to produce shoots than does heading back one-fourth to one-third of the shoot. Very light heading (less than one-fourth of shoot removed), or tipping, may force only a single bud. Heading in pears creates a stiffened, upright branch. Leaders and scaffolds usually need heading, but side branches should seldom be headed unless it is necessary to reduce the number of spurs.

Heading 1-year-old shoots often causes them to grow straight up and create shade, have little fruiting potential, and be costly to remove. Two heading cuts can be employed on 2-year-old wood depending on the number of fruit buds and tree vigor. A ring cut, or fertility cut, can be made at the division between 1- and 2-year-old wood to improve fruit set. Or, the cut can be made closer to the main branch than the ring on the weaker wood to reduce the number of fruit buds (the weaker the wood, the closer the cut should be made to the main branch).

Delayed heading

Delayed heading cuts are heading cuts made after the terminal bud has begun to grow in the spring. This usually causes a greater number of buds to produce shoots below the cut, and these shoots may also have wider angles. Delayed heading on 1-year-old shoots forces lateral breaks and discourages the growth of blind wood. Early delayed heading is done at green tip stage (when the first leaves emerge from the terminal buds), and late delayed-heading is done 2 to 3 weeks later. Later heading causes more but shorter shoots. The response to

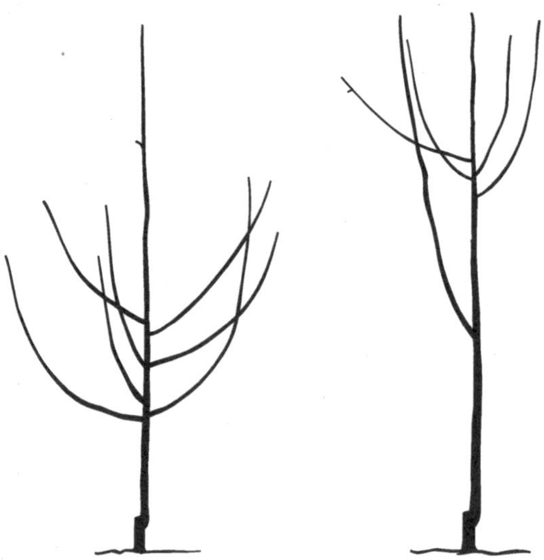

Figure 10.3A. Nursery tree with well-placed "feathers" that can be used as scaffolds. *Source:* After van den Ende 1995, p. 6. **Figure 10.3B.** Nursery tree with poorly located feathers that should not be chosen as scaffolds. *Source:* After van den Ende 1995, p. 7.

delayed heading is restricted to no more than about 16 inches (40.5 cm) below the cut, and the shoot must be at least pencil-thick where the cut is made (fig. 10.4B).

Thinning

Thinning cuts are made at the junction of two branches, removing one branch and leaving the other intact (see fig. 10.2). Thinning removes shoots that would detract from the desired tree structure and also alleviates crowding and shading; it has less effect on crop load than does heading. It is usually best to thin in spring or early summer to direct growth into desirable shoots and limbs. Removal of vigorous, unfruitful shoots that might compete with the leader can be done in the dormant season. Thinning a group of shoots or limbs so that only one remains is known as singulation.

Bench cuts

Bench cuts are thinning cuts that leave a small horizontal "seat." These convert strong, narrow-angled shoots to wider-angled, weaker shoots. Bench cuts are usually made in the winter (see fig. 10.2).

Basic Training Concepts and Practices

Because pruning young trees delays production, minimize severe pruning in the training period and focus on tree development and early production. Techniques that may be used to encourage early fruiting are described below.

Training feathers

Feathers that have wide angles with the main trunk and follow the 3-to-1 rule (see below) should be kept but shortened to 5 inches (12.5 cm) at planting. Spread new shoots to widen branch attachment angles and help produce a balanced scaffold system, especially in varieties that tend to produce very narrow branch angles. Spread the shoots when they are about 6 to 8 inches (15 to 20.5 cm) long using toothpicks, clothespins, or commercially made spreaders that are available in different lengths.

Spreading and bending

In the training period, spreading is preferred to pruning because it enhances early production. Spreading helps control the strong apical dominance of pear and promotes a basal growth habit. Spread lower branches to an angle of 45° from the vertical; spread upper branches to nearly (but not below) horizontal. The more the limb is spread, the more the growth of the terminal shoot is repressed and the more the growth of shoots near the trunk is increased (fig. 10.5).

Spreading scaffolds and branches increases the number and rate of growth of flower buds formed on them; however, fruit on weak limbs can cause the limbs to sag below horizontal. Early fruiting often bends limbs so far toward the horizontal that the vigor of the terminal is lost. When this happens, suckers often rise near the trunk, and the crop load may even cause limbs to break. Excessive spreading often creates the need to tie limbs up for support. It is preferable, however, to spread the branches and develop a taller tree at the same time than to wait many years for initial production on growth that has not been spread. The terminal end of the scaffold or branch should always be higher than the basal end to maintain an upward growth pattern.

Shoot bending, or bowing, can also direct growth. In bowing, a short shoot is bent into an arch or loop; buds near the top of the arch or loop tend to form strong upright shoots, while the terminal shoot loses vigor and buds nearer the trunk are suppressed. Bowing should be employed in specialized cases and avoided in normal spreading.

Limbs may crack or break during or just after spreading or shoot bending. The cambium will normally heal over, although growth may be slowed for a short time.

Wind can bend limbs and even shape whole trees, which makes training a well-shaped symmetrical tree on a windy site very difficult. Few limbs are formed on the upwind side, and staking or anchoring into the wind only partially solves the problem. A small tree with strong lower limbs is preferable in windy locations. This may be achieved by tying scaffolds into the wind

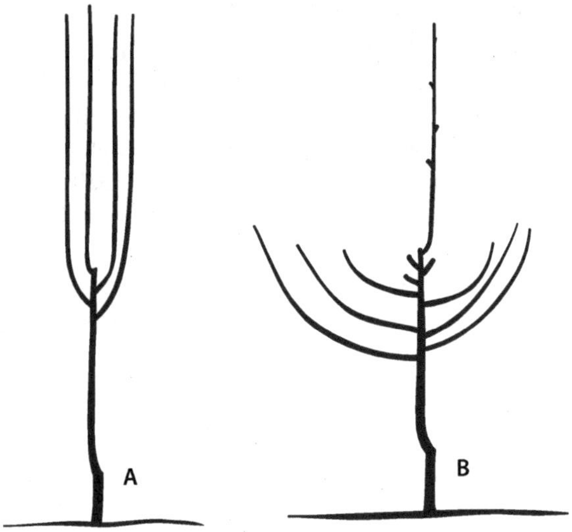

Figure 10.4. The tree on the left (A) was headed in the dormant season, while the one on the right (B) was delayed-headed. Note the difference in branch angles. *Source:* After van den Ende 1990, pp. 21–24.

and anchoring the ties with hop clips or similar anchors in the ground. Pruning into the wind is rarely successful and delays production.

Summer pruning

Summer pruning consists of removing shoot growth during the summer or growing season that does not build the desired structure or encourage early production. It is generally done using thinning cuts in early summer (for example, in mid to late June) when extension growth begins to slow. Excessive summer pruning may devitalize the tree if too much leaf surface is removed. Only experienced personnel should perform summer pruning.

1-2-3 Rule

The "1-2-3 rule" is a system of dormant pruning applied to pear varieties bearing mainly on 2-year-old or older wood that may also be used in training young trees. It is detailed in chapter 11, "Pruning Mature Trees."

3-to-1 Rule

The "3-to-1 rule" refers to the relationship between the diameter of shoots, branches, or scaffolds where they join the leader. It also applies to any fruiting wood higher up in the tree and on the scaffolds. The rule states that a branch must have a diameter at its point of origin that is one-third or less than the diameter of the leader or limb of origin (fig. 10.6). The 3-to-1 rule helps ensure that a strong leader with a balanced scaffold structure develops in the tree. It applies to newly planted trees as well as to older, developing trees.

Notching

In notching, a thin slice of bark is removed above a vegetative bud or short vegetative spur using a piece of a hacksaw blade, pruning saw with fine teeth, or similar tool (fig. 10.7). The notch must extend along the entire width of the bud. Notching increases branching in pear trees by temporarily stopping the flow of the plant hormone auxin from the terminal bud to the buds below it. It is generally done on 2- or 3-year-old wood at bud break when growth begins, encouraging the lower buds to break and form shoots.

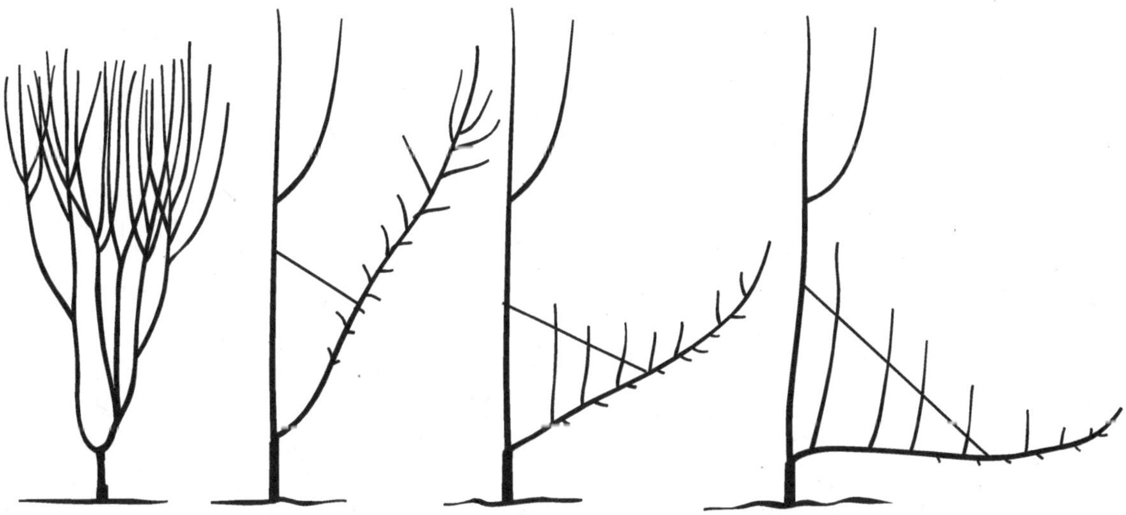

Figure 10.5. Effect of varying degrees of spreading on branch angle and growth pattern. *Source:* After Stebbins 1992, p. 4.

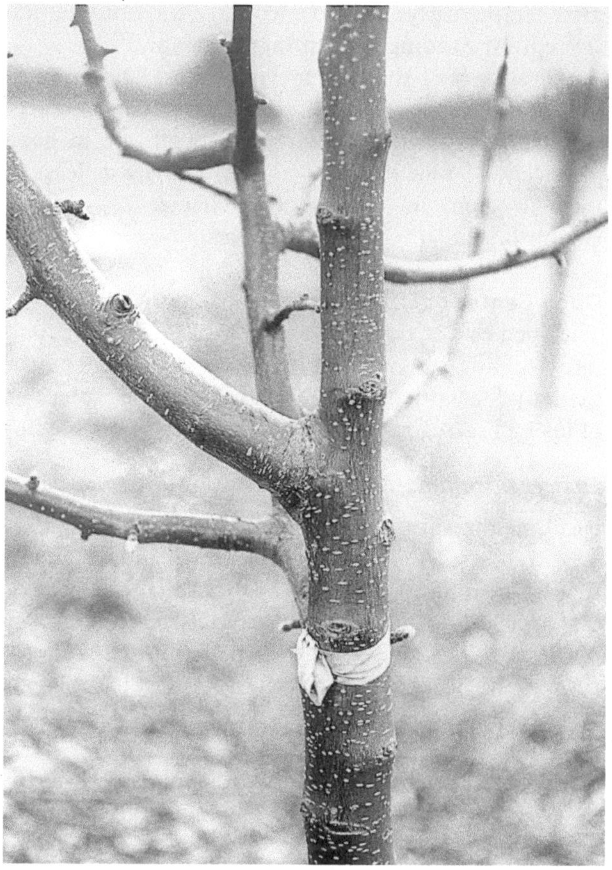

Figure 10.6. The branch on the left should be removed because it exceeds the "3-to-1 rule," that is, it is greater than one-third the diameter of the leader. *Source:* Forshey, Elving, and Stebbins 1992, p. 89 (reprinted with permission).

Scoring

Scoring can be done on unheaded, newly planted, unfeathered trees to encourage early bearing by reducing vigor. To score a branch, make one circular incision all the way around the stem, through the bark only, just above the section where shoot growth is desired (fig. 10.8). A pocketknife or a putty knife that has a small V made in the blade is a good tool to use for this. Keep a constant pressure on the knife. If the incision is made correctly, trees will recover quickly. If the cut is too deep, the leaves above the score can turn yellow or bronze, as in early-fall coloring; the tree can even break off at the score in strong wind. The application of dormant oil in late winter to control pear psylla and mites may also serve to enhance the effect of scoring.

Plucking

Plucking is a mild form of delayed heading that does not remove part of a shoot. It helps develop strong basal scaffolds and also helps restrain vigor in the top of the tree without heading the leader, and it also aids in preventing blind wood. Plucking consists of removing 4 to 6 buds by hand under the tip bud at early emergence to reduce the production of auxin. This encourages buds to break and form new shoots up to about 10 inches (25.5 cm) below the plucked section. It may be necessary to use toothpicks to spread strong shoots that grow after plucking.

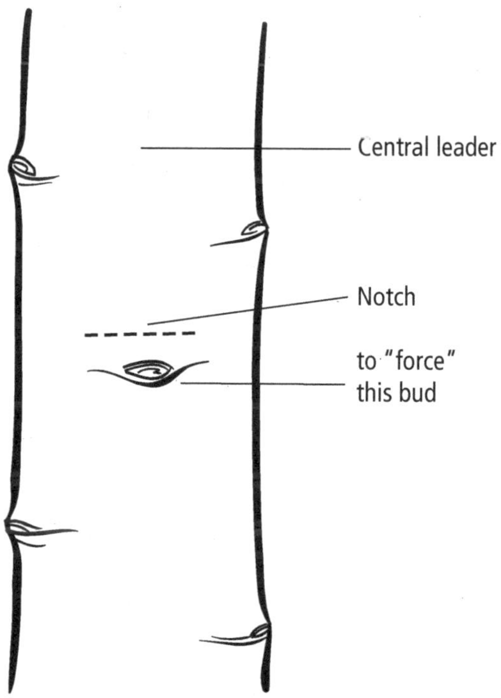

Figure 10.7. Notching above a bud to stimulate new branch formation.

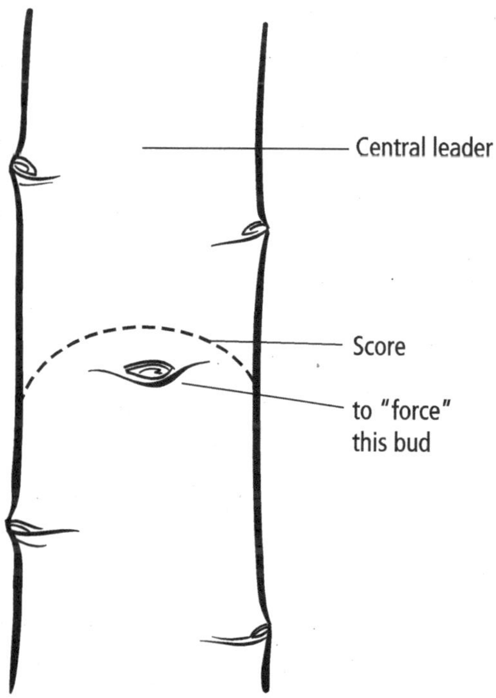

Figure 10.8. Scoring the bark of young trees encourages early bearing. *Source:* After Stebbins 1992, p. 8.

Pinching

Pinching is done on the shoots that grow from the second and third bud under a delayed-headed cut. These new shoots with narrow angles tend to be as strong as the shoot that develops from the topmost bud. Allow the shoot that grows from the topmost bud to extend the leader; pinch back the second and third shoots under the cut to 1 inch (2.5 cm) long when they are about 4 inches (10 cm) long to reduce their growth and encourage more buds with wider angles to break below. Pinching works because the hormones that young shoot tips produce are plentiful in spring in the upper part of the tree, and they control growth lower down in the tree. It may be necessary to pinch again during the growing season.

STANDARD TRAINING SYSTEMS

The four major objectives in training newly planted trees in a standard planting using nondwarfing rootstock and 2-year-old bare root nursery trees are to

- direct growth into a strong tree framework that can support heavy crops
- encourage rapid fruit production
- minimize the risk of fire blight
- develop a short but productive tree that is easy to manage

Because the lack of dwarfing rootstocks to hasten precocity in pears renders the second objective inconsistent with the first and fourth, pear growers often use training systems described in this section to hasten bearing and control tree size.

Most newer orchards in California employ either the open center system or the central leader system or some modification of these systems. Tree spacing in the orchard generally ranges from 9 to 15 feet (2.7 to 4.6 m) between trees and 12 to 20 feet (3.7 to 6.1 m) between rows, but closer spacing may be employed in special situations. Factors that determine spacing include variety, rootstock, and soil type (see chapter 7, "Orchard Establishment"). These factors, as well as climate and growing area, also determine the level and timing of training operations required. For example, trees growing on vigorous rootstocks in excellent soil should be minimally pruned in early years to hasten bearing. Trees grown on less-precocious rootstocks and heavier soil may require more severe early pruning to encourage vegetative growth. Pruning, and particularly heading, however, should generally be minimized in the early years to encourage early production (except for heading at planting; see below).

Regardless of the system you choose, it is critical to plant certified trees grown by a reputable nursery that have been sufficiently hardened off, are in good physical and nutritional condition, and are at least 5/8 inch (16 mm) in diameter (see chapter 7, "Orchard Establishment").

Open Center or Open Vase

The open center, or open vase, system is the most common training system used for pears in California, particularly for Bartlett trees, because it decreases the risk of losing bearing scaffolds or entire trees to fire blight.

First-year training

Head the tops of nursery trees at about 24 to 30 inches (60 to 76 cm) above the ground at planting. Shoots that develop 6 to 10 inches (15 to 25.5 cm) below this cut form the future primary scaffolds. Larger nursery trees may already have well-positioned future scaffolds that can be headed back 6 to 8 inches (15 to 20.5 cm) from the trunk, provided they are of equal length and thickness.

Trees can be lightly pruned in summer during the first growing season. Remove or pinch back branches and suckers that compete with chosen scaffolds. This encourages good growth on branches that are desirably located for future scaffolds and permanent limbs. Walk the first-year orchard two or three times during the summer to prune unwanted growth while it is still small.

Second-year training

Most potential scaffolds grow just below the point where the tree was headed at planting time. If future leaders were not selected during the first growing season, during the dormant season, select 3 to 4 upright shoots of approximately equal size that are distributed evenly around the trunk and spaced vertically 6 to 12 inches (15 to 30.5 cm) apart. These will form the tree's main structure to support fruiting units and the crop load. Retain small lateral branches along these scaffolds to build fruiting units, increase leaf area, and protect the bark from sunburn.

For scaffolding, avoid selecting leaders that are directly above one another. Also avoid excessively dominant shoots that usually have narrow, acute angles from the trunk. The goal should be to attain moderately upright growth; the best angles for scaffold attachments are 30° to 45°. Flat-angled (horizontal) shoots are poor choices because they do not provide enough upward growth. Cut back selected scaffolds to outside lateral branches so as to counteract their natural tendency to grow too upright. In windy areas it is advisable (where possible) to have the lowest scaffold growing toward the prevailing wind. If the trees grew poorly the first year, prune primary scaffolds to 3 or 4 buds to promote vigorous growth the following year. If poor growth continues, investigate and correct the cause or replace the tree, if needed.

Main scaffolds can be summer-pruned in the second growing season to keep them rigid and upright. Walk through the orchard fairly often in late spring and early summer (May and June) to remove vigorous upright watersprouts and rootstock suckers. At this time, however, do not attempt to do detailed pruning and selecting of potential fruitwood because as much leaf surface as possible should be retained to maximize tree growth.

Third-year training

In the dormant season, select two vigorous lateral branches arising 24 to 36 inches (0.6 to 0.9 m) from the head on each primary scaffold to serve as secondary scaffolds. These scaffolds should be positioned around the tree to fill the tree's eventual space. Although two scaffolds are usually selected, some primary scaffolds may bear only one scaffold, and, in a few cases, three may be warranted because of vigor and placement. If more than two secondary scaffolds are left per primary, some of the resulting limbs may be weak, develop less fruitwood, and be apt to break with heavy crops.

Avoid excessive pruning while attempting to form a symmetrical tree. Each tree grows differently, and few provide the opportunity to develop a perfectly shaped tree. Use judgement to train each tree into the best possible framework from the potential scaffolds available. The open center system should provide the expanding fruiting area with adequate structural wood.

Secondary scaffolds can be headed (or delayed-headed) or scored 24 to 36 inches (0.6 to 0.9 m) above the junction with the primary scaffolds, as were the primary scaffolds the previous winter at the junction. When two secondary scaffolds on one primary scaffold are delayed-headed, it is best to leave the bottom one longer than the upper one.

Scaffolds that exceed the 3-to-1 rule should be removed with a bench cut. Completely remove upright vigorous shoots in the center of the tree with a thinning cut to prevent any regrowth. Allow most of the less-vigorous, flatter-angled lateral branches to remain to develop as fruitwood.

Fourth-year training

Continue dormant training by pruning trees to grow upward and outward as scaffolds and fruiting branches fill open spaces. Use delayed heading, scoring, and notching to induce branching and control vigor until the tree reaches its mature size.

At the end of 4 years of growth, a tree should have 3 to 4 primary scaffolds with 5 to 7 secondary scaffolds 4 to 6 feet (1.2 to 1.8 m) above the ground. At 7 to 10 feet (2.1 to 3 m) above the ground, additional branches should fill the periphery of the treetop. Less-vigorous, flatter fruiting wood that produced the first crops should have been thinned out. The center of the tree should be somewhat open to permit light infiltration that will maintain and stimulate further development of fruiting wood (figs. 10.9A and 10.9B).

To increase fruiting area while maintaining structural strength on very upright trees, retain outside lateral branches along major upright limbs or place spreaders horizontally between scaffolds.

Central Leader

The central leader system or some modification of it (e.g., 2-leader, palmette, or slender pyramid) is often used in high-density, or "hedgerow," plantings. A vertical post with a wire trellis may support trees where very high densities are used and where the risk of losing the central leader to fire blight is high.

First-year training

Head newly planted trees to a height of 24 to 30 inches (0.6 to 0.9 m). Train lateral branches as described for open center trees, above. During the following spring and summer, check trees at 4- to 6-week intervals to make sure that the central leader is dominant. Pinch back strongly competing shoots and remove narrow-angled (almost vertical) shoots.

Second-year training

In the dormant season, select 3 to 5 wide-angled scaffolds along the leader, leaving 18 to 24 inches (0.5 to 0.6 m) between the lowest scaffold and the ground. These scaffolds should be distributed as evenly as possible around the tree and should be spaced vertically along the leader 2 to 4 inches (5 to 10 cm) apart. Remove strong unwanted shoots that compete with the leader and scaffolds. Score and pluck the central leader and scaffolds to promote branching and control vigor. Retain some small, flat-angled shoots to produce fruiting wood if the tree has good vigor.

Training during subsequent summers

Remove strong, upright shoots and spread or remove narrow-angled branches that compete with the central leader. Thin branched terminals on each scaffold to a single limb (singulation). Spread scaffolds to a 45° angle with the central leader as the scaffolds approach their desired length (which depends on tree spacing) and

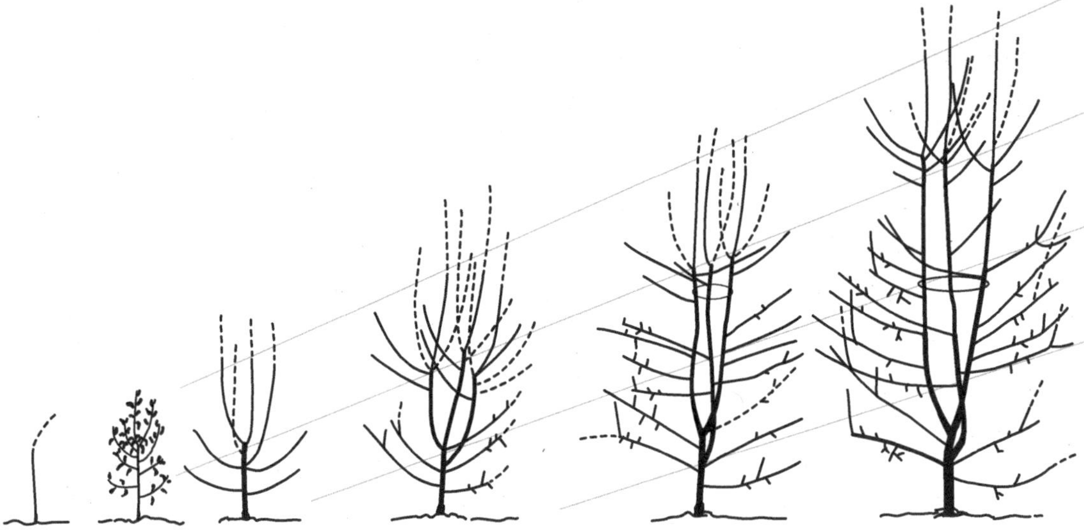

Figure 10.9A. Stages in development of an open center, or vase-shaped, tree. *Source:* After Forshey, Elving, and Stebbins 1992, p. 114.

while they can still be bent easily (are approximately ½ to 1 inch, or 12.5 to 25.5 mm, in diameter at the base). Spreading should be done during the growing season before the wood gets too rigid and while it is less likely to break.

Training during subsequent dormant season

Continue to head the central leader and main scaffolds, removing approximately one-third of the previous season's growth (figs. 10.10A and 10.10B). Spread or remove (using a bench cut) narrow-angled branches and any upright branches from scaffolds. The central leader should be the only upright scaffold in the tree. Create a second whorl of scaffolds 2 to 3 feet (0.6 to 0.9 m) above the first group of scaffolds using wide-angled shoots that arise from the central leader. A third whorl may be formed at an additional 2 to 3 feet (0.6 to 0.9 m) higher in subsequent years, depending on tree vigor. During subsequent growing seasons, spread these higher groups of scaffolds to about 45° to 60°. Maintain at least 2 feet (0.6 m), preferably 3 feet (0.9 m), of space between scaffolds that are directly over each other. Each higher group of scaffolds should be shorter in length than those below it, giving the tree a pyramidal or

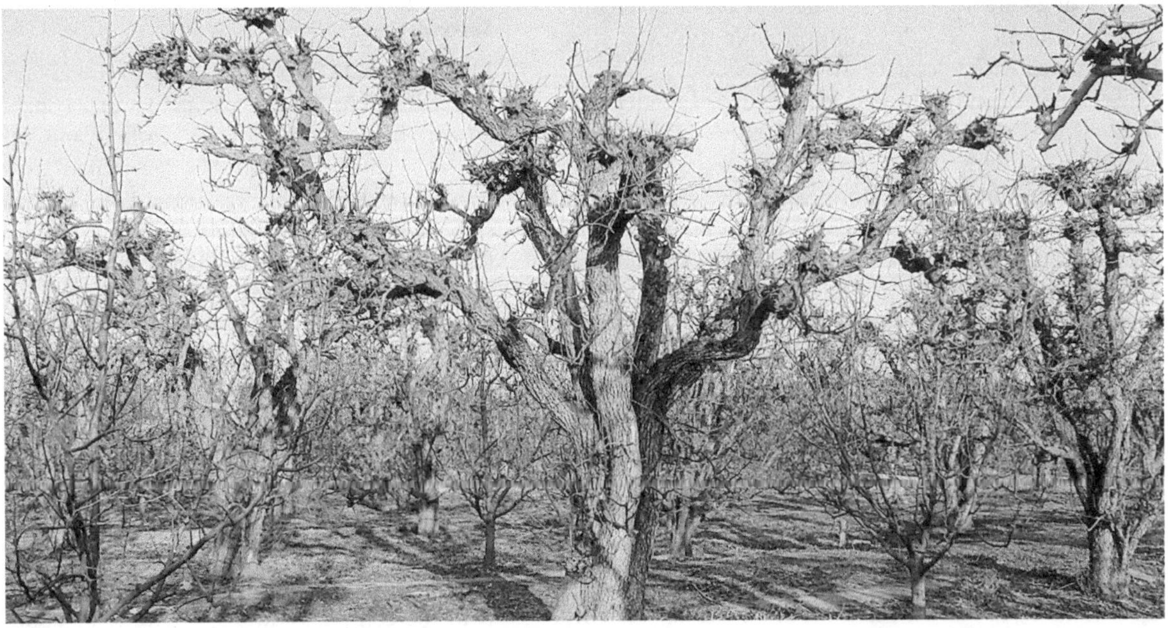

Figure 10.9B. Typical mature open center Bartlett pear tree. *Photo:* R. McClain.

conical shape. Fruiting wood grows on the scaffolds and spurs. Thin laterals to maintain tree vigor and produce high-quality fruit. Singulate the branched terminals. Typical central leader trees are shown in figure 10.10.

HIGH-DENSITY TRAINING SYSTEMS

High-density orchard systems containing more than 300 trees per acre (740 per ha) have been adapted worldwide for pears, but they are used much less in the United States than in other countries. Although this is partly because there are few new pear plantings in the United States, the main reason is the lack of acceptable size-controlling rootstocks adapted to U.S. growing conditions. A major consideration when contemplating a high-density system is how likely the cost of the increased investment in trees, the labor to plant them, and any required support system can be recouped. Varieties that consistently yield higher returns are better candidates for these systems. Fire blight must be strictly managed to prevent tree loss.

Two training systems successfully used worldwide for pears are the palmette and the Tatura trellis. In both systems

- trees should fill the space and reach maximum permissible height after the second year in the orchard
- fruiting should begin in the second year
- trees should produce sufficient flowers to set a crop and influence vigor in the third year
- the full yield potential of 25 to 30 tons per acre (67.2 t/ha) of high-quality fruit should be reached by the sixth year

Both the palmette and Tatura trellis use a post-and-wire support system. While these systems may be required to support pear trees grown on weaker rootstocks, particularly quince, they also optimize available light penetration into the entire canopy, promote uniform scaffold angles, support large amounts of fruit early in the trees' life, reduce scaffold breakage, simplify training and pruning, and maintain short trees to minimize ladder work and reduce labor costs.

For best results, plant unheaded, feathered trees with well-developed tops and roots. Consult a nursery with experience growing such trees, as this type of tree is not customarily provided in California or the western United States and may need special care in the nursery and during planting. Standard dormant whip trees, if used, should be of a large caliper with some side growth in addition to the leader. Using whips will likely delay fruiting beyond the desired intervals for these systems, but pears will still be produced sooner than in a standard planting. The site should be prepared and trees properly planted according to the guidelines presented in chapter 7, "Orchard Establishment."

It is very important to build a balanced tree framework with fruiting units within 3 years in order to achieve early yields. Maximize the number of lateral breaks on the leader using delayed heading, scoring, notching, and other training techniques after planting; encourage maximum growth during spring and summer. Carefully select scaffolds that will form the tree's framework, and prune in the summer to eliminate unwanted growth. Dormant pruning should be minimal; delay corrective pruning using thinning cuts until spring. Regulated deficit irrigation later in the season can slow growth, mature the wood, and form flower buds as long as trees are not overly stressed (see chapter 15, "Irrigation Scheduling"). Good-quality fruiting wood can be maintained by following the 1-2-3 rule of renewal pruning once the full crop potential has been reached (see chapter 11, "Pruning Mature Trees").

Palmette

The palmette system, a modification of the central leader system, is used primarily in Italy and France. In trees in palmette orchards, two scaffolds are trained from the central leader, one on each side parallel to the tree row, to form a hedgerow, or "tree wall." The spacing ranges from 10 to 14 feet (3.0 to 4.3 m) between rows and 6 to 10 feet (1.8 to 3.0 m) between trees within the row, giving about 310 to 725 trees per acre (765 to 1,790 per ha). Very vigorous rootstocks such as *Pyrus betulaefolia* should be avoided, particularly for densities of 400 to 600 trees per acre (990 to 1,480 per ha).

There are many variations of the palmette system. While palmette orchards can be free-standing, most use a wire trellis with 4 to 6 wires to facilitate training. The top wire is generally about 9 feet (2.7 m) above ground, with lower wires spaced evenly at 30-inch (76-cm) intervals. A typical palmette is shown in figure 10.11.

First-year training

After planting, head trees 4 to 6 inches (10 to 15 cm) above the lowest wire and fasten the tree to the wire loosely using vine clips, polyethylene tubing, ties, or a tapening device. Replace the fastener as the limb diameter increases; tape must be replaced sooner than clips or tubing. Two feathers may be retained for initial scaffolds, if available. Remove all growth below 18 inches (0.5 m). When new shoot growth is 2 to 4 inches (5 to 10 cm) long, pinch out any growth that competes with the leader. Attach clothespins to the leader just above the first pair of selected scaffolds to improve their

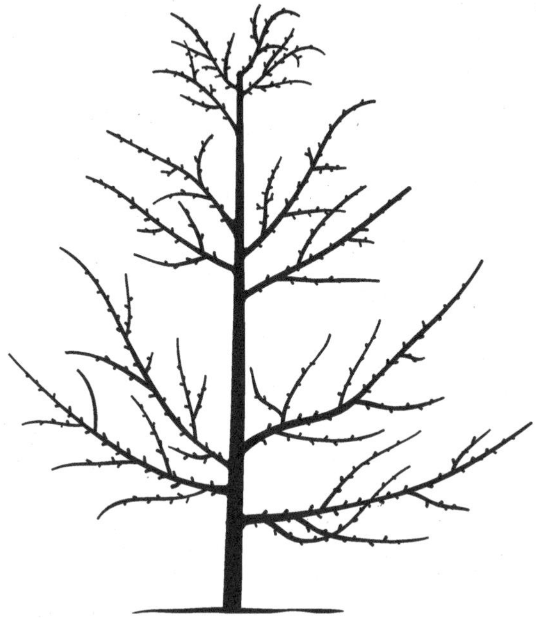

Figure 10.10A. Correctly trained central leader tree. *Source:* After Forshey, Elving, and Stebbins 1992, p. 104.

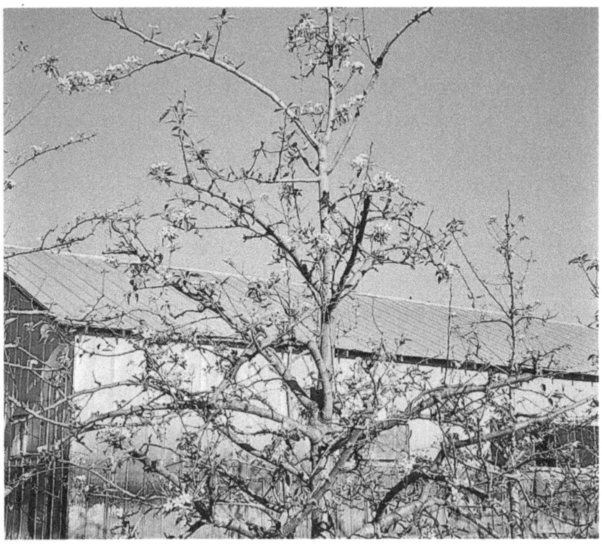

Figure 10.10B. Typical central leader Bosc tree in bloom. *Photo:* R. Elkins.

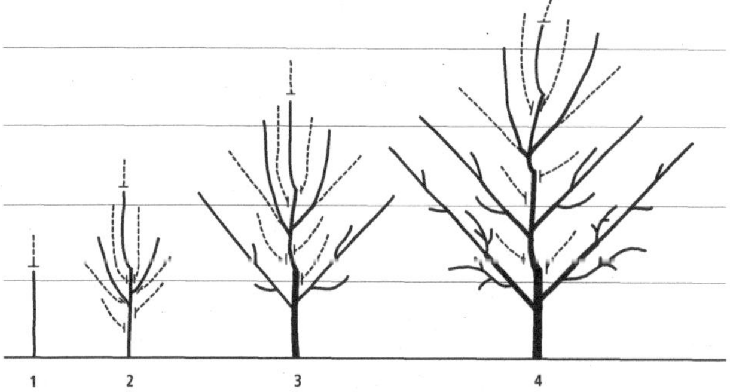

Figure 10.11A. Growth of a palmette-style pear tree from planting through year four. *Source:* After Forshey, Elving, and Stebbins 1992, p. 138.

angles of attachment. Tie the leader when it reaches the second wire.

Second-year training

After the first growing season, head back the dormant leader 4 to 6 inches (10 to 15 cm) above the next wire and remove competing upright growth. If not done at planting, select two scaffolds below the lowest wire and remove all other shoots. Spread these scaffolds and tie them to the lowest wire at an angle of about 30° to 45° (narrower for closer spacings and less-vigorous cultivars).

Repeat these steps annually until the tree fills all wires below the top one (fig. 10.11A) Maintain the proper angle for each set of scaffolds. Once the leader reaches the top wire, either bend it to one side and pair it with a lateral to make a set of scaffolds, or, more ideally, maintain it as a leader and head it back regularly to a weak lateral.

Maintenance pruning

Remove vigorous, upright shoots that arise from the scaffolds and leader. The greatest challenge is to control the vigor in the tops to prevent shading below. The ends of lower branches of adjacent trees can overlap to fill

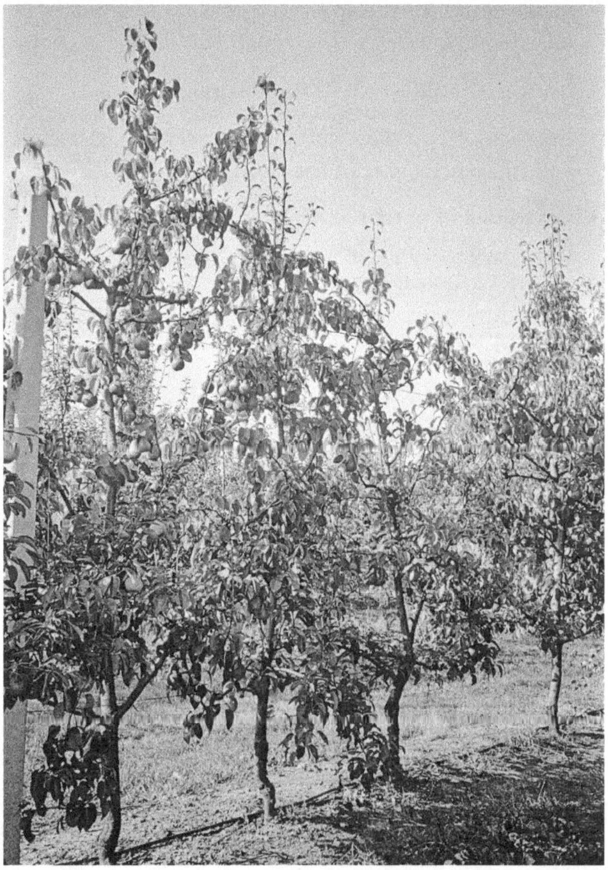

Figure 10.11B. Typical palmette pear orchard in Italy. *Photo:* R. Elkins.

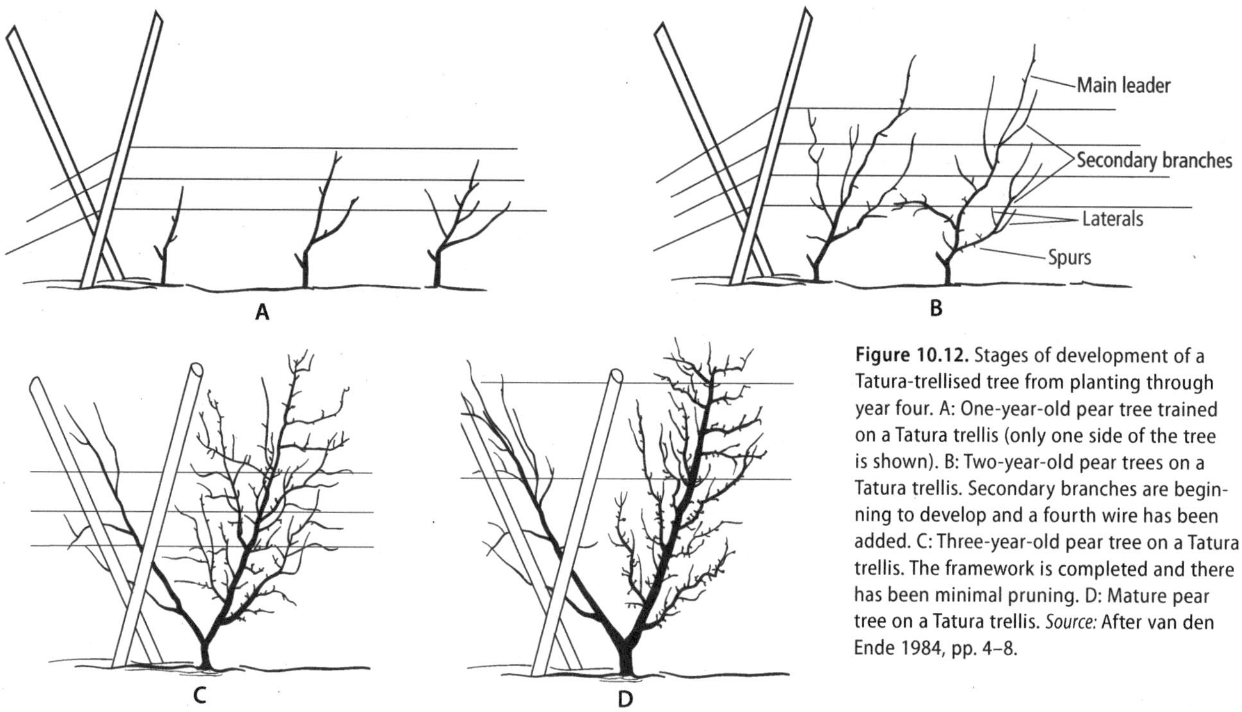

Figure 10.12. Stages of development of a Tatura-trellised tree from planting through year four. A: One-year-old pear tree trained on a Tatura trellis (only one side of the tree is shown). B: Two-year-old pear trees on a Tatura trellis. Secondary branches are beginning to develop and a fourth wire has been added. C: Three-year-old pear tree on a Tatura trellis. The framework is completed and there has been minimal pruning. D: Mature pear tree on a Tatura trellis. *Source:* After van den Ende 1984, pp. 4–8.

in the row space, although judicious heading back and branch renewal may be required to prevent crowding. Remove branches that grow perpendicular to the row.

Tatura Trellis

This tree training system for pears, and the open Tatura system described below, were developed in Australia and are compatible with standard rootstocks if managed properly. The large and uniform surface area of the canopy, which is thin and V-shaped, optimizes the interception and distribution of sunlight. Fruit hangs freely from the V, greatly reducing blemishes from limb rub and contact with other fruit.

Early tree management focuses on training trees to rapidly fill the space on the trellis with fruiting units. Training should be mostly completed in 2 or 3 years (fig. 10.12). Pruning, particularly dormant pruning, should be avoided during the first 2 years. Allow leaders to remain unheaded until they have reached their maximum permissible height. Score them in the spring and pluck the tips to create side branches. Spread shoots that are to be kept for branches when they are 6 to 8 inches (15 to 20.5 cm) long using cocktail-style toothpicks or other spreaders. Remove late bloom from young trees to reduce the threat of fire blight.

In a Tatura trellis orchard, each tree has two equal-sized leaders, one to the left and one to the right, with trees planted 4½ feet (1.4 m) apart. This is preferable to planting trees 6 to 7½ feet (1.8 to 2.3 m) apart with four leaders (two on each side of the trellis) because it is difficult to create four uniform leaders. Allow a spacing of 15 feet (4.6 m) between rows. The angle of the V between the two leaders should be 45° (22.5° to the vertical). The 45° angle

- conforms to the upright growth habit of pear trees
- allows fewer shoots to grow on the backs of the leaders inside the V (and therefore requires less summer pruning)
- fits the 15-foot (4.6-m) row spacing better than a 60° angle
- allows room for orchard personnel to pass through the V during summer pruning

In both the Tatura trellis and open Tatura, the crop load can be calculated according to the number of fruit per square foot of canopy area and expressed as fruit, trays, cartons, bins, or tons per canopy acre. This provides a more accurate measure of canopy and orchard efficiency than calculating tons per acre. For example: a crop load of 28 tons of Bartlett size 100 pears per acre requires 3 pears per square foot of canopy area (28 t × 210 g per fruit = 133,333 pears per acre × 43,560 square feet per acre = 3 fruit per square foot).

First-year training

For pears on standard rootstocks, head trees at a height of 16 inches (40.5 cm) at planting. In the spring, select two shoots of equal length and thickness to become leaders. Attach guide strings to the wires. In the summer, train one of the shoots up one side of the trellis and

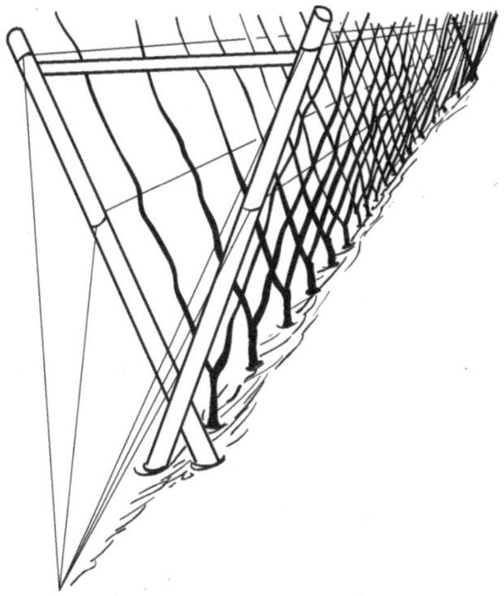

Figure 10.13A. Tatura trellis.

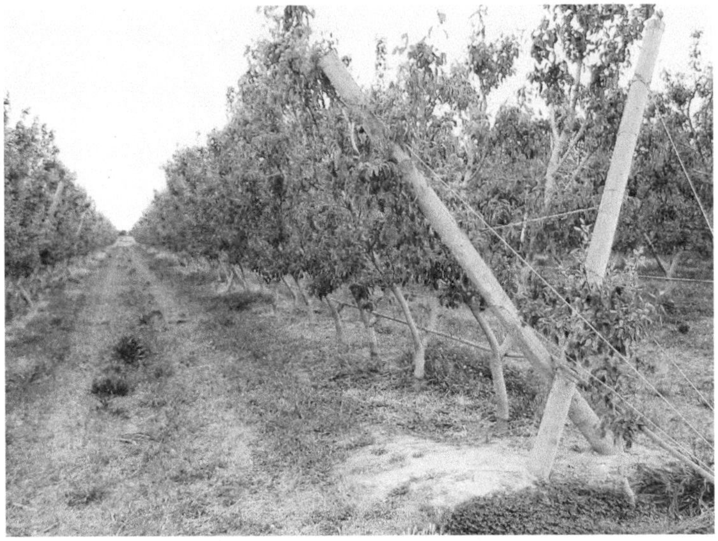

Figure 10.13B. Packham's Triumph trained in a typical Tatura trellis *Source:* Photo courtesy Northern Victoria Fruit Growers.

the other shoot up the other side by tying the shoots to the string and wires using a tapening device. Remove potentially strong side shoots of the same age to maximize the growth of the two leaders. Do not head back the leaders.

Second-year training

It is crucial that uniform branching occur along the leaders. Dormant emulsifiable winter oil at the recommended purity and rate applied for pear psylla and mites about 4 weeks before anticipated bud burst may serve to encourage uniform bud break and branching. Applying a second spray of lighter oil about 2 weeks later may further serve to enhance bud break. Other registered rest-breaking agents that are more potent than winter oil may give better results when applied properly.

Score the leaders 35 inches (89 cm) from the ground at the late green tip stage (when the first leaves emerge from the terminal buds). Pluck the tips of the leaders about 2 weeks later. At that time, make a second score 16 inches (40.5 cm) above the first score if the leaders are longer than 6 feet (1.8 m). Attach the trees tightly to the strings and wire to ensure that buds break and the score(s) can heal. Any rocking of the trees in the wind interferes with the branching process.

Notch individual buds if branching along the leaders is unsatisfactory. Keep the soil moist around the roots and apply 2 ounces (56.7 g) of actual nitrogen per tree every 2 weeks, starting when the new shoots are about 6 inches (15 cm) long.

Allow shoots to grow before positioning them to fill the trellis. Train the bottom shoots up the wires at an angle of about 45°; train branches in the middle section of the fully grown leaders to 30° from the horizontal; and train the shoots near the top wires almost flat. In the summer, remove shoots growing toward the inside of the V. Cut back strong shoots growing anywhere along the trellis to 1-inch (2.5-cm) stubs to develop weaker, more manageable shoots.

Head back the leaders in June or July when their height has reached the equivalent of 60 percent of the width of the rows, measured vertically; allow other shoots to remain unheaded.

Third-year training

At this time, fruiting units begin to develop and vegetative tree growth must be controlled. Evaluate shoot size in the winter and apply the 3-to-1 rule if necessary. Dormant oil applied as in the second year may enhance bud break.

Girdle the trunks to enhance fruit set and fruit growth while controlling tree growth. Use a ⅜-inch (9.5-mm) girdling knife to girdle the trunk of each tree at full bloom. Make two C-shaped incisions halfway between the graft union and where the tree forms the V. Maintain good soil moisture to ensure healing after girdling and apply foliar nitrogen fertilizer to mitigate the temporary effects of arrested root growth (see chapter 18, "Nutrition").

In the summer, remove unwanted growth from inside the V and from the tops of the trees. Head back leaders, long scaffolds, and branches to 60 percent of row width in June and July or even later. Tie up weeping (slender, drooping) branches.

Maintenance

Minimize dormant pruning and focus on repositioning and spacing branches and scaffolds. Fruiting wood and spurs should now be clearly visible. It may be necessary to girdle trees at full bloom and remove unwanted growth by hand in early summer to prevent regrowth. Follow the 1-2-3 rule of pruning for suitable varieties (10.13A and 10.13B) (see chapter 11, "Pruning Mature Trees").

Open Tatura

The open Tatura system differs from the Tatura trellis in that trees alternate left and right, there is one leader per tree, and the angle of the trellis posts forming the V is 17.5° from vertical. It is called open Tatura because the trees in each row are separated by a narrow strip 1½ feet (0.5 m) wide. This system restricts root growth while allowing ample room for the trees to develop simple structures, provides even light distribution, and allows easy access to all parts of the canopy.

Purchase large, unheaded nursery trees if possible. The trees may attain better branching if they are stored 4 to 5 weeks at 34°F (1.1°C) in high-humidity conditions before planting. The trees should be stored in bins filled with damp sawdust and should not be stored with fruit.

Plant trees 2¼ feet (0.7 m) apart within the row and alternate them left and right in a staggered (diamond) fashion so that they are positioned 4½ feet (1.4 m) apart on either side of the trellis. Allow 15 feet (4.6 m) between rows from center to center. The actual width of the row is 13½ feet (4.1 m) because each row consists of two rows of trees separated by a 1½-foot strip. This distance between trees maximizes the surface area of the canopy per acre and still allows narrow tractors, other orchard equipment, and bins to pass through the rows. The maximum height of the canopy is about 8 feet (2.5 m).

The closer the trees are planted together in the rows, the simpler the tree structure and the easier and faster it is to train the trees to intercept a high proportion of sunlight. The 4½-foot (1.4-m) spacing between the trunks of the trees on one side of the trellis allows 2¼ feet (0.7 m) between roots and 1,290 trees per acre (3,186 per ha). This offers a compromise between affordable tree density, filling the space on the trellis quickly with fruiting units, and maintaining good light distribution within the canopy. A typical open Tatura trellis is shown in figures 10.14A and B.

First-year training

After planting, maintain the trees as in second-year Tatura trellis trees. Remove feathers that do not obey the 3-to-1 rule with a bench cut and stub back smaller feathers 3½ inches (9 cm) to a bud that faces downward. Do not head leaders. Dormant winter oil applied as described for second-year Tatura trellis trees may enhance bud break and lateral branching. Keep the roots and soil wet to avoid transplant shock, which can prevent the leaders from forming new shoots.

Second-year training

Manage the trees in the same manner as third-year Tatura trellis trees. Do not girdle trees that have not branched well. However, score the leaders and notch individual buds. At this stage, trees should no longer be subject to transplant shock and should respond well to scoring or notching.

Third-year training and beyond

Tree management procedures are the same as for Tatura trellis in the fourth year and beyond.

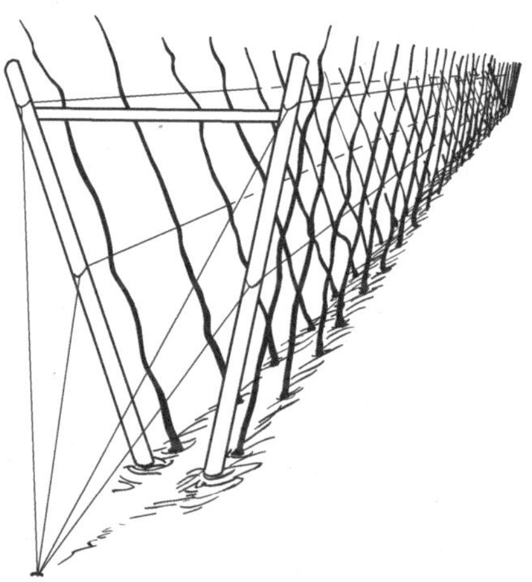

Figure 10.14A. Open Tatura training system.

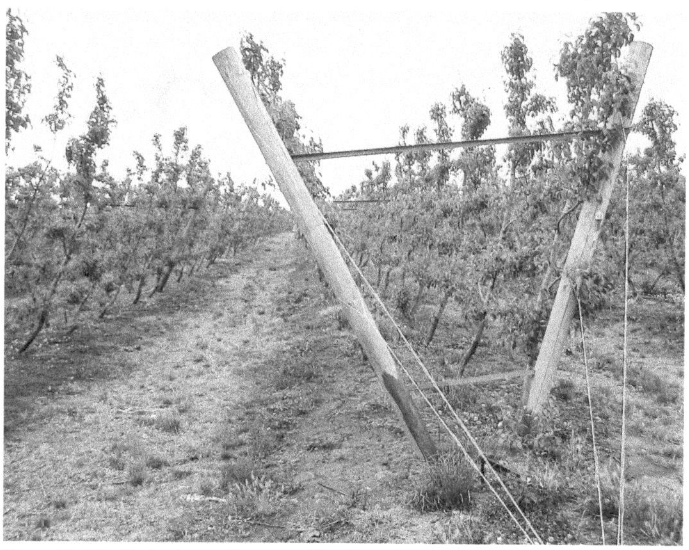

Figure 10.14B. Typical open Tatura trellis. *Photo:* B. van den Ende.

REFERENCES

Barritt, B. H. 1993. Managing sunlight key to high density success. Good Fruit Grower 44(13): 35–38.

Chalmers, D. J., B. van den Ende, and L. van Heek. 1978. Productivity and mechanization of the Tatura trellis orchard. HortScience 13:517–521.

DeJong, T. M. N.d. Physiological principles of pruning and training. Unpublished manuscript.

Dirix, L. 1995. Snoei "light" in de boomgaard. Fruitteelt Nieuws (Jan.): 10.

Elkins, R., and T. M. DeJong. 2002. Effect of training system and rootstock on growth and productivity of 'Golden Russet Bosc' pear trees. Acta Horticulturae 596:603–608.

Forshey, C. G., D. C. Elving, and R. L. Stebbins. 1992. Training and pruning apple and pear trees. Alexandria, VA: American Society for Horticultural Science.

Micke, W., et al. 1980. Pruning fruit and nut trees. Oakland: University of California Division of Agriculture and Natural Resources Leaflet 21171.

Ohlendorf, B. L. 1999. Integrated pest management for apples and pears. 2nd ed. Oakland: University of California Division of Agriculture and Natural Resources Publication 3340.

Robinson, T. L. 1993. Light interception critical to yield in high density blocks. Good Fruit Grower 44(4): 33–36.

Ross, N. W. 1974. Stanislaus orchard handbook. Modesto: University of California Cooperative Extension.

Stebbins, R. 1990. Training and pruning pear trees. In J. Beutel and R. Elkins, eds., Pear growing in the 1990s. Proceedings of a short course held November 27–29, 1990, Kelseyville, CA. Davis: University of California, Davis, Department of Pomology.

———. 1992. Training pear trees in commercial orchards. Pullman: Washington State University Pacific Northwest Extension Publication No. 403.

Strydom, D. K. 1993. Training and pruning of pear trees in countries characterized by long hot summers and where dwarfing rootstocks cannot be used or are not available. Acta Horticulturae 367:13–137.

Tustin, S. D. 1990. The principles and practices of training Slender Pyramid trees for high intensity production. Compact Fruit Tree 23:83–92.

van den Ende, B. 1984. Training, management and productivity of pear trees on Tatura trellis. Victoria, Australia: Victoria Department of Agriculture Ag Note.

———. 1990. Control vigor to optimize productivity. In J. Beutel and R. Elkins, eds., Pear growing in the 1990s. Proceedings of a short course held November 27–29, 1990, Kelseyville, CA. Davis: University of California, Davis, Department of Pomology.

———. 1995. Advanced pear pruning and training. Shepparton, Victoria (Australia): Goulburn Valley Community College.

van den Ende, B., and D. J. Chalmers. 1983. The Tatura trellis as a high-density system for pear trees. HortScience 18:946–947.

11
Pruning Mature Trees

RACHEL B. ELKINS, DAVID J. BURKHART, WARREN C. MICKE, ROBERT STEBBINS, BAS VAN DEN ENDE, AND CHUCK A. INGELS

The long bearing life of pear trees makes it necessary to keep them in a vigorous condition to ensure long-term production of large, quality fruit. Pruning mature trees should renew and maintain 1- to 3-year-old fruiting wood and also maintain adequate penetration of light into all parts of the trees by removing overly vigorous limbs in the upper part of the tree and avoiding an "umbrella" effect (fig. 11.1). Pruning also favors pest control because sprays can more easily penetrate trees that have been pruned for good light distribution. Other reasons for pruning are to adjust potential crop size, remove interfering branches, and contain tree height and spread while maintaining adequate vigor and providing access for workers.

Mature European pear trees in the western United States on standard rootstocks grow from about 12 feet (3.7 m) to over 20 feet (6.1 m) tall, depending on rootstock, soil, and management. It is common to see 14- to 16-foot (4.3- to 4.9-m) ladders being used to prune and harvest pears (fig. 11.2). Tall trees are a problem not only because of labor considerations but also because they reduce fruit production. Pear trees exhibit apical dominance, that is, they tend to grow very upright, especially as young trees (fig. 11.3). Vigor tends to stay close to the trunk; limbs that spread widely under a crop load may break or lose vigor on the ends. Lower limbs tend to lose vigor, and upper limbs tend to grow above lower ones and shade them out. The zone of fruiting tends to move upward from the lower limbs to the treetop and the ends of the upper branches as trees mature (figs. 11.4 and 11.5).

Limbs in poorly lit sections of a tree tend to drop developing fruit soon after bloom. Shading within a tree can be caused by overcrowding limbs, tall higher limbs, and crowding between trees. Arching or horizontal limbs, especially those that are high in the trees, often produce vigorous upright shoots, or watersprouts. When these watersprouts are pruned off the limb, each

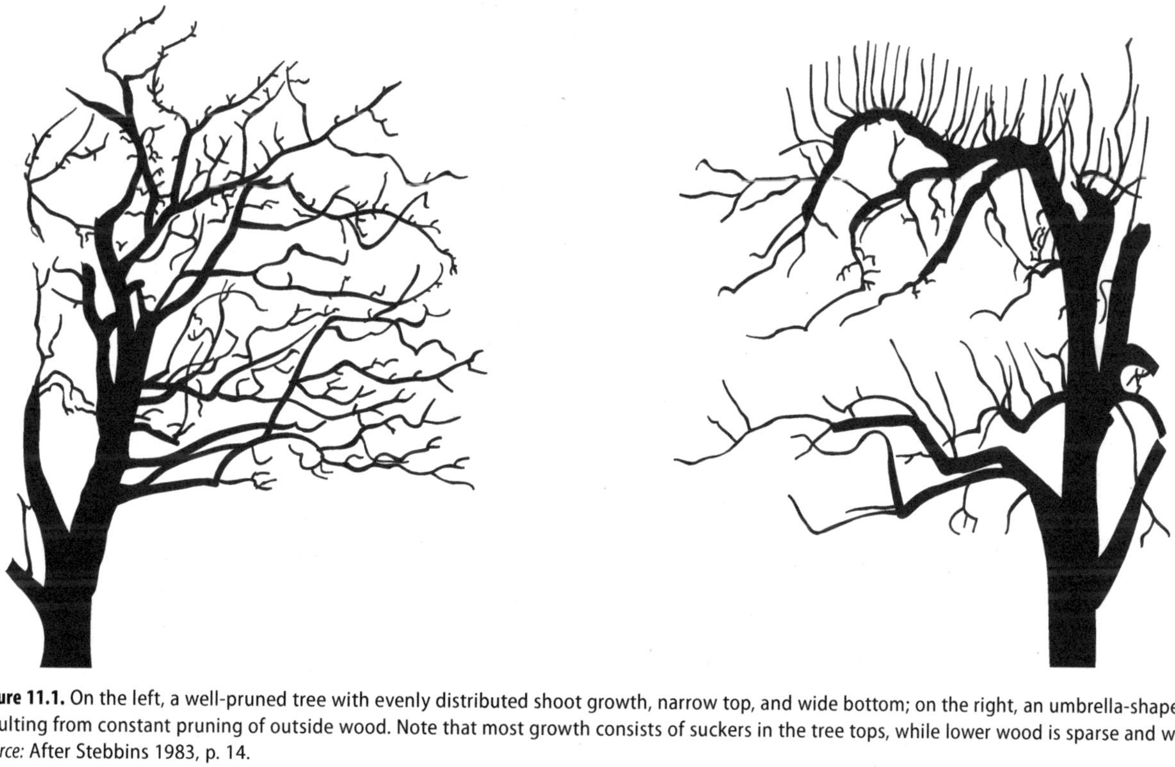

Figure 11.1. On the left, a well-pruned tree with evenly distributed shoot growth, narrow top, and wide bottom; on the right, an umbrella-shaped tree resulting from constant pruning of outside wood. Note that most growth consists of suckers in the tree tops, while lower wood is sparse and weak. *Source:* After Stebbins 1983, p. 14.

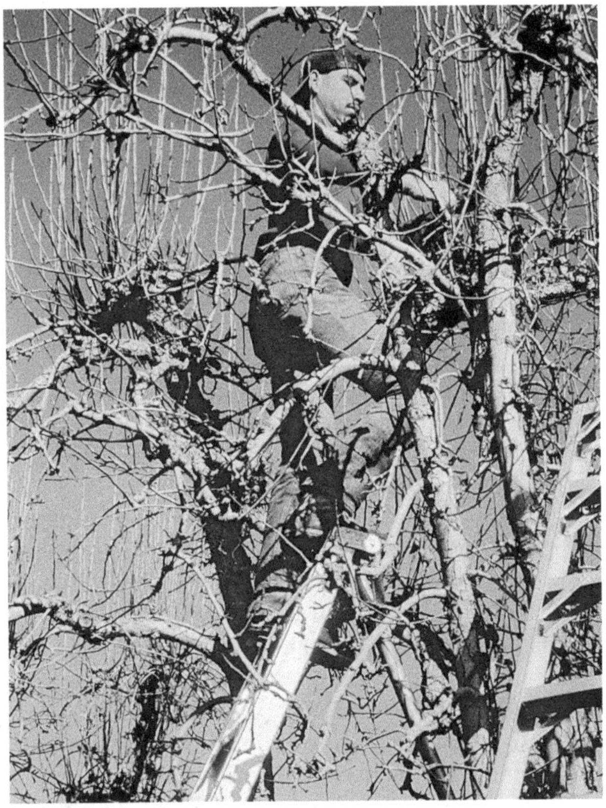

Figure 11.2. Pruning a Bartlett pear tree from a 16-foot (4.9-m) ladder in Lake County, California. *Photo:* J. Fitzgerald.

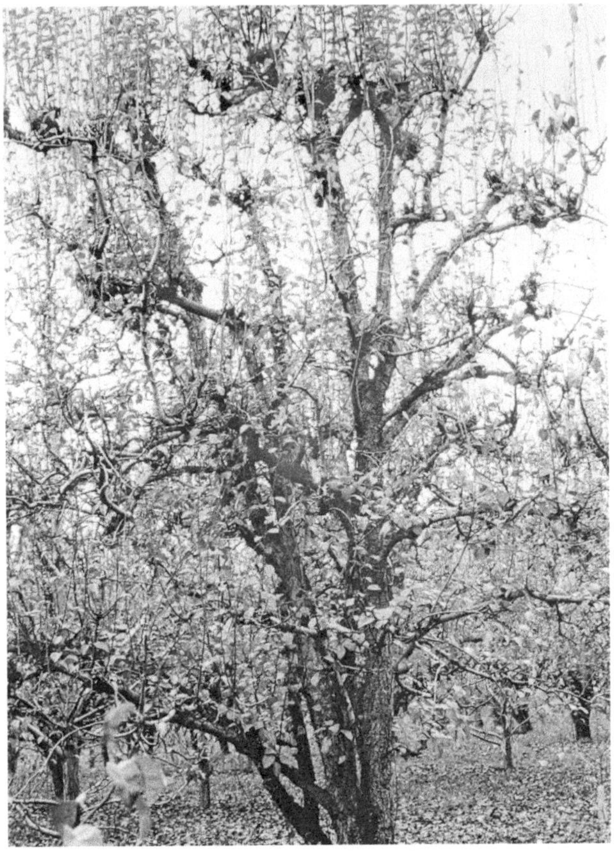

Figure 11.3. A typical upright-growing Bartlett tree that has good production in the lower limbs. The middle limb in the top of the tree needs to be removed for optimal light penetration. Note that overgrown and enlarged stubs have been allowed to grow as a result of yearly pruning of watersprouts. These stubs, which are homes to certain pests, should be removed with a saw cut. *Photo:* D. Burkhart.

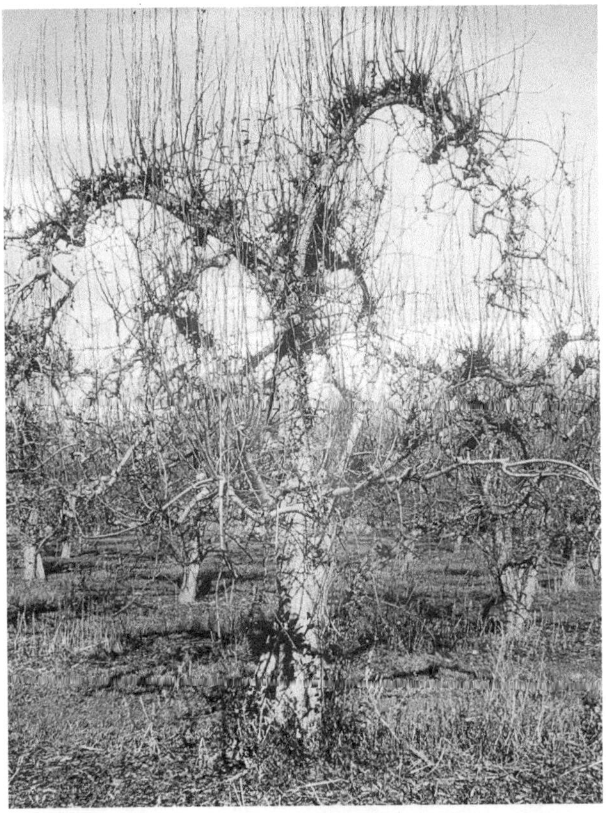

Figure 11.4. Tree shape can influence the amount of shade within a tree. Umbrella-shaped pear trees, like the one shown here, reduce the amount of light that penetrates the interior of the tree and becomes available for the growth of healthy buds and quality fruit. *Photo:* D. Burkhart.

cut can produce two or more additional shoots. Thus, areas within or between trees that have been cleared of water sprouts tend to fill with new growth, sometimes in a single season (fig. 11.6).

Pears produce fruit terminally on long shoots or spurs (see chapter 8, "Vegetative Growth and Fruit Development"). The most productive fruiting limbs and spurs in pear trees are 1 to 3 years old. Flowers on old, weak spurs set very few fruit (fig. 11.7). Trees that are low in vigor and have few new shoots may have an abundance of flowers but set light crops of small fruit.

ANNUAL MAINTENANCE PRUNING

Pruning mature pear trees consists of removing weak wood and upright shoots while thinning branches throughout the tree. Upper branches should be cut back to laterals, but lower branches may be weakened if cut. In addition to thinning shoots and branches, thin or remove the spurs or renew them by removing older parts of some branched spurs in years when fruit spurs are numerous. In years of low fruit spur production (following heavy crops), retain terminal spurs on short shoots and lateral spurs on longer branches. Remove spurs located along the underside of branches (fig. 11.8).

Top pear trees annually by cutting branches back to laterals to maintain a manageable tree height. Trees should be pruned from a ladder of the same height as one intended to be used at harvest. Ideally, tree height should be held to about 80 percent of row width in order to prevent shading caused by overshadowing from trees in the adjacent row. This is particularly crucial where tree rows are oriented parallel to the sun's movement (east to west). Maintain space between leaders of multiple-leader trees and between trees to allow for positioning of ladders. At times, scaffolds must be removed from the trees. This should be done by experienced pruners prior to the main pruning. Each scaffold should be removed completely with a flush cut at the point where the scaffold originates. To avoid splitting the bark below the limb, first undercut the limb about one-third the way through and then finish with a cut from above.

The effect of apical dominance can be used to limit the formation of watersprouts in the tops of trees. At the end of each main scaffold or side limb, retain a strong shoot and head it lightly by removing one-quarter to one-third of the shoot. Completely remove any other vigorous shoots near it. Weaker shoots can be left unheaded to form fruiting wood, but thin them to 8 to 12 inches between shoots along branches. Remove limbs with angles below horizontal unless they end in a shoot that curves upward (fig. 11.9).

Growers often develop different pruning methods for different situations, and no one method is necessarily best. Each orchard or block must be evaluated separately and pruned differently, if necessary. For example, for trees of excessive vegetative vigor it may be appropriate to retain more flower buds than for trees of moderate vigor in order to help control tree growth. There is a point, however, at which fruit size declines with

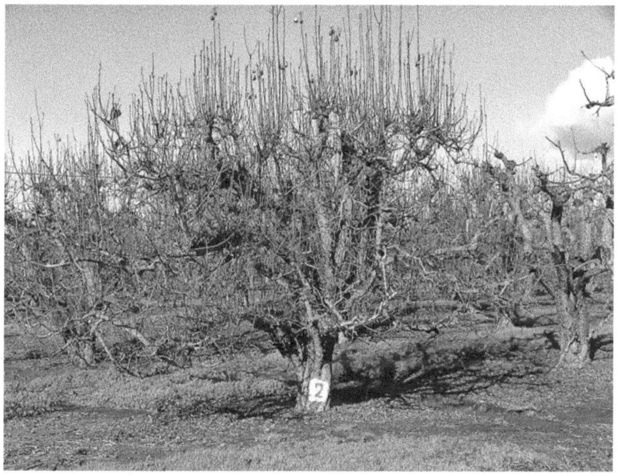

Figure 11.5. Old umbrella-shaped Bartlett pear trees in the Sacramento River Delta before and after pruning. *Photo:* R. McClain.

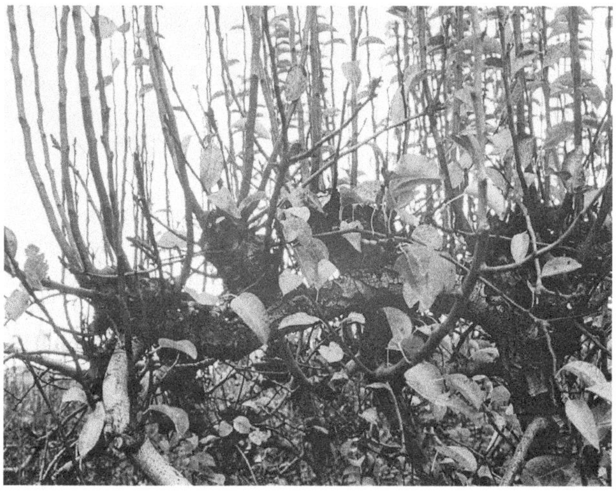

Figure 11.6. Heavy cutting in the top of a tree causes an abundance of strong, vertical-growing watersprouts, or suckers. Remove much of this new growth to settle the tree into annual fruiting, a process that often takes 1 to 3 years. *Photo:* D. Burkhart.

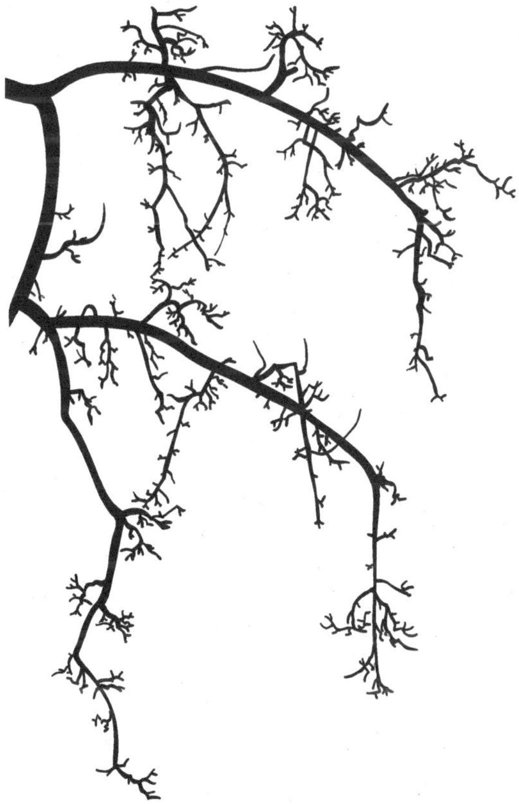

Figure 11.7. Old, weak spurs in umbrella-shaped trees generally produce small crops of mediocre fruit. *Source:* After Stebbins 1990, p. 19-7.

increasing numbers of fruit. If yields are unusually high one year, they will likely be lower the following season, necessitating lighter pruning to retain more buds, and vice versa. This alternate bearing phenomenon is more common with some pear varieties, such as Bosc. Heavier pruning generally results in larger fruit size.

Provide pruners with simple but clear instructions and demonstrate the instructions on one or two trees that they can use as reference trees. Use the same demonstration trees every year so as to comment on the results of previous pruning efforts. Typical instructions may be:

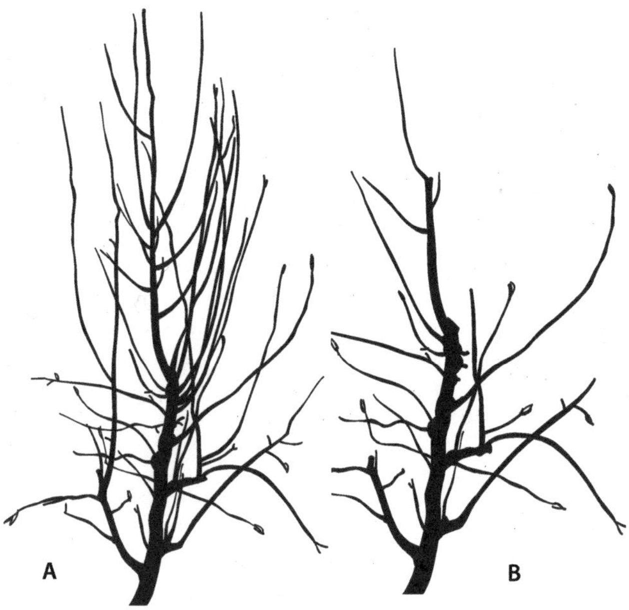

Figure 11.8. Upper scaffold limb before (A) and after (B) pruning. The terminal shoot has been headed and competing shoots have been removed. Weaker shoots left unheaded terminate in flower buds. *Source:* After Stebbins 1990, p. 19-10.

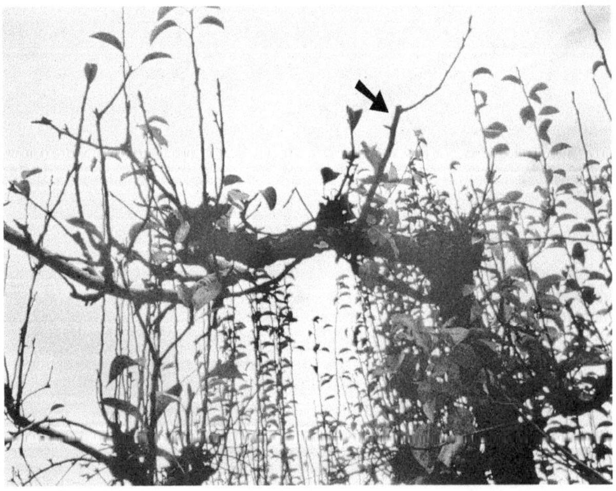

Figure 11.9. Strong, upright growth has been removed from the limb in this picture. Smaller, horizontally inclined growth should be retained and left unpruned. The 2-year-old limb in the middle of this picture should not have been headed, as this delays fruiting (see arrow). *Photo:* D. Burkhart.

- Walk around the tree to see if there are any major scaffolds that need to be removed; remove these first.
- Clean out ladder bays to make setting the ladder easier.
- Stand on the ladder two steps from the top and control tree height by cutting to an upright shoot near the end of each leader; leave this shoot unheaded unless growth extension is desired.
- Prune upper and outer limbs shorter than lower limbs by cutting to upright shoots; head these only if a longer branch is needed.
- Prune to create space between all scaffold limbs.
- Head back old spur systems, especially in shaded areas.
- Remove all long shoots and shoots that are less than pencil-thick. Do not head shoots that are moderately vigorous, pencil-thick, and spaced 8 to 12 inches (20.5 to 30.5 cm) apart.

GUIDELINES FOR PEAR VARIETIES

The majority of Bartlett pear fruit is produced on short laterals with terminal fruit buds. In the winter, select these laterals that are 6 to 12 inches (15 to 30.5 cm) apart (wider spacing for less-vigorous trees). A combination of laterals with fruit buds and young spurs ensures consistent yields. Remove laterals that lack either a terminal fruit bud or spurs. Annually renewing the laterals prevents the growth of long, spurred hangers; shortening spurs generates new laterals and helps produce high-quality fruit.

Varieties such as Anjou, Bosc, Comice, and Packham's Triumph bear primarily on 2-year-old or older wood. They tend to set more fruit near a pruning cut. To stimulate fruit set, leave 1-year-old shoots up to 2 feet (0.6 m) long unheaded, and then head them to a flower bud late the following winter.

1-2-3 Rule

The "1-2-3 rule" of dormant renewal pruning provides a systematic approach to pruning both young and mature trees. The "1" of the 1-2-3 rule refers to 1-year-old laterals, or pencils. These are 12 to 16 inches (30.5 to 40.5 cm) long and a little thinner than a pencil with fruit buds at the tips. These laterals should never be headed; excessive numbers can be thinned to one per 6 to 12 inches (15 to 30.5 cm) along the branch, keeping those that are horizontal and almost pencil-thick. Remove strong, upright shoots and long, thin laterals without fruit buds, sometimes called "wire wood." About one-third of the total renewal wood should be 1 year old (fig. 11.10A).

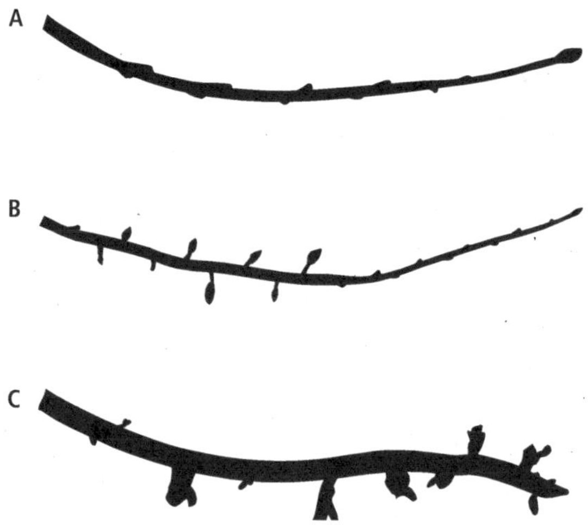

Figure 11.10. These figures illustrate the three stages of shoot growth and how they are treated under the 1-2-3 rule of renewal pruning. (A) 1-year-old shoots no thicker than a pencil, which are left to form fruit and spurs. (B) 2-year-old spurred-up lateral, which generally produces high-quality fruit. (C) 3-year-old wood that has already cropped and has been cut back to varying degrees to generate new shoots. *Source:* After van den Ende 1995, pp. 13–14.

The "2" refers to 2-year-old laterals. These have borne one or more fruit at the tip and spurred up, and the fruit bud at the tip has grown two side, or bourse, shoots. There will often be differences in the fruitfulness of 2-year-old wood on the same tree. Heading cuts back to the ring between 1- and 2-year-old wood promote fruit set on stronger wood; heading deeper than the ring on weaker wood reduces the number of buds (see chapter 10, "Training Young Trees," for details on types of heading cuts). About one-third of the renewal wood should be 2 years old (fig. 11.10B).

The "3" refers to 3-year-old wood that cropped the previous season. Head this wood severely to begin the renewal cycle again. Desirable young fruiting spurs may be left on 3-year-old wood, but the wood should still be cut back to develop new laterals. About one-third of the renewal wood should be 3 years old (fig. 11.10C).

The goal of the 1-2-3 rule is to moderate tree vigor while producing large pears. Vigor is directed into 1- and 2-year-old wood with many large leaves that receive plenty of sunlight. After they have stopped growing in mid to late June, spur leaves contribute to early growth of the fruit, while lateral leaves are important for final fruit growth.

The 1-2-3 rule of renewal pruning is easy for pruners to learn, less expensive than detailed spur pruning, and ensures optimal spur vigor for high fruit quality. Biennial bearing may also be avoided (assuming, where it is employed, that cross-pollination is adequate). The 1-2-3 rule should be complemented by following the 3-to-1 rule of pruning to avoid building up large secondary wood that competes with the leader(s) (see chapter 10, "Training Young Trees").

REDUCING TREE SIZE AND RENEWING THE CANOPY OF OLD TREES

The following guidelines offer techniques for achieving long-term production of large, quality fruit. While all varieties will respond to these techniques, the benefits of heavy pruning and canopy renewal may be less for Bartletts than for winter pears. In all varieties, canopy renewal may take several years to accomplish.

The first phase in renewing the canopy of an older tree is to lower the height in order to open the tree to sunlight. Create light channels, or windows, by making large, well-chosen cuts at various places at or near the

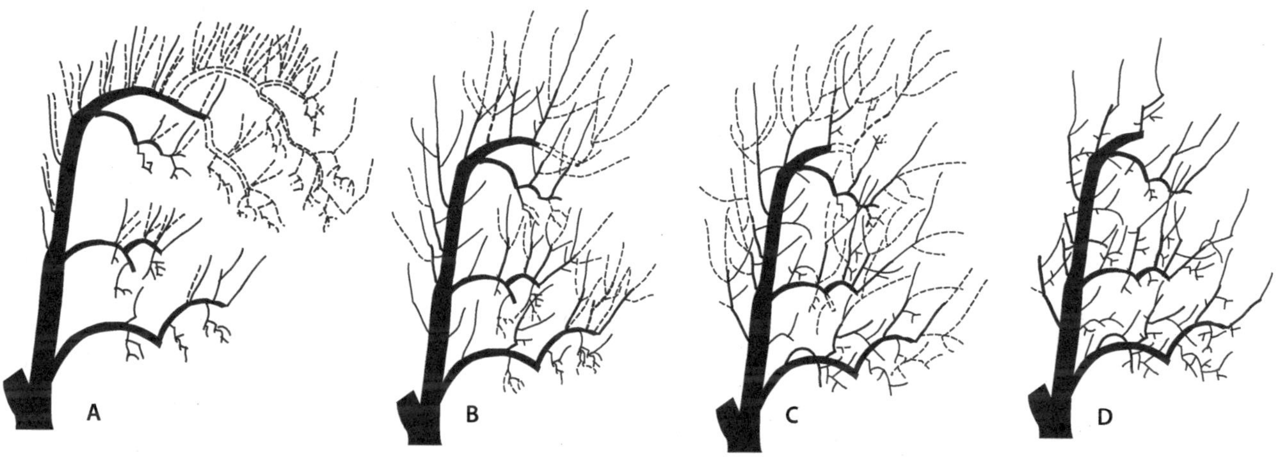

Figure 11.11. The four phases of reshaping and rejuvenating an umbrella-shaped tree. A: Remove excessively overhanging branches and the typical "sucker crowns" at the top. Leave a few weak-angled suckers unheaded to renew fruitwood. B: Thin out new growth, emphasizing fruit wood renewal. C: Continue the process, making sure upper limbs do not overgrow lower ones. Maintain a "Christmas tree" shape. D: Note Christmas-tree shape and only a few upright suckers at top to preserve apical dominance. Fruiting wood is unheaded to form spurs. *Source:* After Stebbins 1983, p. 16.

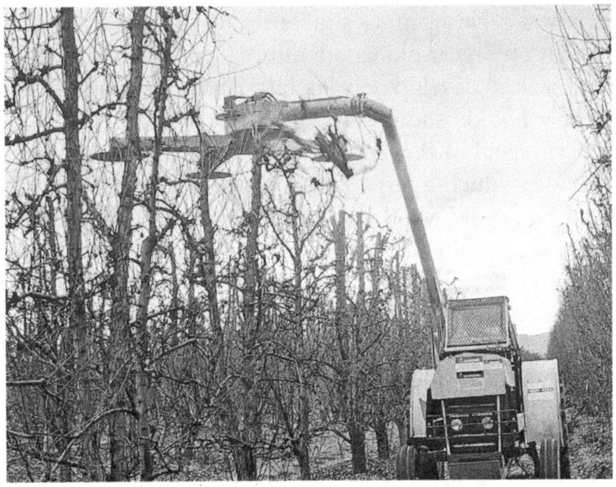

Figure 11.12. Lowering a mature hedgerow Bartlett pear orchard using a machine. Follow-up hand-pruning is necessary to remove stubs, thin out brush, and shape the trees properly. *Photo:* R. Elkins.

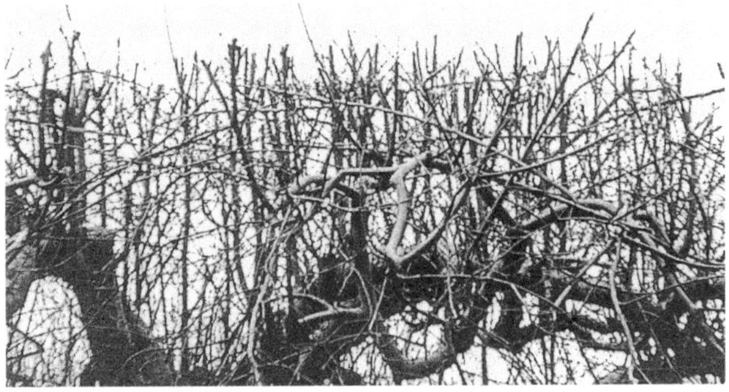

Figure 11.13. Failure to properly follow up dormant mechanical topping creates a thick, dense canopy that interferes with light penetration. *Source:* Forshey, Elving, and Stebbins 1992, p. 31 (reprinted with permission).

top of the tree. With central leader trees, it may be necessary to take out part of the main stem or a whorl of scaffolds; with open center trees it may be necessary to remove a layer of scaffolds. Be careful to preserve a symmetrical tree with sound structural limbs when making these larger cuts (fig. 11.11).

Using mechanical toppers to lower a tree should be approached cautiously. Toppers make indiscriminate cuts, leaving large stubs, scalped limbs, and possibly sites for the entrance of decay-causing organisms (figs. 11.12 and 11.13). If a mechanical topper is used on healthy, vigorous trees, the operation should be performed immediately after harvest in the late summer or early fall when tree vigor is at its lowest. Low-vigor trees should be topped in the winter to stimulate growth. Topping should be followed up in either case with hand pruning in the dormant season to remove stubs and forks, thin out brush, and shape the tree properly. Chainsaws allow cuts to be made more selectively and at a slant. To prevent accidents, chainsaws should be employed from a tree squirrel or attached to poles rather than standing on a ladder. Powered or nonpowered pole saws can also be used, again, not from ladders. Major cuts in old trees should be selected by an experienced pruner who can mark the location of the intended cut from the ground using a long-handled paint roller.

Large canopy openings often appear after big cuts are made. New growth rapidly fills in if a tree is healthy. Remove vigorous, upright new sprouts and thin moderately vigorous ones or leave them unheaded for optimal fruit set. By the second year some fruit will begin to set in the renewed area. Ultimately the new limbs may be selectively headed to direct growth and improve fruit quality. Heavy pruning temporarily reduces subsequent yields; the yield of a healthy tree, however, will recover quite rapidly. Yields are generally reduced the first year, but by the third and fourth years, trees with renewed canopies generally produce more and larger fruit (see chapter 9, "Canopy and Light Management").

The second phase of renewing the canopy of an old pear tree is to modify the typical umbrella shape, which is often responsible for excessively shading lower limbs. A greater proportion of the crop can be grown from scaffolds that are closer to the ground if this shading is prevented. Upper scaffolds must often be headed back severely and forks eliminated in order to accomplish this. Usually the process of narrowing the top and reinvigorating lower weak scaffolds takes several years. It is difficult to replace or invigorate a lower limb in trees with excessive top growth because once higher scaffolds become dominant, lower ones become weak and shaded out. By creating light channels and fostering new growth, however, it may be possible, though difficult, to revive some lower scaffolds (see fig. 11.11B).

The third phase of canopy renewal is to thin out many of the smaller, crowded limbs that may exist within the canopy. Most often these limbs, after repeated fruiting, become pendant hangers with very low vigor. Upright shoots often develop where the limb bends below the horizontal as it loses its apical dominance and heavy fruiting occurs. These upright shoots take over and become new limbs when not removed, leaving the original limb in a shaded area of the tree. The shaded limbs often bear small, poor-quality fruit and should be removed or renewed by removing shading limbs above them (see fig. 11.11C).

The fourth phase of renewal is to stimulate uniform vigor throughout the tree. The goal of this is to stimulate the production of new fruiting wood; thus, the old adage, "A tree is only as old as its fruiting wood."

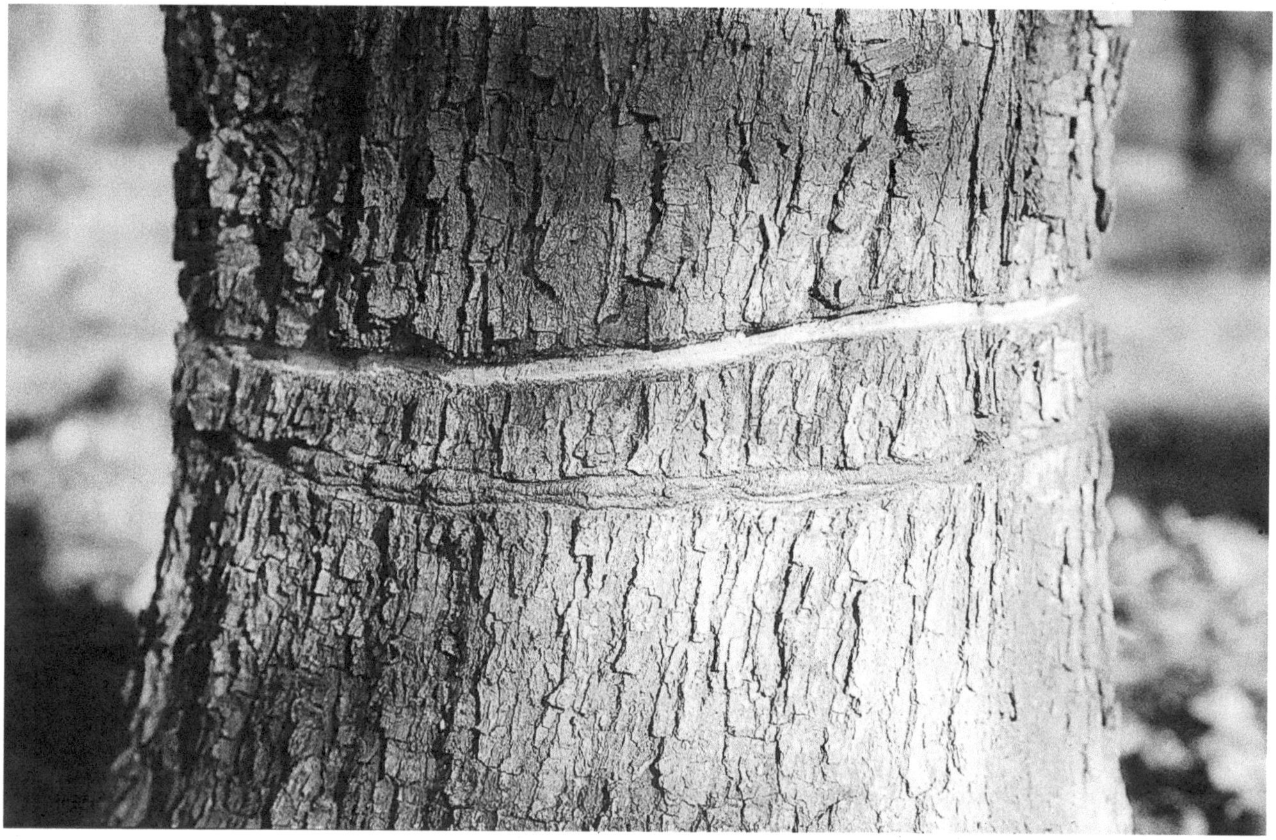

Figure 11.14. Chainsaw girdling cut through bark around entire trunk. Fully healed previous year's cut is below the cut shown here. *Photo:* C. Ingels.

Fruiting spurs under 5 to 6 years of age produce the largest fruit, so old spurs should be removed or reinvigorated. Leaving a stub when removing an old fruit spur will usually, but not always, lead to the development of new growth (see fig. 11.11D).

Watersprouts and other new growth, sometimes referred to as "wire wood," often grow after opening the tree to sunlight. Remove watersprouts but retain limbs that tend to be more horizontal and leave them unheaded. The process of renewing fruiting wood is an ongoing practice. Spur systems that are 2, 3, and 4 years old must be encouraged throughout the tree. If the new growth is long and whippy, the fruit may be damaged by wind; heading cuts will stiffen the new growth.

Though horizontal limbs should be encouraged, main bearing scaffolds may become too flat and have too many watersprouts in the top of the tree. Though it can be a challenge, keeping a main scaffold growing in a more upright position reduces the growth of new shoots. Smaller vertical growth, however, must be removed each year, but flatter, weaker shoots can be kept. These develop into desirable short bearing limbs that can be controlled with fruiting and pruning. A fully renewed Christmas-tree-shaped tree is shown in figure 11.11D.

TRUNK GIRDLING

Trunk girdling can improve fruit set and fruit size, advance maturity, and reduce shoot growth. On pear trees, girdling can reduce excessive vigor with minimal effect on yield or fruit size. This allows more sunlight to reach lower fruiting wood to keep it more productive. However, the practice should be approached cautiously, as trees can be stunted or killed if cuts are made too deep or too wide. Another concern is the possible entry of disease-causing organisms such as *Pseudomonas syringae* (blossom blast) or *Erwinia amylovora* (fire blight) (see chapter 21, "Diseases"). Blossom blast is favored by cold, wet years. Fire blight was reduced on girdled versus nongirdled trees in an experiment in California, despite heavy infection potential, likely due to reduced vigor.

Girdling should be done about 2 weeks before full bloom for best results. The most common method is ring girdling, in which a double cut is made around the trunk through the bark to the cambium layer, which is just under the phloem layer (fig. 11.14). All the phloem tissue must be removed. The cut can be a connected circle, a disconnected but overlapping spiral, or two

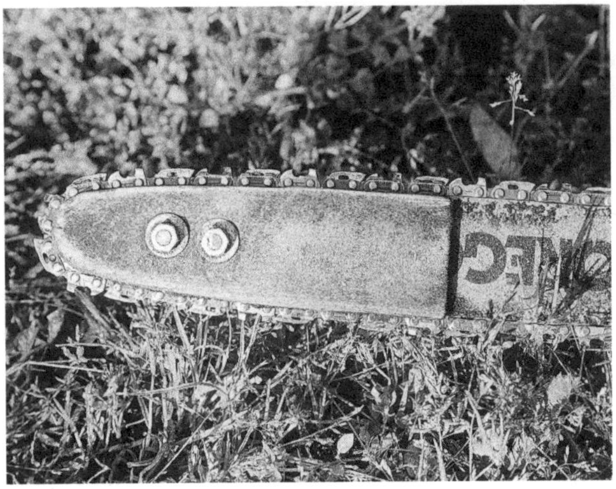

Figure 11.15. Wood guide bolted onto chainsaw bar to prevent cutting too deeply into wood. *Photo:* C. Ingels.

offset C-shaped cuts around the trunk. The cut should be about ¼ to ⅜ inch (6.5 to 9.5 mm) wide, with wider cuts used for larger trunks. On small trees, use a girdling knife. On larger trees, use a small chainsaw with a wooden guide the size and shape of the chainsaw bar bolted onto the bar to prevent cutting too deep (fig. 11.15). Examine the cut surface a short time after girdling to make sure all phloem tissue within the cut area has been removed. Phloem tissue that remains uncut turns brown within a few minutes after cutting.

Callus begins to form along the cut edges within a few weeks, and the phloem connection is reestablished within a month or two. The vigor reduction from trunk girdling generally occurs only in that season, so it must be done every year for consistent results. Girdling over previous year's cuts should be avoided, as the tissues in old cuts may not regrow properly.

REFERENCES

Forshey, C. G., D. C. Elving, and R. L. Stebbins. 1992. Training and pruning apple and pear trees. Alexandria, VA: American Society for Horticultural Science.

Hoying, S. A., and T. L. Robinson. 1992. Effects of chain saw girdling and root pruning of apple trees. Acta Horticulturae 322:167–172.

Ingels, C. 2000. Trunk girdling pear trees for vigor control. Tree Fruit Magazine 32(2): 10–11.

Magness, J. R. 1957. Pear growing in the Pacific Coast states. Washington, DC: USDA Farmers' Bulletin No. 1739.

Micke, W. et al. 1980. Pruning fruit and nut trees. Oakland: University of California Division of Agricultural Sciences Leaflet 21171.

Stebbins, R. L. 1983. Training and pruning apple and pear trees. Oregon, Washington, and Idaho Cooperative Extension Publication PNW 156.

———. 1990. Training and pruning pear trees. In J. Beutel and R. Elkins, eds., Pear growing in the 1990s. Proceedings of a short course held November 27–29, Kelseyville, CA. Davis: University of California, Davis, Department of Pomology.

van den Ende, B. 1995. Advanced pear pruning and training. Shepparton, Victoria (Australia): Goolburn Valley Community College.

12 Plant Bioregulators

STEPHEN M. SOUTHWICK

INTRODUCTION

There have been many attempts to develop commercial uses of plant growth regulators (plant bioregulators, or PBRs) in pear culture. However, although a number of responses after growth regulator application have been documented, very few commercial plant growth regulator products are in wide use throughout the western United States.

This chapter covers PBRs that have been used on pears experimentally and those currently being used in commercial pear production. To obtain a more in-depth review of historical PBR use in apples and pears, see Miller 1988 and Nicotra 1982.

Naturally occurring plant hormones that function as plant growth regulators are generally thought to be molecules that exist at low concentration that facilitate intercellular communication (Cleland 1999). They often interact with one another in plants to elicit responses. Plant hormones are generally grouped into five categories: auxins, gibberellins, cytokinins, abscisic acid, and ethylene.

Indole-3-acetic acid (IAA), which is found in plants, is an example of an auxin. Auxin-like compounds that are used in agriculture but are not found in plants include 1-naphthaleneacetic acid (NAA) and 2,4-D. They are auxin-like because the molecules have structural similarities with IAA and have biological activity. Auxins have been shown to affect cell elongation, apical dominance, tropic responses, abscission of leaves and fruits, ethylene biosynthesis, and bud and root formation (Cleland 1999).

Numerous gibberellins are found in plants. Gibberellins are a large group (more than 100) of related compounds that have some biological activity and share the gibbane ring structure. Gibberellins appear to have a major role in cell elongation, partly by activation of the intercalary meristem in stems and grass leaves (Cleland 1999). Gibberellins have been shown to affect bolting in rosette plants, promote germination of seeds, promote fruit development or alter growth, inhibit and delay flowering, and alter the sexual development of some flowers and control juvenility.

Natural cytokinins are a series of adenine molecules modified by the addition of 5-carbon side chains off the 6 position (Cleland 1999). Two cytokinin groups include the zeatin-related and adenine-related molecules. Cytokinin-like substances include kinetin and 6-benzyladenine. Cytokinins are normally found in root apices, developing embryos, and apical buds. They are thought to affect cell division, apical dominance, senescence of leaves and stems, and root growth. They are most often thought to be produced in root and bud meristems and to stimulate cell division and associated activity in those and other plant organs.

Abscisic acid (ABA) occurs naturally as the cis-(+)-isomer and can be made in all plant parts, but the leaves and root caps appear to be sites of extensive synthesis (Cleland 1999). ABA is made in response to environmental signals like water stress, where it appears to help control stomatal opening and closing. ABA may have a role in dormancy of buds and seeds.

Ethylene is a gaseous molecule that may be produced anywhere in the plant. Maximal synthesis appears to be in apical buds, stem nodes, senescing flowers, and ripening fruit (Cleland 1999). Wounded tissues produce ethylene as well. Ethylene appears to affect plant development by setting in motion a series of events leading to senescence. In fruit ripening, for example, these events include cell wall breakdown, changes in pigmentation, and formation of flavor compounds. These senescence events can lead to abscission of leaves and fruits and death of flower petals. Ethylene also appears to play a role in cell enlargement. There is often a close relationship between auxin and ethylene. In pear, for example, NAA has been thought to stimulate the production of ethylene and can hasten fruit maturation and softening. Other molecules currently thought to have plant hormone characteristics include oligosaccharins, jasmonic acid, methyl jasmonate, salicylic acid, systemin, and brassinosteroids (Cleland 1999).

AUXINS AND AUXIN-LIKE COMPOUNDS

Preharvest Drop

Naphthaleneacetic acid and 2,4-dichlorophenoxyacetic acid

Since 1940, spraying pears with 1-naphthaleneacetic acid (NAA) in the western United States for the control of preharvest drop has been a commercial practice (Strickland et al. 1941). NAA formulations include 1-NAA, ammonium salt; 1-NAA, potassium salt; 1-NAA, sodium salt; 1-NAA, ethyl ester; and 1-NAA, amide. These hormone sprays were justified in California's Lake and Santa Clara Counties even with a moderate amount of drop (Davey and Hesse 1942). Hormone sprays were often added to the last cover spray, and they became especially used where light crops, light soils, or weak trees were thought to contribute to excessive fruit drop (Gerhardt and Allmendinger 1946).

Confronted with large crops and labor shortages, growers have sometimes used hormone sprays to lengthen the normal picking season or to hang fruit on the tree where fruit color or size has been slow to develop. In the early 1940s, canners and shippers experienced losses of some Bartlett lots upon arrival in eastern markets or after a brief storage period, suggesting that hormone sprays were detrimental to the keeping quality of pears (Allen and Davey 1945). Severe watery breakdown developed in Lake County pears stored for a relatively short period of 6 weeks. The problem appeared to be associated with overmaturity. When this normal dropping is delayed or prevented by the NAA sprays, Bartlett pears continue to ripen rapidly. A 7- to 10-day delay in harvesting may mean that there are some overmature fruit and some yellow, soft pears on the tree. The undue delay in harvesting greatly impairs the keeping quality. The point is that NAA can reduce the dropping of pears, especially when a slight delay in harvest is required, but increased softening may be associated with the use of NAA. When fruit are harvested at the proper maturity, the likelihood of storage problems is reduced. Although the use of NAA generally reduces preharvest drop, in some instances there is no effect on drop from NAA use. These nonresponses to treatment are not well understood. See Allen 1953; Allen and Davey 1945; ARS Special Report 1956; Gerhardt and Allmendinger 1946.

In experiments conducted by Allen and Davey (1945), they found that early-picked lots of NAA-sprayed and unsprayed pears ripened fairly evenly. Later-picked, sprayed fruit ripened ahead of unsprayed fruit, and the ripening was uneven. Pears treated with spray deteriorated rapidly as a result of delayed harvest. Overall, sprayed trees had less drop but varied from 50 percent less drop to no differences with control. The PBR spray seemed to be effective throughout the picking period in reducing drop, but its influence seemed to diminish after midseason picking in the Sacramento River and Santa Clara Valley areas. Because the spray tended to hold the fruit on the spur, pears that otherwise would drop and become straw pears remained attached and continued to ripen. For this reason, certain pears on sprayed trees became more advanced in maturity than most others. Watery breakdown occurred in pears that were sprayed and appeared relatively green but were actually more mature than the rest.

Gerhardt and Allmendinger (1946) indicated that problems developed in pears treated with NAA when growers attempted to delay harvest to improve size. Although fruit drop was generally reduced, NAA-treated fruit softened more rapidly, and with the delay in harvest to improve size, there was increased fruit softening during storage. They suggested that fruit sprayed with NAA should be harvested at the normal maturity (1 to 2 weeks after spray) to avoid these postharvest overmaturity problems.

Work by Batjer, Thompson, and Gerhardt (1948) showed that treatments of either NAA or 2,4-dichlorophenoxyacetic acid (2,4-D) sprayed 14 days before harvest at either 2.5, 5, 10, or 20 ppm were equally effective in preventing preharvest drop. 2,4-D above 2.5 ppm caused phytotoxicity, with 20 ppm killing leaves and buds. Doses of NAA above 20 ppm led to uneven fruit ripening. NAA at 20 ppm decreased return bloom by more than 50 percent, and some phytotoxicity was seen. In fact, flower and vegetative buds were killed by the treatment. Even when using NAA at label rates these effects as well as later and more erratic bloom have been observed by growers, and they seem to vary from season to season. It has been suggested that the return bloom problems are more acute when the preceding season was cooler than normal. At concentrations of 2.5 ppm, fruit drop was reduced without an effect on fruit maturity or breakdown compared to untreated fruit.

The NAA spray becomes effective within about 3 days after application (ARS Special Report 1956). If an NAA spray actually hastens ripening, then the earlier it is applied, the more ripening may occur. Therefore, with an early application, losses from overmaturity might actually outweigh the advantage of protection from wind and drop. Application of the PBR might best be delayed until after the first picking. Instead of lengthening the harvest period with NAA sprays, the treatment may actually shorten the period because the treatment hastens the ripening process.

There are different use labels for different NAA formulations (potassium salt, ammonium salt, etc.) A common NAA product is the potassium salt of NAA (K-salt, Fruit Fix 800), and the use label indicates that NAA should be applied at 12.5 to 25.0 grams per acre 5 to 7 days prior to anticipated harvest or when there is evidence of the first normal fruit drop (AMVAC Chemical Corporation International Label Guide). Growers may normally apply NAA at 25 to 37.5 grams of active ingredient per acre 7 to 10 days before harvest or at 21 to 22 pounds fruit firmness. Use of varying spray volumes per acre are outlined on the label. NAA spray should be applied when temperatures are moderate (65° to 80°F, or 18.3° to 26.7°C) and skies are clear for more consistent responses. Pick with little or no delay and reduce fruit core temperature as quickly as possible. NAA can stimulate endogenous ethylene production (Miller 1988), accelerate fruit softening, and reduce storage life. On trees sprayed with NAA, there may be a number of overmature fruit that did not drop, making it necessary for the overmature fruit to be sorted out either by the picker or by the packing crew. NAA is thought to have only a short period (up to 2 weeks) of effective control of preharvest drop. The use of NAA could be avoided in orchard blocks that are harvested early in the harvest period with no harvest delay. Fruit drop is minimal early in the harvest period. Nevertheless, most growers use NAA to insure against preharvest fruit drop.

Aminoethoxyvinylglycine

Aminoethoxyvinylglycine (AVG) is a specific inhibitor of an enzyme important in the biosynthesis (production) of ethylene gas in plant tissues. Recently, a formulated product containing AVG has been developed by Valent Corporation called ReTain. The product was developed as a harvest management tool, where it can be used to extend the normal harvest period. It can be applied approximately 4 weeks prior to harvest at 50 grams active ingredient per acre in 100 gallons of spray volume and generally improves the firmness of Bartlett pears during the normal and extended harvest period. Fruit drop is reduced as harvest is delayed (7 to 14 days) compared to untreated trees; however, the reduction in drop is not as pronounced as that observed with the use of NAA. As a result of extending or delaying the normal harvest period by using ReTain, fruit size (diameter and weight) may be increased while maintaining fruit firmness. ReTain has not been fully developed for use on pears in California and is not labelled for use there, but it appears that it could be useful for harvest and storage management. Currently, limits to the widespread use of ReTain for pears include variable performance from season to season with regard to improvements in fruit firmness, size, and preharvest fruit drop control. Application timing, spray combinations with NAA, and the use of penetrating surfactants may improve product performance and are generally recommended.

Thinning

In some seasons, Bartlett and other pear cultivars (e.g., Seckel, Red Sensation) may produce high numbers of fruit, leading to reduced fruit size at harvest. Both NAA and NAA AMIDE reduce fruit set when applied approximately 15 to 20 days after bloom (Williams et al. 1978). The potassium salt of NAA (Fruit Fix 200) is labeled for use at a rate of 15 to 20 ppm (about 13 to 19 grams/acre or 8 to 12 fl oz/acre) in a dilute spray in up to 400 gallons of water per acre. Fruit Fix 200 should be applied 15 to 28 days after full bloom with a surfactant (wetting agent). In some seasons on Bartlett there may be little if any thinning effect. NAA AMIDE (AMID THINW) is also labeled for use on Bartlett and Bosc pear and should be sprayed at a rate of 10 to 50 ppm at petal fall or within 5 to 7 days after petal fall. NAA AMIDE may, at times, cause overthinning. Less variation in thinning response (fruit set reduction) has been suggested to occur when temperatures range from 70° to 75°F (21.1° to 23.9°C) during the application period. Although increases in fruit size may approach 18 percent of untreated fruit, the fruit size increases obtained have been relatively modest. In general, pears do not show an increase in fruit size relative to crop load from chemical or hand-thinning to the same extent as do apples (Williams et al. 1978).

Flowering and Fruiting

Auxins

DeTar, Griggs, and Crane (1950) showed that sprays of 2,4,5-trichlorphenoxypropionic acid (2,4,5-TP) could increase parthenocarpic fruit set of Bartlett pear. The spray was applied in the fall (September) and increased fruit set by approximately 40 percent, but there was decreased fruit size and firmness and increased seedlessness. Subsequent studies (Griggs and Iwakiri 1961) indicated that both fall (September 25) and spring (March 20; at pink bud, 2 to 3 days prior to anthesis) sprays of 2,4,5-TP at 40 ppm increased parthenocarpic fruit set of Bartlett. Sprays applied at 20 ppm increased fruit set only when applied in the fall. The spring sprays also hastened fruit softening. Sprays of 2,4,5-TP were most effective on Bartlett, while they were less or not effective on Anjou, Hardy, or Winter Nelis. Spraying 2,4,5-TP is not allowed in California.

Daminozide

Daminozide (Alar-85; butanedioic acid mono [2,2-dimethyl-hydrazide]) stimulates flowering of pear (Dennis 1968; Griggs and Iwakiri, 1968) when applied at 1000 ppm. However, Alar is not available for use in the western United States at this time.

Watersprout Suppression

Vigorous upright shoots produced from latent buds on the trunk, on scaffolds, or around pruning wounds, especially in the upper part of tree canopies, reduce light and spray penetration. Additionally, these succulent shoots and shady areas in tree canopies are ideal feeding sites for harmful insects such as aphids and psylla. Experiments have shown that applying a mixture of 1.0 percent NAA ethyl ester in a 20 percent white latex paint solution to the trunk and scaffold limbs of mature pear trees in the early spring gave sprout control in the season of application. Effective control was achieved in the second season as well (see AMVAC Chemical Corporation International Label Guide).

Root Sucker Control

Some pear rootstocks produce undesirable shoots from adventitious buds below the soil surface, making orchard management more difficult. These root suckers can be controlled by NAA at a 0.5 percent concentration when the suckers are 6 to 12 inches (15 to 30.5 cm) tall. NAA (Tre-Hold) should be applied only to vigorous, established, nonstressed trees (see label or AMVAC Chemical Corporation International Label Guide).

GIBBERELLINS AND CYTOKININS

Flowering and Fruit Set

Increasing fruit set in some seasons would appear to have value for pear growers, especially where pears are produced through parthenocarpic fruit set (Gorter and Visser 1958). Work by Varga (1969) has demonstrated that gibberellins and kinetin (cytokinin; 6-furfurylaminopurine) were applied to 4-year-old Beurré Hardy (BH) and Doyenne du Comice (DC) trees growing in 50-liter (13.2-gal) containers. The trees were placed in greenhouses where the minimum temperature was 18°C (65°F). The application of 25 ppm kinetin at full bloom increased fruit set of both BH and DC varieties with or without pollination. Application of 25 ppm GA_3 at full bloom increased set of BH and DC. Additionally, a 25 ppm full bloom spray of GA_{4+7} increased fruit set of DC. Gibberellins stimulated seed abortion.

Kinetin sprays applied at 25 ppm 21 days after pollination reduced "June drop" in BH and DC pears. Similarly, gibberellins reduced drop in DC but not in BH. GA_3 sprays of 25 ppm applied at full bloom to BH increased fruit set, and the resulting crop was threefold higher than the untreated controls (Varga 1969). It has been shown in a number of studies that GA_3 improves fruit set in pears, and the effect is especially pronounced in varieties that produce parthenocarpic fruits, such as Conference pear (Gil, Griggs, and Martin 1972; Luckwill 1960; Modlibowska 1963; Varga 1969). Gibberellic acid is most useful in situations where fruit set is presumed to be low, for example, where spring frost or other weather conditions compromise fruit set. Gibberellin helps to promote the development of parthenocarpic fruit set. Side effects from GA use, such as misshapen fruit (protrusions at calyx ends and longer) and reduced flowering (Griggs and Iwakiri 1961) in the season after application can be reduced or eliminated by adjusting the concentration and timing of sprays (Nicotra 1982; Tromp and Wertheim 1980; Turner 1973). Gibberellic acid can inhibit flowering in the "off" year in Seckel pear (Miller 1988).

Growth Control

Chlormequat

Chlormequat (2-chloro-ethyl trimethyl ammonium chloride; Cycocel) inhibits gibberellin synthesis in certain fungi. It has been shown to be effective for controlling growth and initiating flower buds (promoting flowering) on pear in Europe. Suppression of vegetative growth (through shorter internodes) and enhanced fruiting is achieved by using two sprays at 1,000 ppm, one at petal fall (PF) and one 2 to 3 weeks later (Miller 1988; Nicotra 1982; Wertheim 1985). Cultivars such as Comice and Beurré Hardy respond well to multiple applications, but the effects of Cycocel on tree growth pervade across most cultivars tested. In the season following Cycocel application, an increase in flowering in the following season has been observed. On young trees coming into bearing, Cycocel has been show to increase early production. Reduction in vegetative growth by Cycocel leads to a reduction in the need for pruning. The use of Cycocel has been suspended in several European countries, and it has never been sold in the United States.

Ethylene

Ethylene (2-chloroethyl phosphonic acid; ethephon; Ethrel), a compound naturally produced by pear fruit during ripening, is one of several breakdown products released from ethephon spray solutions (pH above 4.0). Ethephon has been used to promote abscission of fruits (thinning) in stone fruit and apple. Knight (1982) thinned Conference pear treated with GA_3 once between bloom and petal fall. Ethephon has been used to control shoot growth in Bartlett, as a defoliant in *Pyrus calleryana* nursery trees, and to promote ripening and fruit softening in Anjou pears (Nicotra 1982).

Currently, there is no commercial use of ethephon in California pear culture, but late-ripening, rat-tail fruit can be reduced or eliminated with postharvest sprays of ethephon. The removal of these rat-tail fruit appear to aid in the management of codling moth. Commercial use of ethephon is expected in 2005 in California for this purpose.

Prohexidione calcium

Prohexidione calcium (Apogee) is a gibberellin biosynthesis inhibitor. Growth control of shoots and reduction in fire blight has been shown in European pear when the product is sprayed on mature trees at 1 to 3 inches (2.5 to 7.5 cm) of initial shoot growth. Multiple applications are recommended at 2- to 3-week intervals and spray concentrations range from 125 to 250 ppm with 100 to 200 gallons of water per acre (Southwick et al. 2004). Apogee has a registration for use in California pear production but does not have a label. The product is under development and shows promise.

REFERENCES

Allen, F. W. 1953. The influence of growth regulator sprays on the growth, respiration, and ripening of Bartlett pears. Proceedings of the American Society for Horticultural Science 62:279–298.

Allen, F. W., and A. E. Davey. 1945. Hormone sprays and their effect upon the keeping quality of pears. Berkeley: University of California Division of Agricultural Sciences Bulletin 692.

AMVAC Chemical Corporation International Label Guide. 1994. Newport Beach, CA: AMVAC.

ARS Special Report. 1956. Use of stop-drop sprays on apples and pears. Washington, DC: USDA Agricultural Research Service USDA-ARS 22-28.

Batjer, L. P., A. H. Thompson, and F. Gerhardt. 1948. A comparison of naphthaleneacetic acid and 2,4-dichlorophenoxyacetic acid sprays for controlling pre-harvest drop of Bartlett pears. Proceedings of the American Society for Horticultural Science 51:71–74.

Cleland, R. E. 1999. Introduction: Nature, occurrence and functioning of plant hormones. In P. J. J. Hooykaas, M. A. Hall, and K. R. Libbenga, eds., Biochemistry and molecular biology of plant hormones. Amsterdam: Elsevier Science. 3–22.

Davey, A. E., and C. O. Hesse. 1942. Experiments with sprays in the control of preharvest drop of Bartlett pears in California. Proceedings of the American Society for Horticultural Science 40:49–53.

Dennis, F. G. 1968. Growth and flowering responses of apple and pear seedlings to growth retardants and scoring. Proceedings of the American Society for Horticultural Science 93:53–61.

DeTar, J. E., W. H. Griggs, and J. C. Crane. 1950. The effect of growth regulating chemicals applied during the bloom period on the subsequent set of 'Bartlett' pears. Proceedings of the American Society for Horticultural Science 55:137–139.

Gerhardt, F., and D. F. Allmendinger. 1946. The influence of α-naphthaleneacetic acid spray on the maturity and storage physiology of apples, pears, and sweet cherries. Journal of Agricultural Research 73:189–207.

Gil, G. F., W. H. Griggs, and G. C. Martin. 1972. Gibberellin-induced parthenocarpy in 'Winter Nelis' pear. HortScience 7:559–561.

Gorter, C. J., and T. Visser. 1958. Parthenocarpy of pears and apples. Journal of Horticultural Science 33:217–227.

Griggs, W. H., and B. T. Iwakiri. 1961. Effects of gibberellin and 2,4,5 trichlorophenoxypropionic acid sprays on Bartlett pear trees. Proceedings of the American Society for Horticultural Science 77:73–89.

———. 1968. Effects of succinic acid 2,2-dimethyl hydrazide (Alar) sprays used to control growth in 'Bartlett' pear trees planted in hedgerows. Proceedings of the American Society for Horticultural Science 92:155–166.

Knight, J. N. 1982. Regulation of cropping and fruit quality of Conference pear by the use of gibberellic acid and thinning. II. The effect of ethephon as a flower thinner when used in conjunction with gibberellic acid application. Journal of Horticultural Science 57:61–67.

Luckwill, L. C. 1960. The effects of gibberellic acid on fruit set in apples and pears. Report of the Long Ashton Research Station 1959:59–64.

Miller, S. S. 1988. Plant bioregulators in apple and pear culture. HortReviews 10:309–401.

Modlibowska, I. 1963. Effect of gibberellic acid on fruit development of frost damaged Conference pears. Report of the East Malling Research Station 1962:64–67.

Nicotra, A. 1982. Growth regulators in pear production. Acta Horticulturae 124:131–148.

Southwick, S. M., C. Ingels, R. Hansen, and K. Glozer. 2004. The effects of Apogee on shoot growth, secondary flowering, fireblight, fruit quality, yield, and return bloom in 'Bartlett' pear growing in California. Journal of Horticultural Science and Biotechnology 79(3): 380–389.

Strickland, A. G., H. K. Kemp, and J. A. Beare. 1941. Spraying with plant growth substances. Fruit World and Market Grower 42:7–10.

Tromp, J., and S. J. Wertheim. 1980. Synthetic growth regulators: Mode of action and application in fruit production. Physiological aspects of crop production. Bern, Switzerland: Proceedings of the 15th Colloquium of the International Potash Institute. 137–150.

Turner, J. N. 1973. Gibberellic acid for controlling fruit production of pears. Acta Horticulturae 34:287–297.

Varga, A. 1969. Effects of growth regulators on fruit set and June drop of pears and apples. Netherlands Journal of Agricultural Science 17:229–233.

Wertheim, S. J. 1985. Developments in pear production in the Netherlands. Compact Fruit Tree 18:111–118.

Williams, M. W., H. M. Couey, H. Moffitt, and D. L. Coyier. 1978. Pear production. Washington, DC: USDA Agricultural Handbook No. 526.

13

Orchard Floor Management

GLENN McGOURTY AND CLYDE L. ELMORE

The amount and type of vegetation growing on the orchard floor can influence tree growth and development, fruit finish, frost protection, and yield. There are many approaches to managing this vegetation. Growers must consider the age of the trees, the time of year, aesthetic standards, the type of irrigation system, and the potential for erosion. Frequently, growers must combine and adapt various strategies to achieve their management objectives.

NEWLY PLANTED AND NON-BEARING ORCHARDS

During the first few years of a tree's life it is susceptible to competition from vegetation growing 3 to 6 feet (0.9 to 1.8 m) around the base of the tree. Weeds compete for water and nutrients and may suppress tree growth to the point where they reduce the crop. Perennial weeds are particularly detrimental, since they compete for these limiting factors during the summer. Control weeds from at least 2 to 4 feet (0.6 to 1.2 m) around the base of the tree; additional open area is usually beneficial and can result in improved tree growth.

Clean culture, or maintaining bare soil, is the best way to manage orchard floors beneath young trees to promote rapid root and tree growth. In clean culture, no weeds or cover crops compete with tree roots while the trees are actively growing. This reduced competition can be accomplished through tillage, directed sprays of nonselective postemergence and preemergence herbicides, or combinations of both. Ideally, in the fall prior to planting, the orchard site should be cultivated and weeds controlled in the area where trees will be planted. Apply preemergence sprays in a strip about 2 to 3 feet (0.6 to 0.9 m) wide on either side of the newly planted trees, protecting the trees with tree wrap (see table 13.1 and *UC IPM Pest Management Guidelines for Pear*). Oxyfluorfen, oryzalin, and napropamide are frequently used as preemergence herbicides, as they are quite safe even around newly planted trees. All preemergence herbicides require incorporation by rainfall or irrigation after application. Follow all label directions when using herbicides.

Research in apple and peach showed that improved tree growth occurs if the strip of herbicide control is increased to 6 to 8 feet (1.8 to 2.4 m) wide. The percentage of tree roots in the upper 6 inches (15 cm) of soil and overall root biomass is greater beneath herbicide-treated soil compared to grass-cover-cropped plots. As trees mature and grow more extensive root systems, they are better able to compete with cover crops and weeds, and the effect of orchard floor management generally becomes less pronounced on tree vigor and growth (Elmore 1990).

Weeds can be removed by hoeing or shoveling them away from the trunk area, mulching to reduce weed growth, or close cultivation. Plant-derived mulches (plant products) should not be placed close to the base of young trees. This is especially the case if flaming will be used for weed control. Landscape fabrics, also known as synthetic ground cover, may promote more tree growth than mechanical weed control (hand-hoeing) and application of postemergence herbicides. They also increase soil moisture content and decrease soil temperature, maintaining both at a moderated level without excessive highs or lows. Additionally, they prevent competition from weeds. Fabric costs range from $800 to $1,000 per treated acre ($2,000 to $2,500 per ha), and the estimated payback based on potential herbicide savings is 8 to 10 years (Facteau 1991).

Table 13.1. Major preemergence and postemergence herbicides for use in pear orchards

Preemergence		Postemergence	
Chemical name	**Trade name**	**Chemical name**	**Trade name**
diuron	Karmex	paraquat	Gramoxone
pendimethalin	Prowl	glyphosate	Roundup
oryzalin	Surflan	sulfosate	Touchdown
oxyfluorfen	Goal	2,4-D*	2,4-D Envy
napropamide	Devrinol		
norflurazon	Solicam		
simazine	Princep		

Notes: *Low-volume formulations preferred. Numerous trade names.

Do not disk the soil close to the trees. Disking cuts young roots and can provide an entrance for disease. If a tractor driver is not careful with the equipment, cultivators may hit and damage the trees.

ESTABLISHED ORCHARDS

In many California pear orchards, growers choose to not till the orchard floor, but mow the resident vegetation of annuals and perennials instead. This can be an option in areas where water supplies are adequate, but it requires additional irrigation compared to clean-tilled orchards.

Weed Control in the Tree Rows

Total weed control in strips 2 to 4 feet (0.6 to 1.2 m) wide beneath the trees reduces weed competition; assists in mite control by reducing perennial morningglory (field bindweed, *Convolvulus arvensis*), which harbors two-spotted mites; helps prevent rodent damage from voles (*Microtus* spp.) since it allows predators to find the animals and does not give them a hiding place near the tree trunk where they can girdle the trees; and improves the irrigation uniformity of solid-set sprinklers. Also, many growers prefer the appearance of a clean tree row.

Timing options for orchard floor management can be found in table 13.2. If a preemergence herbicide is to be used alone, it must be applied before weeds germinate in the fall. If weeds are present, apply a postemergence herbicide along with a preemergence herbicide in mid to late winter to control vegetation in the tree row. If the herbicides are applied too late in the spring, the weeds may get too large and the postemergence herbicide will not control them. A fairly broad array of selective and nonselective preemergence and postemergence herbicides are available. Sometimes it is feasible to use only postemergence herbicides by using low rates in the winter and early spring with an increased rate in the early summer (chemical mowing), or two applications in late winter and early summer. Before starting any weed management program, it is imperative to correctly identify the target species that you hope to control and choose the right program to accomplish this. For help in identifying weeds, consult the sources listed in the references at the end of this chapter, especially *The Grower's Weed Identification Handbook* and *Weeds of the West*.

The efficacy of herbicides varies with the timing and season of application. Preemergence herbicides must be applied before weeds emerge. If preemergence herbicides are applied in late fall, it is often necessary to split the application so the later treatment will have enough residual effect to control weeds into the summer. If preemergence herbicides are applied in late winter, existing weeds may be so tall that a postemergence herbicide must be applied to kill them. Also, enough moisture must be available to properly incorporate preemergence herbicides into the soil. If there is inadequate rainfall after application, the soil should be irrigated. Herbicides persist from 7 days to several weeks before rainfall must be available to incorporate them into the soil. For more information, see the label of the material being applied.

In high-rainfall areas, delaying herbicide applications until later in the winter may work better, as preemergence herbicides become diluted in the soil; they degrade and are less effective after 30 inches (76 cm) of rainfall. Additionally, soil erosion is less likely if vegetation is growing on the orchard floor during the peak of the winter rains. Some growers apply preemergence herbicides in the fall and treat the orchard again in the spring.

Most growers set up their sprayers to treat the area beneath the trees in strips that are 2 to 3 feet (0.6 to 0.9 m) wide on either side of the tree trunks. Low rates of glyphosate or paraquat are often used in combination with preemergence herbicides to control emerged weeds. These materials should be applied before bud break to prevent drift from damaging lower limbs of the trees. Shields, low sprayer pressure (less than 40 psi, or

Table 13.2. Herbicide application timing

Area of orchard floor	Herbicide type	Fall	Winter	Spring	Summer
strip	postemergence	X	X		
	preemergence (split)	X	X		
	preemergence + postemergence		X		
	preemergence + postemergence (split)	X	X		
middles (frost protection and russet management)	postemergence			X (Feb)	
middles (chemical mowing)	postemergence		X	X	X

276 kPa) and larger nozzles can help reduce the risk of drift if spraying must be done after bud break. Keep glyphosate off green bark or foliage at all times.

Low rates of postemergence herbicides can be used to reduce weed growth without killing most weeds, a practice known as chemical mowing. This leaves some plant cover in the orchard, reducing erosion, but also reduces the competition from weeds. The plant species must be identified so the rates can be properly adjusted. Too little herbicide may not be effective enough, and too much may kill the species, leaving bare soil. In February and early March, chemical mowing between the rows is often done using applications of postemergent herbicides such as glyphosate (Roundup) or paraquat to suppress young winter weeds. As an alternative, some growers mechanically mow and disk at this time to suppress vegetation between the rows. It is still beneficial to maintain weed-free or reduced vegetation in the tree row so that equipment does not have to get close to the tree trunks. Also, by not cultivating, you do not cut or disturb the shallow roots of the tree, which absorb nutrients from the surface area of the soil.

Continued use of the same management practice or herbicide season after season causes tolerant weeds or weeds that are not present when herbicides are applied to grow on the orchard floor. Repeated mechanical mowing shifts the weed population to annual grasses and perennial weeds such as bermudagrass *(Cynodon dactylon)*, annual ryegrass *(Lolium multiflorum)*, and foxtail fescue *(Festuca megalura)*. Repeated chemical mowing using low rates of glyphosate shifts the population to perennial weeds such as bermudagrass, nutsedge (*Cyperus* spp.), and field bindweed *(Convolvulus arvensis)*. In some almond orchards where glyphosate has been used for many years, an annual ryegrass has become resistant to this herbicide. Weed management practices must be rotated (since the crop cannot be rotated) to reduce this risk of resistance. Competition from tall-growing cover crops, such as vetch (*Vicia* spp.) or Berseem clover *(Trifolium alexandrinum)*, help suppress these weeds. Periodic tillage, changes in herbicide use patterns, mowing, and planting cover crops can help reduce the risk that weeds will become a problem.

APPLICATION OF HERBICIDES

Herbicides are applied with specialized equipment either to bare ground or to young weeds. At no time should herbicides be applied in a way that drift can get onto the foliage of trees or other crops. You are legally required to follow all directions on the pesticide label. For more information on the legal requirements and other aspects of pesticide use, including pesticide formulations and hazards and equipment use and calibration, see *The Safe and Effective Use of Pesticides* (O'Connor-Marer 2000). Consult manuals supplied by equipment manufacturers for details on equipment configuration and use.

Preemergence herbicides are applied with small, lightweight sprayers that place the specified amount of herbicide on a given area of the soil. This differs from most other pesticide applications, which apply a specified volume amount to foliage or limbs. Spray equipment typically is set up to have a boom with one or more flat fan nozzles (depending upon strip width) and an off-center (OC) nozzle at the end of the boom to spray into the tree row. This equipment can spray both sides of the tree row to give a treated strip. Another newer piece of equipment is a spinning-disk, ultra-low-volume (ULV) applicator that can be hooded to reduce spray contact with the tree trunk. In established orchards, booms should be constructed to spray below hanging tree limbs. Because preemergence herbicide is usually sprayed onto berms, the nozzles should be positioned to apply pesticide uniformly to the soil.

Postemergence herbicides are applied with similar equipment, except the booms should be high enough to spray over the weed foliage but not into the tree foliage. A contact herbicide such as paraquat will burn green leaves or young bark; a systemic herbicide such as glyphosate must not be allowed to touch any green plant tissue. For good spray coverage, applications of paraquat typically need 20 to 60 gallons of water per acre (187 to 560 l/ha), depending on the size of the weeds, but glyphosate needs less water, only 10 to 40 gallons per acre (94 to 375 l/ha), since it is translocated in the plant and does not have to touch every part of the leaf to be effective.

TILLAGE

If water is in short supply, soil moisture availability is limited, or flood irrigation is practiced, the orchard floor can be disked several times during the growing season to suppress weeds. In the spring, maintaining clean, tilled, firmed soil on the orchard floor is one way to reduce the risk of frost damage. Drawbacks to this practice include the expense of operating and maintaining tillage equipment, dust (which can promote mite activity), poor footing for spray equipment during wet periods, and soil compaction from tillage and wheel traffic. Soil erosion can also be a problem if heavy rains occur on exposed soils. Unless there is a specific need for this practice, such as danger of frost, few growers in California choose it as their primary approach to orchard floor management.

COVER CROPS

Maintaining a cover crop on the orchard floor can help reduce erosion, which can be a significant problem in orchards on slopes and adjacent to waterways, especially early in the orchard's life. Cover crops also help increase water penetration and reduce soil compaction. Certain cover crops can add nitrogen to the soil. Recommended cool-season annuals include Zorro fescue *(Vulpia myuros)*, Blando brome *(Bromus mollis)*, burr medic *(Medicago truncatula)*, rose clover *(Trifolium hirtum)*, and subterranean clover *(Trifolium subterraneum)*. These covers are generally seeded in the fall, mowed through the spring, and allowed to dry in the early summer. They often reseed themselves for the following season's growth. During these species' period of active growth, pear trees are mostly dormant or just beginning to grow. Consequently, these cover crops do not compete very much with young trees, and they can be grown between tree rows and have little effect on established trees.

On sites with deep soils, high fertility, and abundant water, perennial cover crops can be grown, including perennial ryegrass *(Lolium perenne)*, turf-type tall fescue *(Festuca arundinaceae)*, red fescue *(Festuca rubra)*, strawberry clover *(Trifolium fragiferum)*, and white clover *(Triofolium repens)*. Mixes of these species are often grown. These species must be managed carefully to prevent them from competing with young trees. Attributes of some cover crops and their use in orchards are given in table 13.3; see also the publications listed in the references, such as Ingels 1998 and McGourty 1994.

Cover crops can have disadvantages in an orchard. If water costs are high or water is in short supply, perennial cover crops should not be used. Annual cover crops should always be kept away from the base of the trees. Certain cover crops such as clovers increase the potential for damage from rodents (voles) or other pests (nematodes and mites). Preparing the soil and planting and maintaining a cover crop can be expensive and increase water use and overall management in an orchard.

Cover Crops, Weeds, Frost, and Fruit Russetting

Strong evidence exists for a relationship between weeds, cover crops, frost, and russetting of fruit (see Lindow et al. 1996, 1998). Many grasses and certain hairy-leafed weeds may have large numbers of the bacteria *Pseudomonas syringae* on the foliage. When these bacteria are blown from the floor vegetation to pear foliage just as the foliage is breaking dormancy, the risk of frost is increased. Controlling vegetation on the orchard floor can significantly reduce the risk of damage by reducing the number of these bacteria. Other studies have shown that bacteria such as *Erwinia herbicola* can cause a reaction on the fruit surface that leads to corkiness (suberization) of these cells, giving the fruit a rusty appearance at harvest. Using chemical mowing to control grass populations can reduce these bacteria, but planting cover crops that do not host the bacteria can also reduce the potential for russetting.

Table 13.3. Selected winter annual cover crops grown in pear orchards

Scientific name	Common name	Growth habit	Seeding	Flowering	Maturity
Avena sativa	oat	upright	Oct–Dec	Apr–May	May–Jun
Bromus hordeaceus ssp. *molliformis*	Blando brome, soft chess	upright	Sep–Oct	Mar–Apr	Apr–May
Lolium multiflorum	annual ryegrass	upright	Sep–Oct	Apr–May	May–Aug
Pisum sativum	field pea	upright	Sep–Nov	Mar–May	May–Jun
Secale cereale	cereal rye	upright	Oct–Dec	Feb–Apr	Apr–Jun
Vicia faba	bell bean, fava bean	upright	Sep–Nov	Mar–May	May–Jun
Vicia villosa	hairy vetch	vine	Sep–Nov	Apr–May	May–Jun
Vulpia myuros var. *hirsuta*	Zorro fescue, foxtail fescue	upright	Sep–Oct	Mar–Apr	Apr

Sources: Ingels 1998; McGourty 1994; Miller et al. 1989.

REFERENCES

DiTomaso, J.D., and E.A. Healy. 2007. Weeds of California and other western states. 2 vols. Oakland: University of California Agriculture and Natural Resources Publication 3488.

Elmore, C. 1990. Vegetation management in pear orchards. In J. Beutel and R. Elkins, eds., Pear growing in the 1990s. Proceedings of a short course held November 27–29, 1990, Kelseyville, CA. Davis: University of California, Davis, Department of Pomology.

Facteau, T. J., and E. Milke. 1992. Reducing herbicide usage in orchards. In 1991–92 Progress reports to the Agriculutural Research Foundation for the Hood River Grower-Shipper Association. Corvallis: Oregon State University Agricultural Research Foundation. 102–108.

Ingels, C., ed. 1998. Cover cropping in vineyards: A grower's handbook. Oakland: University of California Division of Agriculture and Natural Resources Publication 3338.

Lindow, S. E., G. McGourty, and R. Elkins. 1996. Interactions of antibiotics with *Pseudomonas fluorescens* strain A506 in the control of fireblight and frost injury to pear. Phytopathology 86:841–848.

Lindow, S. E., C. Desurmont, R. Elkins, G. McGourty, E. Clark, and M. Brandl. 1998. Occurrence of indole-3-acetic-acid-producing bacteria on pear trees and their association with fruit russetting. Phytopathology 88:1149–1157.

McGourty, G. T. 1994. Covercrops for North Coast vineyards. Practical Winery and Vineyard 15:2 (July/Aug.): 1–6.

Miller, P., W. L. Graves, W. A. Williams, and B. A. Madson. 1989. Covercrops for California agriculture. Oakland: University of California Division of Agriculture and Natural Resources Publication 21471.

O'Connor-Marer, P. 2000. The safe and effective use of pesticides. 2nd ed. Oakland: University of California Division of Agriculture and Natural Resources Publication 3324.

Ohlendorf, B. 1999. Integrated pest management for apples and pears. 2nd ed. Oakland: University of California Division of Agriculture and Natural Resources Publication 3340.

UC IPM pest management guidelines for pear. Oakland: University of California Division of Agriculture and Natural Resources Publication 3455. UC Statewide IPM Web site, http://www.ipm.ucdavis.edu/PMG/selectnewpest.pears.html.

Whitson, D., ed. 2000. Weeds of the West. Oakland: University of California Division of Agriculture and Natural Resources Publication 3350.

14
Water Relations

KENNETH A. SHACKEL

GENERAL PRINCIPLES

Pear trees require a healthy root system and a good supply of available soil water below ground in order to be healthy and productive above ground. This is because the above- and below-ground parts of a pear tree are completely dependent on one another: roots depend on shoots for carbohydrates and other organic nutrients, and shoots depend on roots for water and mineral nutrients. Any factor that has a direct impact on either of these systems in the short term can indirectly cause changes in the entire tree, and in its productivity, over the long term.

Water plays a key role in almost all plant processes, beginning with carbon fixation, the most basic function of plant life. Plants have developed many physiological responses to help them survive periods of water stress because lack of water is a commonly occurring condition in nature. Most of these responses involve reductions in growth of shoots and fruit. Finding the appropriate balance in pears is a challenge because reduced shoot growth may be desirable for many reasons in a mature pear tree, but maximizing fruit size is also usually important.

While it is impossible to discuss all of the possible physiological responses to water stress and their consequences for fruit production in pears, this chapter presents the general principles of how trees respond to reduced water availability and gives examples how pear trees respond to water stress. Further information on irrigation management in pears can be found in chapter 15, "Irrigation Scheduling."

WATER USE AND TREE PRODUCTIVITY

During daylight hours, a tree looses (evaporates) a substantial quantity of water into the air from its leaves. This process of water loss is called transpiration (fig.

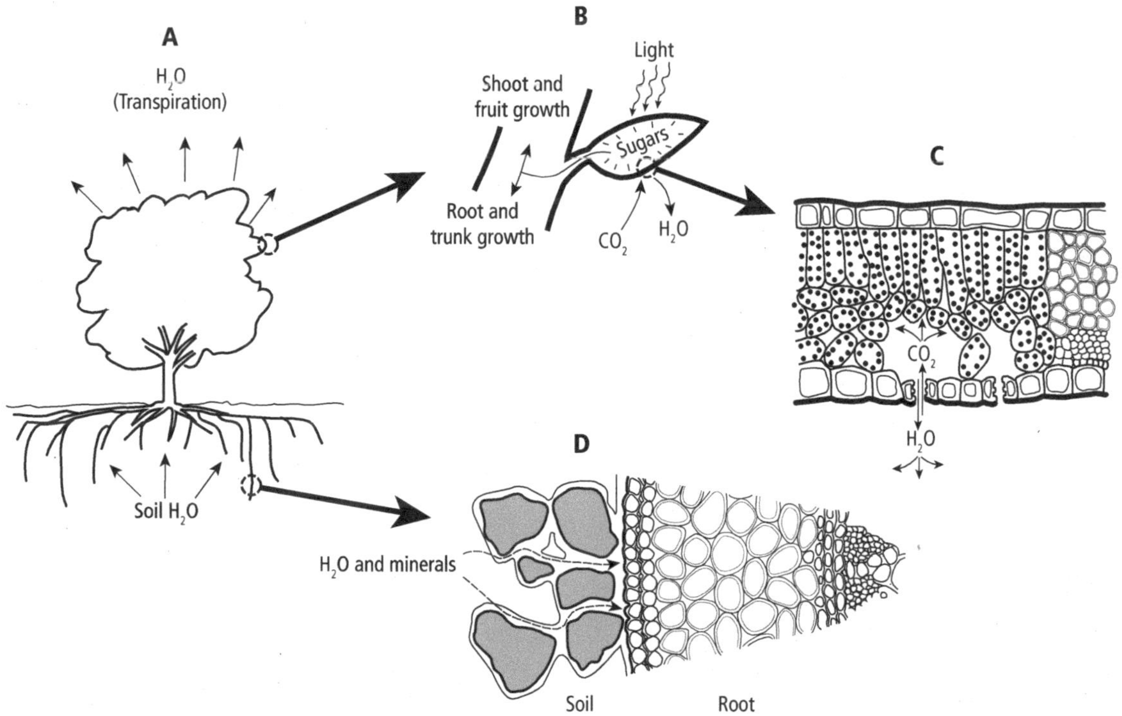

Figure 14.1. Water flow through the soil-plant-atmosphere continuum (SPAC). Water evaporates through the same stomatal pores in the leaf (C) that allow the uptake of atmospheric CO_2 for the production of sugars in the process of photosynthesis (B). The water leaving the tree canopy (A) must be continuously replaced by water uptake from the soil (D).

14.1A). Water can also evaporate into the air directly from a wet soil surface. The combined quantity of water lost by the tree plus the water lost by the soil is called orchard or crop evapotranspiration (ETc). Transpiration usually accounts for most of ETc, and under hot, dry conditions a tree can transpire its entire fresh weight in water each day. The water lost by each leaf is continuously being replaced by water that the leaf "pulls" from its stem, much like a wick or a siphon pulls water from a reservoir. The stem, in turn, pulls water from the branch, and so on, through the plant to the roots, which must ultimately pull water from the soil (fig. 14.1D). Most of the transport of water within the plant is through the xylem, which is the woody tissue made up of cells that essentially act like pipes carrying this water under tension.

Transpiration can be considered a "necessary evil" in plants, since the rates of transpiration and photosynthesis are closely related. All plant life is base on photosynthesis, the biochemical process of using sunlight to convert atmospheric carbon dioxide (CO_2) into sugars (fig. 14.1B). Photosynthesis does not itself require the plant to lose water through transpiration. However, in order to allow carbon dioxide into the leaf, specialized cells on the leaf surface (stomatal guard cells) must press open a gap called the stomatal pore. This gap gives the cells within the leaf access to the carbon dioxide in the atmosphere. Unfortunately, the atmosphere also has access to the water vapor within the leaf through the same opening (fig. 14.1C), and this starts the first step in the transpiration process. The stomatal pore can be thought of as a main door to the photosynthetic factory in the leaf, and this door must remain open for the factory to receive its most important raw material (CO_2). The fact that water and CO_2 share this doorway is the basic reason why tree productivity and tree water use are so closely related. It is also the reason why any treatments that cause the tree to reduce its water use can be expected to reduce some aspect of tree productivity. Whether reducing water use influences economic production depends on the processes that are occurring at the time productivity is reduced, how important these processes are to fruit yield or quality, and whether the process relies heavily on the current level of photosynthesis or can use stored carbon (like starch) to compensate for a reduction in the quantity of sugars being generated by the leaves.

PHYSIOLOGICAL RESPONSES TO WATER STRESS

The flow of water through the plant from the soil to the atmosphere, is a fundamental concept in plant physiology called the soil-plant-atmosphere continuum. It is a key concept for understanding tree water use and the physiology of water stress. The driving force for transpiration begins with water evaporation from the leaves. Thus, environmental factors such as high temperature, low relative humidity, or high wind speed, which would increase evaporation from any wet surface, can also increase tree transpiration, and therefore orchard ETc. In fact, this is the primary reason why ETc- based methods of irrigation scheduling are useful: these methods use weather information to predict tree water requirements. An increase in transpiration however, is also associated with an increase in the pulling force experienced by the water in the plant. This pulling force (the water potential) is best understood as a tension that exists in the xylem and in the cell walls throughout the tree. As transpiration increases, more tension is needed to pull water through the plant, just as increasing the flow of water through irrigation pipes requires more pressure.

The tension of the water in the tree can be measured using a device called a pressure chamber (also known as a pressure bomb; see chapter 15, "Irrigation Scheduling"). Pear trees experience a certain amount of tension as a normal part of daily and seasonal variation in ETc, with the greatest tension occurring at midday, when the rate of water loss from the tree is at its highest (typically 1:00 to 3:00 P.M. daylight savings time). Measuring midday values of tree water potential is analogous to measuring a person's blood pressure during a treadmill test: it depends partly on how hard the person is working, so it might vary from one test to the next, but it is also a very good way to test how well the person's circulatory system works as a whole. In the case of a pear tree, the measurement indicates the ability of the soil and roots together to supply the tree with enough water to match the tree water demands.

A standard method of sampling that has been developed for pears and a number of other trees is to sample covered leaves at midday, which gives a value known as midday stem water potential. Midday stem water potential is measured with pressure, but since it is actually a tension, it is given a negative sign. The more negative the value, the greater the tension, and for the tree, the greater the stress. An easy way to remember this is to think of water stress as a "deficit," that is, the greater the stress, the more the plant is experiencing a deficit of water. The most standard unit at present is the megapascal (MPa), although many devices still use the unit of bars. The conversion between these units is straightforward: 10 bars = 1 MPa (10 bars is also equivalent to 145 psi). For fully irrigated pear trees, we know what the midday stem water potential would be under various conditions of air temperature and relative humidity (table 14.1), so any value more negative than this would indicate the occurrence of some water stress.

It is interesting to note that the soil tension measured by a tensiometer is measured in centibar units (100 centibars = 1 bar); most water-filled tensiometers cannot measure tensions below about –80 centibars. Therefore, you can see that the tension that would be experienced in a pear tree even under wet soil and mild weather conditions (e.g., –7.0 bars) corresponds to an extremely dry soil (–700 centibars). This is because the plant needs to pull the water through the root system, which appears to offer a significant resistance to water uptake. As the soil dries, the tree must exert an increasing strength of pull to extract the water held by the soil particles, so additional water stress develops as the soil dries.

Water stress has a number of effects on plants. For instance, as water potential decreases (becomes more negative), the pressure within the cells of the plant (turgor pressure) is also reduced. Cell turgor is the primary force that keeps the soft parts of the plant erect. The most obvious sign of a loss in cell turgor is wilting, such as of growing shoot tips, or flagging of leaves. A less obvious but probably more important result of a loss in turgor is a reduced growth rate. Turgor is an important part of the cell expansion process in the growth of essentially all plant parts, including fruit. If turgor is low, we can expect growth to be reduced. In pears we have found a very strong reduction in fruit size as the level of water stress increases (fig. 14.2). Along with the decrease in size, however, was an increase in fruit soluble solids (fig. 14.3), which may be considered a positive effect on fruit quality. In addition to these effects there was also a reduction in pruning weights with stress, similar to the reduction of fruit size shown in figure 14.2. This illustrates the multiple effects that water stress can have, which must be balanced by the grower as part of any irrigation strategy.

From a survival standpoint, a reduction in tree growth is a beneficial physiological response of all plants to water

Table 14.1. Values of midday stem water potential (in bars) to expect for fully irrigated pear trees under selected conditions of air temperature and relative humidity

Temperature		Relative humidity (RH, %)				
(°F)	(°C)	10	20	40	60	80
60	15.6	-6.6	-6.6	-6.4	-6.2	-6.0
70	21.1	-7.0	-6.8	-6.6	-6.4	-6.1
80	26.7	-7.4	-7.2	-6.9	-6.6	-6.2
90	32.2	-8.0	-7.8	-7.3	-6.8	-6.3
100	37.8	-8.8	-8.4	-7.8	-7.2	-6.5
110	43.3	-9.8	-9.3	-8.5	-7.6	-6.7

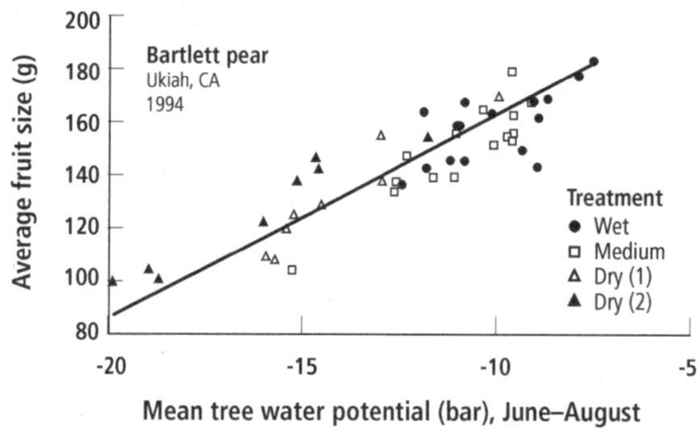

Figure 14.2. The relation of average fruit size and average water potential for Bartlett pear trees under four different irrigation treatments. Each point represents the average size of all fruit at harvest and the average midday stem water potential during the growing season (June to August) for an individual tree. *Note*: 10 bars = 1 MPa.

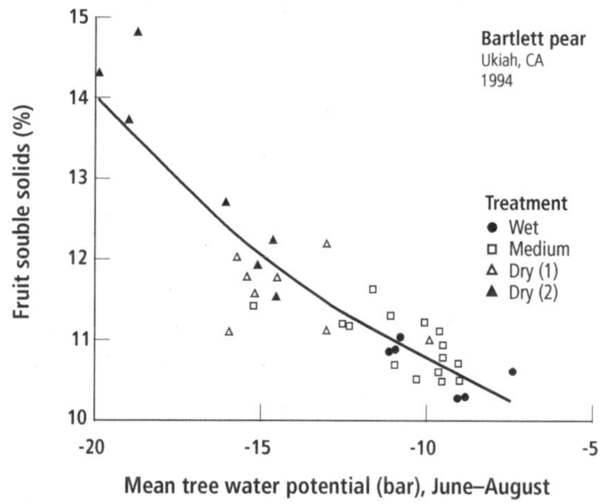

Figure 14.3. The relation of fruit soluble solids at harvest to average water potential during the growing season (June to August) for Bartlett pear trees under different irrigation treatments, as in figure 14.2. *Note:* 10 bars = 1 MPa.

stress, because a small plant requires less water to survive than a large plant. A smaller pear tree may also require less labor to prune and harvest and may also allow better light penetration into the canopy. These effects may be considered positive effects from a production standpoint in a mature orchard as long as yearly growth is sufficient to maintain active fruiting wood. Of course in the early years of orchard establishment, reductions in growth slow down the development of the canopy and may reduce early orchard productivity.

Regulating the size of the stomatal pores themselves is another physiological response that allows the plant to partially control leaf transpiration and tree water use. Guard cells are very responsive to many environmental conditions and can close completely under severe water stress conditions (fig. 14.4). Stomatal closure is ben-

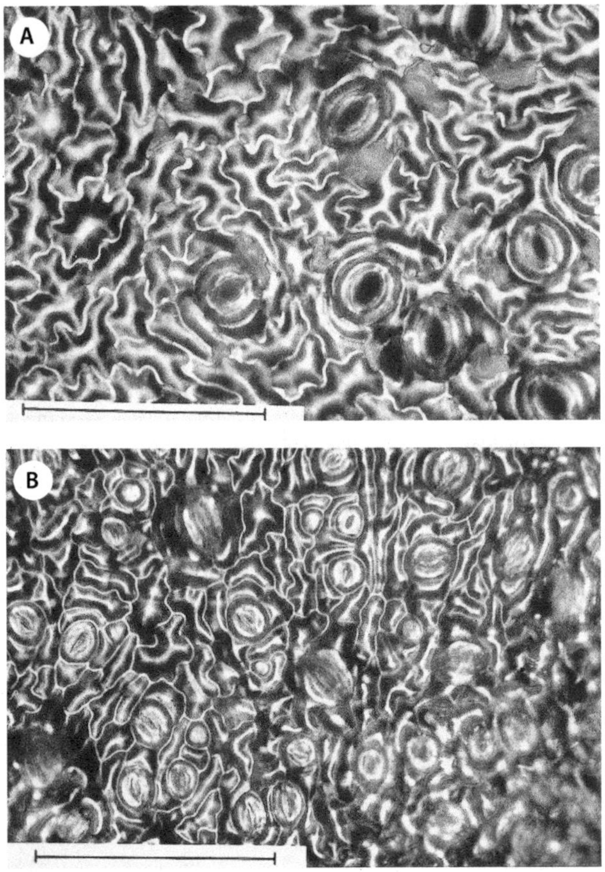

Figure 14.4. Surfaces of apple leaves, observed under the microscope, showing open (A) and closed (B) stomatal pores. The bar on both micrographs equals 100 μm. *Photo:* K. Shackel.

eficial for the tree from a survival standpoint because it reduces water use. Just as with growth reductions, the closing of stomata may ultimately reduce tree productivity by reducing CO_2 uptake and photosynthesis, but it is difficult to say whether a reduction in photosynthesis caused by partial stomatal closure will have any economic impact on tree productivity. When stomatal closure reduces leaf transpiration, it also reduces evaporative cooling of leaves, which can cause leaf temperatures to increase. This is the reason why canopy temperature measurements with infrared "guns" or other remote sensing equipment can help detect stress in crops. One drawback of these methods, however, is that substantial stomatal closure may be required before a significantly warmer leaf or canopy temperature can be detected. These methods may detect stress only after it has already had a negative impact on crop growth.

Roots are the point of water entry into the tree, so any factor that reduces the health of the root system and the roots' ability to absorb water can cause symptoms of water stress. Diseased root systems do not function properly, so trees affected by root diseases may show leaf flagging, reduced growth, and closed stomata, even though the soil is wet. Similar water stress symptoms may occur if the soil is flooded and roots are deprived of the oxygen they require to live. From the standpoint of tree irrigation, either too much or too little water can have similar effects on the physiology of the tree and can cause similar responses. Pears are considered relatively tolerant of flooding, but the data shown in figure 14.2 indicate that even when the soil was adequately irrigated (the "wet" irrigation treatment), some trees experienced stress. The reasons for this stress have yet to be identified, but it appears not to be due to any known root disease, pest problem, or rootstock effect. This may be an example of poor overall root health causing tree water stress even though irrigation is adequate.

REFERENCES

Bradford, K. J., and T. C. Hsiao. 1982. Physiological responses to moderate water stress. Encyclopedia of Plant Physiology 12B:263–324.

Fereres, E., and D. A. Goldhamer. 1990. Deciduous fruit and nut trees. In B.A. Stewart and D.R. Nielsen, eds., Irrigation of agricultural crops. American Society of Agronomy/Crop Science Society of America/Soil Science Society of America Monograph 30. 987–1017.

Jones, H. G., A. N. Lakso, and J. P. Syvertsen. 1985. Physiological control of water status in temperate and subtropical fruit trees. Horticultural Reviews 7:301–344.

Kramer, P. J. 1983. Water relations of plants. New York: Academic Press.

Ramos, D. E., S. A. Weinbaum, K. A. Shackel, L. J. Schwankl, E. J. Mitcham, F. G. Mitchell, R. G. Snyder, G. Mayer, and G. McGourty. 1994. Influence of tree water status and canopy position on fruit size and quality of Bartlett pears. Acta Horticulturae 367:192–200.

Shackel, K. A., H. Ahmadi, W. Biasi, et al. 1997. Plant water status as an index of irrigation need in deciduous fruit trees. HortTechnology 7(1): 23–29.

15
Irrigation Scheduling

LAWRENCE J. SCHWANKL, KENNETH A. SHACKEL, AND STEPHEN M. SOUTHWICK

An important aspect of irrigation water management is irrigation scheduling, that is, determining when and how much to irrigate. Underirrigation of pears can reduce yield and size; overirrigation can encourage root disease, cause excess vigor and additional pruning, and may waste water and energy. There are numerous ways to determine an irrigation schedule, and many growers use combinations of irrigation scheduling techniques to achieve the results they want.

Methods of irrigation scheduling can be grouped into weather-based methods, soil-based methods, and plant-based methods. Most modern irrigation research has been based on the assumption that fruit production will be maximized only if the water used by the tree is fully replaced by irrigation on a regular basis. For both weather- and soil-based methods, the grower must determine how much of the soil water reservoir can be safely depleted by the tree before irrigation is required. For soil-based methods this depletion is estimated by measuring soil wetness directly, whereas for weather-based methods this depletion is calculated using information about the age and condition of the orchard and either real-time or historical average weather information. Reliable plant-based methods for irrigation scheduling are a more recent development. If a grower is already using a particular soil- or weather-based scheduling method, plant-based measurements can be used to check whether this method is having the desired effect on the pear tree. Plant-based methods may also allow the grower to use a deficit irrigation strategy, which can conserve irrigation water and achieve horticulturally beneficial tree responses, such as controlling tree vigor or reducing the incidence of root diseases.

WEATHER-BASED IRRIGATION SCHEDULING

A common method of irrigation scheduling is based on monitoring weather conditions to determine the water use (evapotranspiration, or ET) of the pear trees. The ET of a tree is related to solar radiation, wind speed, and relative humidity. By carefully monitoring weather parameters and using that information in empirically developed and calibrated models, orchard, or crop, ET (ETc) can be estimated. Adding in determinations of the soil's water-holding capacities and the trees' root zone depth allows water budget irrigation scheduling to be used.

Water Budget Irrigation Scheduling

Water budget irrigation scheduling uses the concept that the soil in the crop's root zone stores moisture in a "bank account." Each day, the tree withdraws water from the account as part of its normal evapotranspiration activities. A certain level of soil moisture depletion, prior to irrigation, is allowed. This allowable depletion is chosen so that the tree is not unduly stressed and yield or quality is not reduced. Once the allowable depletion level is reached, irrigation water must be "deposited" back into the root zone's soil moisture bank account. If the irrigation is efficient, only enough irrigation water is applied to refill the trees' root zone. Irrigation water applied in excess of this amount will drain below the tree's roots and be lost.

The water budget system of irrigation scheduling assumes that the trees are healthy. Unhealthy trees often use less water (have less evapotranspiration) than healthy trees. A step-by-step process of water budget scheduling is as follows.

Step 1: Determine your soil type and use table 15.1 to determine the available soil moisture per foot of root zone depth.

If you do not already know the soil type(s) in your orchard, refer to the soil survey for your area, which is available from your local UC Cooperative Extension office. Using table 15.1, choose the appropriate available soil moisture (inches of water per foot of rooting depth). Additional soil moisture information is available in Bowers et al. 1989.

Step 2: Determine the root zone depth of your trees.

While pears may root as deep as 4 to 5 feet (1.2 to 1.5 m), the majority of soil water uptake occurs in the top 3 feet (0.9 m). It is best to determine the actual rooting depth for your orchard. Using an auger, or better yet a backhoe trench, check the rooting depth of your

Table 15.1. Field water capacity, 15-bar soil moisture content (often considered to be the wilting point), and available soil moisture (in/ft) for various soil types

Soil texture	Field capacity (in/ft)	15 bars (in/ft)	Available moisture content (in/ft)
Sand	1.2	0.5	0.7
Loamy sand	1.7	0.7	1.0
Sandy loam	2.3	1.0	1.3
Sandy clay	2.9	1.8	1.1
Sandy clay loam	3.0	1.8	1.2
Loam	3.2	1.4	1.8
Silt loam	3.6	1.6	2.0
Clay loam	3.8	2.3	1.5
Silty clay loam	4.3	2.4	1.9
Clay	4.7	3.1	1.6
Silty clay	4.9	3.2	1.7

Source: Hanson, Schwankl, and Fulton 1999, p. 193.

trees. Young pear trees have a smaller root zone than do mature trees. This should be accounted for in choosing a root zone depth.

Step 3: Determine the tree root zone's total available soil moisture.

Combine the information from Step 1 (available soil moisture per foot of depth) and Step 2 (root zone depth) to determine the total available soil moisture in your trees' root zone:

$$\left[\begin{array}{c}\text{total available}\\\text{soil moisture}\\\text{in the root}\\\text{zone (in)}\end{array}\right] = \left[\begin{array}{c}\text{available soil}\\\text{moisture per}\\\text{foot of soil}\\\text{(in/ft)}\end{array}\right] \times \left[\begin{array}{c}\text{root zone}\\\text{depth (ft)}\end{array}\right]$$

Step 4: Determine the allowable soil moisture depletion.

To minimize water stress on the trees and its negative impact on yield and quality, only a portion of the total available soil moisture should be depleted prior to irrigation. A rule of thumb is to allow 50 percent of the available soil moisture to be depleted; this percentage must be customized to fit your orchard. Observation of the tree, or better yet, tree-based stress measurements, should be used, and the amount of irrigation water that can be applied within your irrigation practices must also be considered. For example, if excessive runoff begins to occur after a certain number of hours of sprinkling, your allowable soil moisture depletion may be set by the amount of water applied prior to onset of the excessive runoff:

$$\left[\begin{array}{c}\text{allowable}\\\text{soil moisture}\\\text{depletion}\\\text{(in)}\end{array}\right] = \left[\begin{array}{c}\text{total available}\\\text{soil moisture}\\\text{in the root}\\\text{zone (in)}\end{array}\right] \times \left[\begin{array}{c}\text{allowable}\\\text{depletion}\\\text{(\%)}\end{array}\right] \div \left[\begin{array}{c}100\end{array}\right]$$

Step 5: Determine the daily tree water use (in/day).

A tree's daily water use (evapotranspiration, or ET) is available from a number of sources. Table 15.2 lists the historical average ET values for pear trees at a number of California locations. A more refined estimate is available by using the California Department of Water Resources CIMIS (California Irrigation Management Information System) network. The CIMIS network makes use of weather station(s) in your area to measure daily ET. This "real-time" information is more accurate than historical ET estimates.

Step 6: Determine the irrigation interval (days) between irrigations.

Using the allowable soil moisture depletion (Step 4) and the daily tree water use (Step 5), determine the interval (days) between irrigations:

$$\text{irrigation interval (days)} = \text{allowable soil moisture depletion (in)} \div \text{daily tree water use (in/day)}$$

Step 7: Determine the irrigation set time (number of hours).

Using the allowable soil moisture depletion (Step 4) and the application rate (in/hr) of your irrigation sys-

Table 15.2. Historical average evapotranspiration for mature pear orchards in Courtland (Sacramento Co.), Lakeport (Lake Co.), and Ukiah (Mendocino Co.), California

	Pear evapotranspiration (in)		
Date	**Courtland**	**Lakeport**	**Ukiah**
Mar 1–15	—	—	—
Mar 16–31	0.83	—	0.25
April 1–15	1.28	1.65	0.86
April 16–31	1.77	1.44	1.46
May 1–15	2.22	1.90	1.83
May 16–31	2.68	2.28	2.21
June 1–15	2.90	2.52	2.44
June 16–30	3.31	3.14	2.87
July 1–15	3.33	3.05	2.91
July 16–31	3.31	3.14	2.87
Aug 1–15	3.05	2.79	2.64
Aug 16–31	2.77	2.52	2.40
Sept 1–15	2.46	2.20	2.11
Sept 16–30	1.79	1.61	1.53
Oct 1–15	1.26	1.16	1.13
Oct 16–31	0.88	0.82	0.79

Note: If cover crops are present in the orchard, the evapotranspiration values should be increased by approximately 35%. If the orchard is young, see table 15.3 to determine the percent reduction of evapotranspiration values.

Table 15.3. Reduction of mature pear tree evapotranspiration for immature trees (see table 15.2)

Ground shading, %	% of mature tree water use
10	25
20	48
30	65
40	80
50	90
60	100
>60	100

tem, determine the irrigation set time (hr). If you do not know your irrigation system's application rate, refer to chapter 16, "Irrigation Systems." Since no irrigation system is 100 percent efficient, additional water must be applied to account for water lost to runoff or deep percolation. If you do not know the irrigation efficiency of your system, use an estimate of 75 to 80 percent efficiency for a well-designed, well-maintained solid-set sprinkler system. Hand-move sprinklers would be less efficient, and micro-sprinklers would be slightly more efficient.

$$\text{irrigation set time (hr)} = \text{allowable soil moisture depletion (in)} \div \text{irrigation system application rate (in/hr)} \div \text{irrigation efficiency (\%)} \times 100$$

Example for steps 1 through 7

Assume that your orchard has a clay loam soil, a root zone of 3 feet, and you irrigate it with a solid-set sprinkler system with a spacing of 20 feet × 47 feet and 3⁄32-inch nozzles operated at 40 psi. What should be your irrigation interval and irrigation set time?

Step 1: Determine your available soil moisture per foot of soil (in/ft).
From table 15.1, the available soil moisture is 1.5 in/ft of soil.

Step 2: Determine your tree's root zone depth (ft).

$$\text{Rooting depth} = 3 \text{ ft (given)}$$

Step 3: Determine the tree's total available soil moisture in the root zone (in).

$$\text{total available soil moisture} = 1.5 \text{ in} \times 3 \text{ in} = 4.5 \text{ in}$$

Step 4: Determine the allowable soil moisture depletion (in).
Assume that you want to allow 50 percent of the total available soil moisture to be depleted.

$$\text{allowable soil moisture depletion} = 4.5 \text{ in} \times 50 \div 100 = 2.25 \text{ in}$$

Step 5: Determine the daily tree water use (in/day).
From historical pear ET records (table 15.2), determine the daily crop water use. Assume this to be 0.20 in/day during midseason.

Step 6: Determine the irrigation interval (days).

$$\text{irrigation interval} = 2.25 \text{ in} \div 0.20 \text{ in/day} = 11 \text{ days}$$

Step 7: Determine the irrigation set time (hr).
For a 20-foot × 47-foot solid-set sprinkler system with 3⁄32-inch nozzles operated at 40 psi, the application rate is 0.17 in/hr (see chapter 16, "Irrigation Systems"). Assume an irrigation efficiency of 80 percent.

$$\text{Irrigation set time} = 2.25 \text{ in} \div 0.17 \text{ in/hr} \div 80 \times 100 = 16.5 \text{ hr}$$

SOIL-BASED IRRIGATION SCHEDULING

Soil-based methods of irrigation scheduling assume that the soil moisture content largely determines the moisture status of the plant. If there is less water in the soil, and the water is held at a greater tension, it will be more difficult for the plant's roots to take up that water. The result is that the plant will be under greater stress. Irrigation scheduling can be done by monitoring the soil moisture and maintaining the soil moisture within acceptable limits that do not adversely affect plant growth.

The use of soil moisture monitoring devices such as tensiometers, gypsum blocks, neutron probes, or other devices is an excellent way to check the water budget irrigation scheduling method described above. Monitoring the soil moisture before and after irrigation provides you with additional information on the timing of your irrigations as well as the depth to which irrigation water penetrated.

The most basic method of monitoring soil moisture is to take soil samples at various depths (e.g., 1-foot increments) within the root zone. Using a soil auger, sampling tube, or a shovel requires significant time and effort to sample throughout the root zone. If a volumetric soil moisture measurement (e.g., inches per foot of soil moisture) is desired, an undisturbed soil sample must be collected, which can be a difficult task. In order to make monitoring of soil moisture easier and quicker than collecting soil samples, a number of soil moisture monitoring tools are available. Some of these soil moisture monitoring tools are described below.

Tensiometers

A tensiometer measures soil water tension, which is a measure of the surface tension force holding water in the soil. Tensiometers consist of a cylindrical pipe about 1 inch (2.5 cm) in diameter with a porous ceramic cup attached to one end and a vacuum gauge attached to the other (fig. 15.1).

Tensiometers measure soil moisture tension indirectly. They are installed for the entire season or

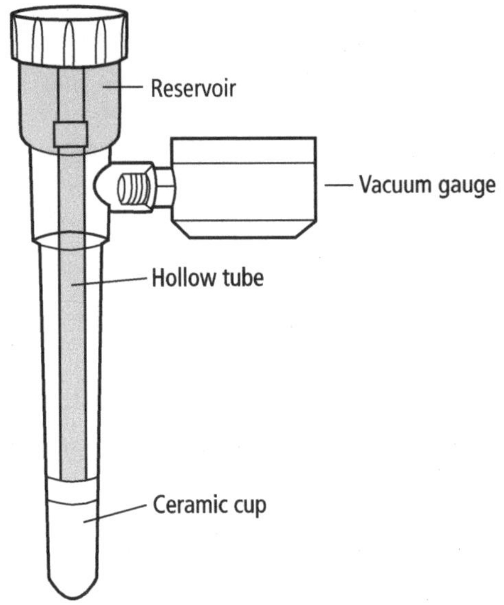

Figure 15.1. Tensiometer.

longer and provide readings at the same location over an extended period of time. Tensiometer readings are easily interpreted and indicate the soil water conditions experienced by the trees' roots. Soil salinity does not affect the readings.

In an unsaturated soil, soil water tension, frequently called the "suction," falls below atmospheric pressure. As wet soil dries, the soil water suction increases, causing water to flow out of the tensiometer through the porous cup. The small pores of the saturated cup prevent air from entering the tensiometer. This outflow of water creates a vacuum inside the tensiometer and increases the reading on the vacuum gauge. When the soil is wetted by irrigation, water is drawn back into the tensiometer, reducing the vacuum inside, and the reading on the tensiometer gauge decreases. The vacuum gauge measures the suction in centibars (100 centibars = 1 bar = 1 atm), with a range of 0 to 100. A reading of 0 indicates a saturated soil in which plant roots will probably suffer from poor aeration. A reading of 10 to 25 reflects a soil at field capacity. The lower reading is for sandy soils at field capacity, and the higher reading is for finer-textured soils. Readings of 70 to 80 indicate a dry soil. Tensiometers will not read above 85 centibars.

Tensiometers do not provide information on the amount of water depleted from the soil unless they have been calibrated for the particular soil type. They indicate when to irrigate, but do not indicate how much to irrigate.

Although tensiometers are used most frequently for monitoring soil moisture, they can also be incorporated into automated irrigation systems. Tensiometers with solenoids can be used to control an irrigation system, and tensiometers equipped with transducers can be used with computerized irrigation systems.

Installation

To install a tensiometer, insert the end of the tensiometer with the porous cup at the depth desired through a pilot hole made with a soil probe in the soil. The porous cup should be soaked in water for several hours before installation. After installation, fill the tensiometer with water and allow it to equilibrate with the soil water for about 24 hours. Install the tensiometer in the zone of greatest root density, at about one-quarter to one-third of the maximum root depth. A tensiometer at this depth can be used to schedule irrigations. For sprinkler- or flood-irrigated orchards, a tensiometer reading of approximately 60 indicates a need for irrigation in the near future. Refinements based on soil sampling will be required to adjust for site-specific conditions.

A second tensiometer should also be installed in the bottom one-third of the root depth to assure that the moisture extends to an adequate depth. Frequently, the shallower tensiometer will begin to show an increased reading (indicating drier soil) before the deeper tensiometer begins to change. Trees tend to take up water from shallower soil first because they have more actively growing roots closer to the surface. Some water managers who use tensiometers use the beginning of moisture extraction from the deeper depths, indicated by a change in the lower tensiometer's readings, as an indicator of when to irrigate. This practice seems to work well as long as irrigation water can penetrate to the lower depths to replace the soil moisture. If the tensiometer reading at the lower root depth remains unchanged following an irrigation or continues to rise during the growing season, irrigation applications may be insufficient to penetrate to the lower root zone.

The number of tensiometer stations required depends on the irrigation system and on soil uniformity and management. For areas up to 40 acres (16 ha), at least two stations should be established. Stations should be located in areas representative of overall moisture status, with separate stations for problem areas or for areas having different soil conditions. In sprinkler-irrigated orchards, tensiometers should not be placed in areas where the sprinkler application pattern is "shaded" by a tree.

Maintenance

Tensiometers must be properly maintained. This requires periodically filling the pipe with water and replacing the porous cups. If the soil becomes too dry (tensiometer readings greater than 85), the porous cup will break tension and air will enter the tensiometer. A cracked cup also prevents a vacuum from developing in the tensiometer and causes the instrument to always read 0. In locations where temperatures fall below freezing, the tensiometers should be protected or removed from the field during cold periods. Finally, the porous cup of a tensiometer filled with water should not be exposed to the atmosphere for long periods of time. Such exposure causes evaporation of water from the cup's surface, which may cause salt buildup and clogging of the cup.

Electrical Resistance Blocks

Electrical resistance blocks, commonly referred to as gypsum blocks, are an inexpensive and simple soil moisture measurement tool. They are useful for timing irrigations but do not provide information on the amount of irrigation water required. Electrical resistance blocks are installed permanently at a site and have the advantages of allowing the user to quickly monitor soil moisture at the same location throughout the season.

Resistance blocks monitor soil moisture indirectly by measuring the electrical resistance between two electrodes attached to a small cast block of gypsum buried in the soil. Some electrical resistance blocks have electrodes mounted in fiberglass or in plastic or metal-encased blocks containing a sand-gypsum mixture. Wires, available in various lengths, lead from the blocks to the soil surface. The electrical resistance should be read on a portable conductance meter. Measuring conductance (1 ÷ resistance) requires the use of alternating current (AC), which the resistance block meter produces from its direct current (DC) battery. Using a simple resistance meter or ohmmeter to read the electrical resistance blocks does not give stable or reliable readings. Electrical resistance blocks and meters vary in cost: blocks range from $6 to $25 each, and meters cost approximately $250. Since the meter is portable, only a single meter is required to monitor all the installation sites. At a normal site, a block may last 2 to 3 years, but in areas of frequent irrigation or high water table it may require annual replacement.

One manufacturer produces a block and meter that reads directly in centibars of soil water tension, while most other manufacturers provide charts or tables to correlate the meter readings to soil water tension values. Soil water tension information can help you determine the timing of irrigations but does not tell the amount of irrigation water required. The irrigation amount can be arrived at indirectly through a trial-and-error procedure of monitoring the resistance blocks before and after irrigations. Blocks whose readings indicate wetter conditions after an irrigation show that irrigation water has reached their depth. If a block does not indicate wetter soil conditions following an irrigation, the irrigation was insufficient to reach the block's depth. Monitoring the changes in resistance block readings after a number of irrigations allows you to gauge the irrigation to match the soil moisture deficit.

Data loggers, which can be used with soil moisture blocks, record moisture block readings on a frequent and automated schedule. This "continuous" measurement of soil moisture can assist you in improving irrigation water management while minimizing the labor associated with reading the moisture blocks.

Gypsum blocks should be placed at several soil depths. At a minimum, blocks should be placed in the top one-third and bottom one-third of the root zone. Consultants who install soil moisture monitoring using resistance blocks often place blocks at 1-foot (0.3-m) intervals throughout the root zone. Soil moisture conditions are monitored only in the soil closely surrounding a resistance block. Wires from the blocks are brought to the soil surface to facilitate periodic connection to the portable meter for taking measurements. The matric potential of the blocks is assumed to be in equilibrium with the surrounding soil, so the blocks act much like the surrounding soil, taking up and releasing water as the soil wets and dries. The electrical conductance between the electrodes varies according to the water content of the block: the higher the water content, the higher the conductance and the lower the electrical resistance.

The placement of resistance block stations in an orchard depends primarily on the soils. Generally, monitoring stations should be representative of what is happening in the surrounding area. At a minimum, two stations per 40 acres (16 ha) are recommended, with additional stations in problem areas or in areas with different soil conditions. Since gypsum is soluble, gypsum blocks slowly dissolve. The life of a block can be extended by 1 or 2 years by the addition of a small quantity of gypsum to the backfill soil at the time of installation. A small quantity of lime ($CaCO_3$) may also prolong the life of a block in acidic soil.

Most electrical resistance blocks monitor soil moisture more effectively in the drier range of soil water tensions (in excess of 0.33 bar). Consequently, blocks are more useful in medium- to heavy-textured soils (loams and clays), which retain more available water in the higher soil water tension range. Sandy and coarse-textured soils tend to release much of their water at low soil water tensions, where the blocks are not as accurate or responsive. Blocks are also affected by highly saline or alkali soil conditions and may not be appropriate for use under such conditions.

In pear orchards, moisture resistance blocks can often be installed in the tree row to protect the blocks from traffic or cultivation damage. They should be installed within the drip line of a tree. When sprinkler irrigation is used, avoid placing the resistance blocks in the drier "shadow" area caused by the tree blocking the sprinkler's spray.

At a station, resistance blocks to be placed at various depths can either be sequentially stacked in the same hole or placed in separate holes. Placing blocks in separate holes has the advantage of locating the block in the soil environment closest to an undisturbed condition, but it has the disadvantage of requiring more work than installing multiple blocks in a single hole. Users of resistance blocks have employed both techniques successfully.

Prior to installation, test each block by soaking it in water and hooking it up to its companion conductance meter. The installation hole should be only slightly larger than the block and can be made using a soil probe or small auger. Dig the hole to the depth at which you want to install the block. Before inserting the block, add about 2 ounces (60 ml) of water to the hole and allow it to soak into the soil at the bottom of the hole.

To place the block in the hole, run the wire lead through an appropriate length of ½-inch-diameter PVC pipe and put tension on the wire to hold the block on the end of the pipe. One manufacturer's block has a special collar that fits into the end of ½-inch PVC pipe to facilitate installation. Lower the block into the hole and push it firmly into the moist soil in the bottom of the hole. Remove the PVC pipe and fill the hole for several inches with the soil removed in making the hole. Pack the soil firmly, preferably using a wooden rod such as a broom handle to avoid damaging the blocks or wires. Continue refilling the hole in stages, making sure that the soil is firmly packed at each stage in an effort to simulate undisturbed conditions. Refilling the hole properly is important to ensure that water and roots do not penetrate the hole more easily than the surrounding soil.

When installing blocks in trafficked or cultivated areas, trench or bury the lead wires to prevent damage. Identify the blocks by knotting the wire, color coding, or by tagging. If possible, protect the bare ends of the wire leads since they tend to become corroded. Scraping the wire leads with a knife blade or fingernail file prior to attaching the meter facilitates a better connection.

How to use information from the blocks

Resistance block information is best used by evaluating changes in block readings. A block that shows that the surrounding soil is becoming drier between irrigations indicates that roots are taking up water at that depth. Most frequently, trees preferentially take up water from the shallower depths, drying the shallow areas before taking up water from deeper areas. The shallower-placed resistance block shows a change first. Some block users consider this transition of water uptake from the drier, shallow areas to the deeper areas, along with a particular meter reading of the shallow block, as a signal to irrigate. These signals to irrigate are often arrived at over time by watching and comparing the tree response and block readings.

As mentioned earlier, the blocks provide information only on when to irrigate, not how much to irrigate. Irrigation amounts can be estimated by trial and error. Reading the blocks just prior to irrigation and then reading them 2 to 3 days after irrigation (when drainage has nearly stopped) can provide information on the depths to which irrigation water penetrated. If a block's readings remain unchanged (assuming the block is not reading completely wet), water did not reach that depth. If you wanted to wet the soil to that depth, more water (a longer irrigation) would need to be applied.

If blocks at deeper areas never change readings, you may be able to schedule more time between irrigations, assuming that there is root activity at that soil depth. A lack of root activity can be due to saturated soil conditions resulting from overirrigation, or a high water table, disease, or nematode problems. In addition, if a deeper block continually reads wet, you may be overirrigating and losing water to deep percolation.

If a block's readings remain unchanged when you think it should be changing, check a nearby location's soil moisture using a soil auger and the "feel" method to gauge the accuracy of the readings.

Neutron Soil Moisture Meters

A neutron soil moisture meter (neutron probe) consists of an electronic gauge, a connecting cable, and a probe cylinder that has a nuclear source and a detector tube (fig. 15.2).

The nuclear source is an americium 241–beryllium capsule, which emits high-energy neutrons. When lowered down an access tube into the soil, the high-speed neutrons collide with surrounding soil particles. Collisions with hydrogen atoms in the soil cause the neutrons to lose energy, creating low-energy, or thermal, neutrons. Some of these neutrons are reflected back to the probe cylinder and are counted by the thermal neutron detector tube. These "counts" are transmitted through the cable back to the gauge and displayed to the user. The neutron probe is a sophisticated hydrogen counter. Since water is the primary source of hydrogen atoms in the soil, the neutron probe indirectly measures the soil moisture level.

Neutron probe measurements are taken by setting the gauge atop the access tube and lowering the probe cylinder into the access tube to the desired depth. The neutron probe measures a large (volleyball-sized) sample. This is a larger sample than most other soil moisture sensors measure, one advantage of using the neutron probe. The probe cylinder can be lowered to the same depth each time, allowing repeat measurements at the same location. The neutron probe allows you to obtain very accurate soil moisture measurements in a matter of minutes.

The neutron probe has some disadvantages, however. Soil moisture measurements at shallow depths (6 inches, or 15 cm, or less) are often unreliable because neutrons escape through the soil surface. Initial calibration of the neutron probe at a site can be time-consuming. Since the neutron probe uses a nuclear source, the operator must be trained in its handling, storage, and use. A license to operate the gauge is required, as are periodic inspections. The initial cost of the neutron probe is also quite high, approximately $4,000.

Installation

Establishing a neutron probe monitoring site entails installing an access tube to the desired depth. Access tubing can be aluminum pipe, PVC pipe, or electrical metal tubing (EMT). Since the probe readings are affected by the access tubing material, only one type of material should be used for all the tubing in the orchard. Aluminum has the least effect on the readings; PVC pipe gives lower readings because the chlorine in the PVC absorbs the low-energy neutrons.

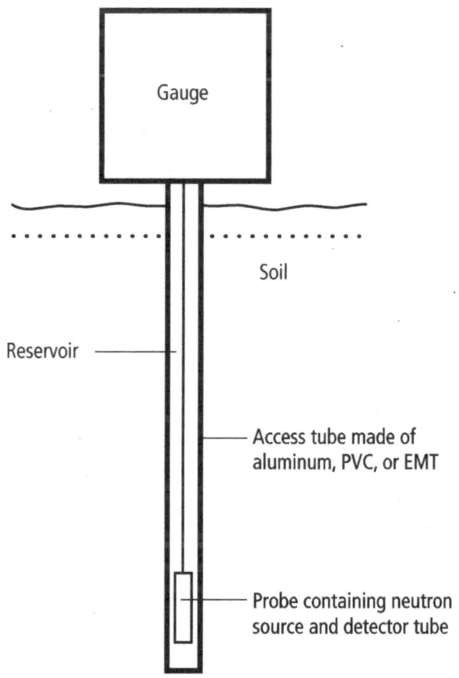

Figure 15.2. Neutron soil moisture gauge.

Access tubing should be approximately 2 inches (5 cm) in diameter. The inside diameter of the access tube should be nearly the same diameter as the probe cylinder. In areas with a high water table, the bottom of the access tubing should be sealed.

Auger a hole the same diameter as the outside diameter of the access tubing into the soil to a depth slightly greater than the maximum measurement depth. The access tubing should fit snugly into the augered hole. Cover the top of the access tube when not in use to prevent water or other foreign material from entering.

Placement of monitoring sites

The number of required monitoring sites depends on soil uniformity and management. For areas up to 40 acres (16 ha), at least two sites should be established. Stations should be in areas representative of overall moisture conditions, with separate sites for problem areas or for areas with different soil or management conditions.

The maximum depth at which the access tube is installed should be at least the depth of the active root zone. For established pear trees with no restricting soil layers or high water table, measuring to a depth of 5 feet (1.5 m) monitors most water usage by the tree. Installing to a slightly greater depth is recommended for monitoring deep percolation losses below the root zone.

Place the access tube in the tree row. For flood- or sprinkler-irrigated orchards, install the access tube within the drip line of the tree, but not directly next to the trunk. In sprinkler-irrigated orchards, do not place the access tube where trees "shade" the sprinkler distribution pattern.

Evaluating the readings

The readings of a neutron probe are simply counts of returning neutrons, so the instrument must be calibrated to translate the readings to soil moisture measurements. For scheduling irrigations, it may be sufficient to identify the neutron probe reading at which the soil is "full" of water (is at approximately field capacity) and the reading at which irrigation should begin (refill point). A more thorough calibration, in which neutron probe readings are compared to actual soil moisture as determined by volumetric soil sampling, allows you to determine not only the timing of irrigations but also the required irrigation amounts. This calibration technique is time-consuming but need be done only once for a given monitoring site. Sites with different soil types must be calibrated separately. Once calibrated, the neutron probe is very accurate.

Calibrations should relate soil water content to a relative count. The relative count is the actual gauge count divided by a standard count. This procedure is recommended because the actual count can change with time due to deterioration of the radioactive source and because of electronic drift. The standard count is made by placing the probe at least 2 feet (0.6 m) above the soil and 2 feet away from any source of hydrogen ions, such as a human body. Make the count with the source tube locked in the gauge.

Maintenance

Neutron probes require little maintenance beyond checking to ensure that the probe is operating properly and that the access tubes are free of water or foreign material. When repairs are required, they can be quite costly and must be done by the gauge manufacturer.

Dielectric Constant Soil Measuring Devices

A relatively new technology for measuring soil moisture uses the dielectric constant, a physical property of materials. The dielectric constant of a material affects the travel time of a high-frequency electromagnetic pulse passing through the material. The dielectric constant of soil and water are significantly different: 3 to 7 for soil and 80 for water. Two soil moisture measurement techniques take advantage of these differences in dielectric constant. Time domain reflectometry (TDR) measures the travel time of an electromagnetic wave along two or more waveguides of known length. Frequency domain reflectometry (FDR), also known as capacitance measurement, measures the soil capacitance of a high-frequency radio wave pulsed through the instrument and the soil. Most of these devices require permanent installation of sensors in the soil. For TDR, the most common soil-installed sensor has been a pair of parallel stainless steel rods spaced a known distance apart, usually about 5 centimeters (2 in). The length of the rods determines the monitored soil volume, and the device measures a cylinder the length of the probes with a diameter slightly greater than the spacing of the rods. Monitoring various depths requires installation of a number of pairs of parallel rods, each pair of different lengths. One manufacturer of TDR devices has combined the parallel rods, with the remote diode shorting into a single segmented probe that can measure various depths.

A number of FDR, or capacitance, probes are currently available. They require the permanent installation of one or more sensors or the installation of a PVC access tube down which a probe is lowered. One company makes a portable instrument that can be pushed into the soil as a soil probe.

If you choose a TDR or capacitance device, work closely with a consultant or other professional familiar with their use.

Advantages

A major advantage of most TDR and capacitance devices is that they can be automated to take continuous measurements. This allows you to have an accurate picture of how soil moisture changes with time. For example, you can measure how the soil moisture changes when the tree is taking up moisture, or how the soil moisture changes as a result of an irrigation. The electromagnetic, wave-generating component of these TDR and FDR systems can monitor and store information from multiple sensors using a multiplexing system and a data logger. These capabilities are unique to TDR and FDR soil moisture monitoring devices. The TDR and capacitance probes can also be accurate once calibrated to the site.

Disadvantages

There are some significant disadvantages to TDR and capacitance devices when used for field applications. First, both types of devices monitor a relatively small volume a soil, only a slight distance out from the probes. For TDR devices using pairs of parallel metal rods, you must install pairs of different lengths if you want soil moisture readings for various depths. Excellent contact between the soil and the TDR or capacitance soil sensor is very important. Some of the greatest difficulties in using the devices have resulted from poor contact following installation. It is often difficult to attain this good contact due to installation difficulties, particularly in gravely or heavy soils. Dielectric constant devices are difficult to calibrate to achieve the desired accuracy. Finally, the cost of TDR and FDR devices is high. Often the systems, including the wave generators, multiplexer, and data storage units, are $7,000 or more. The portable capacitance probe is approximately $600.

Placement of monitoring sites

As with other soil moisture measurement devices, the soil, irrigation system, tree type and age, and overall orchard management often determine the number of sites necessary in an orchard. A minimum of two sites per 40 acres (16 ha) is recommended. The multiple-station monitoring capability of TDR and FDR devices, with their necessary wiring connections, limits the distance between monitoring sites. At each site, it is useful to take measurements at 1-foot (0.3-m) intervals to monitor root uptake and irrigation refill and penetration.

Monitoring sites should be chosen with the objective of attaining the information desired. If the information is to be used for scheduling irrigations for the orchard, choose a location typical of the surrounding area. It is best to choose a site in the tree row, within the drip line of the tree and a few feet from the trunk. If a problem area of the orchard is to be investigated, siting should be chosen accordingly.

PLANT-BASED IRRIGATION SCHEDULING

Even under wet soil conditions, pear trees must balance the water demand of their environment with the ability of their roots to take up water from the soil. Plants use many different responses to keep their water supply and demand in balance, and most plant-based methods for irrigation management are based on the measurement of one or more of these responses. The pressure chamber method for measuring the tension of water within the plant has shown some promise as a physiological measure of water stress in pear trees (see chapter 14, "Water Relations"), but many other plant-based methods have been proposed in the past and new ones are being proposed almost as fast as new technologies are developed. One advantage of a good plant-based method is that it can be used to check whether the trees are responding favorably to the weather- or soil-based methods of irrigation scheduling. The weather-based methods tell you how much water the trees are using, and the soil-based methods tell you how much irrigation is required to recharge your soil reservoir, but neither reliably tells you when to irrigate. In order to determine when to irrigate using these methods, you must make a number of assumptions about the available soil moisture and the root zone depth of your orchard. Plant-based methods are very complimentary to weather- and soil-based methods: plant-based methods indicate when to irrigate but not how much, and weather- and soil-based methods indicate how much to irrigate but not when.

Temperature-based Methods

During the daytime, leaf canopy temperatures reflect the balance between the heating effects of the sunlight and the cooling effects of water loss (transpiration). One of the most common plant responses to water stress is to reduce the rate of water loss by closing their stomata (see chapter 14, "Water Relations"). This benefits the plant because it reduces water demand and brings the demand into balance with water supply; but in addition to reducing water loss, the closing of stomata also reduces sugar production by the leaves and may ultimately reduce the growth of fruit. Another side effect of reducing water loss is that leaf temperature increases because there is less evaporative cooling. This is the response that is measured by temperature-based methods.

A number of temperature-based methods are available, ranging from hand-held infrared "guns" that can be pointed at an individual tree to satellite-based remote sensing systems that can measure the average tree canopy temperature over a much larger area. Unfortunately, there are many drawbacks with these methods:

- Leaf or tree canopy temperature increase only after the stomata reduce water loss by closing, and by that time some negative effects of the water stress may have already occurred, so the information may not be timely.
- Remote sensors give an average temperature value over a large area, so any inclusion of sky (pointing from the ground up into the tree canopy), or of soil (pointing down from above), can cause substantial measurement errors.
- In order to interpret the measured value of canopy temperatures, a number of additional environmental variables such as air temperature, relative humidity, wind speed, and sunlight intensity must be obtained and combined to form an index of stress. This index must then be compared to a previously determined reference value. These reference values have not yet been determined for pears.

Pressure Chamber Method

The pressure chamber method measures the tension on the water in the water-conducting tissue of the plant (xylem). It can be both a sensitive measure of plant water stress and potentially useful in determining the effectiveness of a particular irrigation schedule or method. The pressure chamber method involves cutting a leaf and its attached petiole from the plant, protecting it from any water loss, and sealing the leaf inside a pressurization chamber with a small amount of the cut end of the petiole exposed to the outside. The pressure

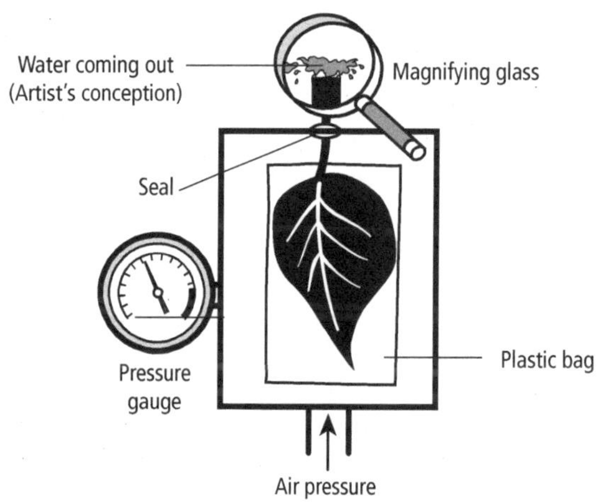

Figure 15.3. Pressure chamber.

within the chamber is increased until water is observed at the cut end of the petiole (fig. 15.3). The tension within the xylem is believed to be equivalent to the pressure in the chamber at the first appearance of water, and this can be used as a measure of the physiological "dryness" of the plant, which is called the plant water potential (see chapter 14, "Water Relations").

The units of water potential are negative pressures, with the most commonly used unit being the bar, which is equivalent to about 14.5 pounds per square inch (psi), or 6.9 kilopascals (kPa). In the scientific literature, the megapascal (MPa) unit, which is simply 10 bars, is preferred. Commercial pressure chambers are used to schedule irrigations in crops such as cotton. All pressure chambers operate on the same principle, and commercially available devices only differ in how the sample is sealed into the chamber; whether a supply of pressurized gas (usually industrial nitrogen) or hand-pumping is required to generate the pressure; the size, shape, and maximum operating pressure of the pressure vessel; and whether an industrial or scientific accuracy gauge is used. For irrigation management in pears, a gauge with an industrial level of accuracy (about 5 percent) is sufficient. As soils dry, plant water stress increases. In addition to soil drying, however, other factors, such as root disease, can restrict water availability to the plant and can also cause water stress. If increased irrigation does not alleviate plant water stress, there may be a problem related to root health.

Use in pears

To date, the most successful diagnostic use of the pressure chamber in pear trees is to measure midday stem water potential. For this method, shaded, inner-canopy leaves that are close to the trunk or main scaffolds are enclosed in a reflective plastic envelope and left attached to the tree for a minimum period of time, usually 10 minutes. The envelope stops the process of water loss from the leaf, and the 10-minute period is usually sufficient to allow the water potential of the leaf to come into balance with the water potential of the stem where the leaf is attached. Longer time periods (2 hours) can also be used if there is any doubt that the leaf has not come into balance with the stem, but in almonds, walnuts, and prunes no difference has been found between leaves covered for 10 minutes and leaves covered for over 1 hour, indicating that 10 minutes is sufficient.

As long as the leaf is physically undamaged, other physiological characteristics related to leaf nutrition or leaf type (shoot or spur) have not been found to influence the measurement of stem water potential. Also, leaves from the same tree are typically within ± 0.2 bar, so a single leaf usually represents an adequate sample. The number of trees that represent an adequate sample for any given orchard or irrigation section within an orchard depends on how much tree-to-tree variation occurs in that orchard. Usually a sample of 10 trees over an area of about 50 acres (20 ha) is sufficient to indicate whether substantial tree-to-tree variation occurs and a larger sample of trees is required, or whether the trees are relatively uniform and the sample of 10 is adequate.

Under typical field conditions, stem water potential is lowest (i.e., at its driest) at midday, because tree water use is the most rapid at this time. For diagnostic purposes, stem water potential is best measured over a period of about 2 hours beginning with solar noon (1:00 to 3:00 Pacific Daylight Savings time). Under fully irrigated conditions, the expected value of stem water potential in pears under different midday temperature and humidity conditions is given in table 14.1 in chapter 14. If the day is unusually cold or overcast, midday stem water potentials may indicate that the trees are under little or no moisture stress even if the soil is dry. This is an accurate reflection of the status of the trees under those conditions; but for the purposes of irrigation scheduling, measurements of midday stem water potential made under unusually cold or overcast days should not be interpreted as indicative of the typical day.

DEFICIT IRRIGATION SCHEDULING

Regulated deficit irrigation (RDI) is the intentional withholding of irrigation water during specific periods of crop development to achieve particular horticultural goals. In pears, RDI during the period of rapid shoot growth in the spring has been proposed to reduce excessive shoot growth in the current year, maintain or improve fruit size in the current year, and increase return bloom in the subsequent year. A reduction in excessive shoot growth may have a number of important direct horticultural benefits, such as reduced pruning and reduced susceptibility to fire blight.

In order to use RDI in pears, the timing and degree of water stress that is necessary to accomplish RDI goals must be determined. One goal is usually that the water stress should not irreversibly reduce fruit growth, and some researchers outside the United States have proposed that the period of stress should end about 60 days after full bloom, when the growth rate of the fruit increases from about 1.5 cm^3/day to about 2.0 cm^3/day. In an irrigation experiment performed on Bartlett pear in Ukiah, California, we found a relatively gradual increase in fruit volume over the season (fig. 15.4A), although some trees exhibited a marked increase in fruit growth rate around mid-June (fig. 15.4B), about 80

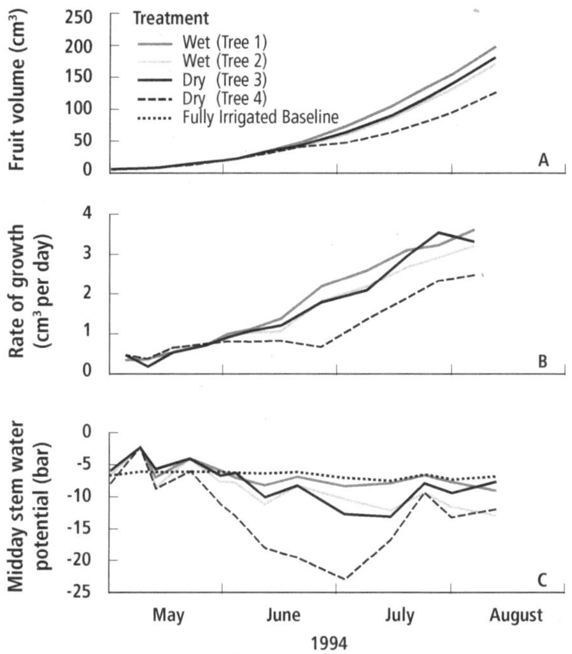

Figure 15.4.Fruit volume (A) and rate of volume growth (B) for ten tagged fruit on each of four trees, two of which were irrigated frequently (wet treatment) and two of which were not irrigated until early July. Midday stem water potentials measured on each tree over the growing season, along with the value expected for a fully irrigated tree (baseline) are shown in C.

days after full bloom. The midday stem water potential of individual trees was also monitored throughout the season in this experiment (fig. 15.4C), and over most of the season, the fruit growth rate (fig. 15.4B) was lowest for trees with the lowest stem water potential (i.e., trees that were under the most stress), and highest for trees with the highest stem water potential (for the relation of final fruit size to seasonal average stem water potential for this experiment, see fig. 14.2 in chapter 14).

Water was withheld from all of the trees in the dry treatment until July 5, 1994, well after the recommended RDI period, but the effect of this treatment on fruit growth was not the same for all trees. Some trees exhibited large and rapid reductions in midday stem water potential and fruit growth (dry treatment tree B, fig. 15.4), and even though the fruit growth rate increased for these trees following irrigation, the final fruit size was substantially reduced. In the same treatment, however, other trees exhibited more moderate and gradual reductions in midday stem water potential and fruit growth (dry treatment tree A, fig. 15.4), and the final fruit size was relatively unaffected. These results suggest that the response of fruit size to RDI depend on how RDI influences the degree of tree stress (stem water potential), and that this influence may vary from tree to tree within an orchard.

More extensive experiments were carried out with Bartlett pear in 1999 to further define how early-season RDI might be used in pear irrigation management programs. Large plots, approximately ½ acre (0.2 ha) or more, were developed in two locations near Ukiah and Hopland, California. At each location either an irrigated or an RDI plot was established (two treatments). The irrigated trees were irrigated as normal by the grower; in the RDI plot, trees were not irrigated until midday stem water potential reached approximately −17 to −19 bars. The unirrigated trees had water withheld until either July 1 (Ukiah location; average stem water potential −18.8 bars) or July 8 (Hopland location; average stem water potential −17.3 bars), and then were irrigated the same as the irrigated controls. Regular irrigation commenced on June 4 at the Ukiah location and June 16 at the Hopland location for the irrigated controls. The irrigated controls had stem water potentials different from the RDI trees and average stem water potentials of – 9.0 to −11.8 bars during the treatment period at either location. The fruit size (diameter) of RDI-treated trees was slightly smaller at only the Hopland location on July 7, 1 day prior to initiation of irrigation. At harvest there were no differences in fruit size or yield at either location as a function of irrigation treatment. Fruit firmness at harvest was improved at the Hopland location by RDI. These results suggested that water could be saved by reducing early-season irrigation in Bartlett pear without reducing fruit size or yield. Additional benefits from reducing irrigation might include tree vigor control with improved light penetration in tree canopies, which benefits fruit quality and disease control.

These data indicate that monitoring the degree of water stress experienced in the orchard should be an important component of imposing RDI in pears, and they suggest that achieving good final fruit size may be possible even with irrigation deficits well into the rapid phase of fruit growth, as long as plant water stress is not severe. However, since some experiments have shown that reduced fruit size is closely associated with the occurrence of even mild plant water stress in pear, more research will be required to determine how irrigation deficit is related to plant water stress (midday stem water potential) under different soil and environmental conditions.

REFERENCES

Bowers, W. O., R. L. Snyder, S. B. Southard, and B. J. Lanini. 1989. Water-holding characteristics of California soils. Oakland: University of California Division of Agriculture and Natural Resources Leaflet 21463.

Hanson, B., L. Schwankl, and A. Fulton. 1999. Scheduling irrigations: When and how much water to apply. Oakland: University of California Division of Agriculture and Natural Resources Publication 3396.

PART 3

IRRIGATION AND FERTILIZATION OF PEARS

16

Irrigation Systems

LAWRENCE J. SCHWANKL

The critical components of good irrigation water management are knowing how much water is applied during an irrigation (application amount or rate) and how uniformly or evenly the water is applied (irrigation uniformity). This chapter discusses the application rate and irrigation uniformity issues associated with sprinkler and flood irrigation, the most common methods of irrigating pears.

SPRINKLER IRRIGATION

Application Rate

The application rate, usually measured in inches per hour (in/hr), of a sprinkler system is determined from the sprinkler discharge rate and the spacing of the sprinklers. The discharge rate (gallons per minute, gpm) of a sprinkler head is controlled by the sprinkler nozzle size and the operating pressure at the sprinkler.

Table 16.1 lists the sprinkler discharge (gpm) for various nozzle sizes and operating pressures. The greater the nozzle size at a given pressure, the greater the sprinkler discharge rate. The nozzle size is frequently engraved on the side of the nozzle, or it can be determined by testing the nozzle opening with drill bits of known size. The operating pressure can be determined by placing a pressure gauge fitted with a pitot tube (available at most irrigation supply stores) into the water stream just outside the nozzle opening.

The sprinkler discharge rate can also be quickly determined by using approximately 4 feet (1.2m) of garden hose, a 5-gallon (20-l) bucket, and a stopwatch. By placing the hose over the nozzle and directing the water into the bucket and timing how long it takes water to fill the bucket, the discharge rate from the sprinkler can be determined. The following formula can be used to calculate the sprinkler head discharge rate:

$$\left[300 \right] \div \left[\begin{array}{c} \text{time to fill} \\ \text{5-gal bucket} \\ \text{(sec)} \end{array} \right] = \left[\begin{array}{c} \text{sprinkler} \\ \text{discharge} \\ \text{rate (gpm)} \end{array} \right]$$

Note: The 300 in the above equation is a unit conversion constant.

For example, if it takes 150 seconds to fill the 5-gallon bucket, the sprinkler discharge rate is 2 gallons per minute: 300 ÷ 150 sec = 2 gpm.

Whether the method used to determine the sprinkler discharge rate is the sprinkler operating pressure and table 16.1 or the bucket and stopwatch method, measurements should be taken at various locations within the orchard. Sprinkler discharge rates vary with the pressure differences in the orchard. It is a management decision whether to select irrigation set times to adequately irrigate the orchard section(s) receiving the

Table 16.1. Sprinkler discharge for selected nozzle sizes and pressures

Pressure	Nozzle size (in)										
(psi)	3/32	7/64	1/8	9/64	5/32	11/64	3/16	13/64	7/32	15/64	1/4
20	1.17	1.60	2.09	2.65	3.26	3.92	4.69	5.51	6.37	7.32	8.34
25	1.31	1.78	2.34	2.96	3.64	4.38	5.25	6.16	7.13	8.19	9.32
30	1.44	1.95	2.56	3.26	4.01	4.83	5.75	6.80	7.86	8.97	10.21
35	1.55	2.11	2.77	3.50	4.31	5.18	6.21	7.30	8.43	9.69	11.03
40	1.66	2.26	2.96	3.74	4.61	5.54	6.64	7.80	9.02	10.35	11.79
45	1.76	2.39	3.13	3.99	4.91	5.91	7.03	8.30	9.60	10.99	12.50
50	1.85	2.52	3.30	4.18	5.15	6.19	7.41	8.71	10.10	11.58	13.18
55	1.94	2.64	3.46	4.37	5.39	6.48	7.77	9.12	10.50	12.15	13.82
60	2.03	2.76	3.62	4.50	5.65	6.80	8.12	9.56	11.05	12.68	14.44
65	2.11	2.88	3.77	4.76	5.87	7.06	8.45	9.92	11.45	13.21	15.03
70	2.19	2.99	3.91	4.96	6.10	7.34	8.78	10.32	11.95	13.70	15.59
75	2.27	3.09	4.05	5.12	6.30	7.58	9.08	10.66	12.32	14.19	16.14

least water, or to irrigate based on the orchard's average sprinkler discharge rate.

Determining the sprinkler application rate (in/hr) is simple once the discharge rate of the sprinkler and the sprinkler spacing are known. The following equation can be used:

$$\left[\text{application rate (in/hr)}\right] = \left[96.3 \times \text{sprinkler discharge rate (gpm)}\right] \div \left[\text{sprinkler spacing along row (ft)} \times \text{sprinkler spacing across row (ft)}\right]$$

For example, if the sprinkler discharge rate is 1.0 gpm and the sprinkler spacing is 48 feet along the tree row and 20 feet across the tree row, the application rate is 0.1 in/hr:

$$\left[\text{application rate}\right] = \left[96.3 \times 1.0 \text{ gpm}\right] \div \left[48 \text{ ft} \times 20 \text{ ft}\right] = \left[0.1 \text{ in/hr}\right]$$

Table 16.2 can also be used to determine the sprinkler system application rate. The sprinkler area in square feet (ft^2) is determined by multiplying the sprinkler spacing along the row (ft) times the sprinkler spacing between rows (ft). In the previous example, the sprinkler area would be 20 feet × 48 feet = 960 ft^2. Looking in table 16.2 under 960 ft^2 and a sprinkler discharge rate of 1.0 gpm gives the same sprinkler application rate of 0.1 in/hr.

Table 16.2. Average application rate for various sprinkler discharge rates and areas of coverage

Sprinkler area (ft^2)	Average application at sprinkler discharge rate (gpm)							
	0.75	1.00	1.25	1.5	1.75	2.00	2.25	2.5
500	0.14	0.19	0.24	0.29	0.34	0.39	0.43	0.48
550	0.13	0.18	0.22	0.26	0.31	0.35	0.39	0.44
600	0.12	0.16	0.20	0.24	0.28	0.32	0.36	0.40
650	0.11	0.15	0.19	0.22	0.26	0.30	0.33	0.37
700	0.10	0.14	0.17	0.21	0.24	0.28	0.31	0.34
750	0.10	0.13	0.16	0.19	0.22	0.26	0.29	0.32
800	0.09	0.12	0.15	0.18	0.21	0.24	0.27	0.30
850	0.08	0.11	0.14	0.17	0.20	0.23	0.25	0.28
900	0.08	0.11	0.13	0.16	0.19	0.21	0.24	0.27
950	0.08	0.10	0.13	0.15	0.18	0.20	0.23	0.25
1,000	0.07	0.10	0.12	0.14	0.17	0.19	0.22	0.24
1,050	0.07	0.09	0.11	0.14	0.16	0.18	0.21	0.23
1,100	0.07	0.09	0.11	0.13	0.15	0.18	0.20	0.22
1,150	0.06	0.08	0.10	0.13	0.15	0.17	0.19	0.21
1,200	0.06	0.08	0.10	0.12	0.14	0.16	0.18	0.20
1,250	0.06	0.08	0.10	0.12	0.13	0.15	0.17	0.19
1,300	0.06	0.07	0.09	0.11	0.13	0.15	0.17	0.19
1,350	0.05	0.07	0.09	0.11	0.12	0.14	0.16	0.18
1,400	0.05	0.07	0.09	0.10	0.12	0.14	0.15	0.17
1,450	0.05	0.07	0.08	0.10	0.12	0.13	0.15	0.17
1,500	0.05	0.06	0.08	0.10	0.11	0.13	0.14	0.16
1,550	0.05	0.06	0.08	0.09	0.11	0.12	0.14	0.16
1,600	0.05	0.06	0.08	0.09	0.11	0.12	0.14	0.15

An alternative method of determining the amount of applied water (inches) during an irrigation is to monitor the flow meter (if available) at the pump. The following formula can then be used:

$$\left[\text{application rate (in/hr)}\right] = \left[\text{pump discharge (gpm)}\right] \div \left[449 \times \text{irrigation area (acres)}\right]$$

Note: 1 cubic foot per second (cfs) = 449 gallons per minute (gpm).

It is often difficult to choose an application rate when designing a sprinkler irrigation system. A major design objective is to minimize runoff from the orchard. Ideally, irrigation water should soak in where it lands. The infiltration rate of the soil is difficult to determine since it changes during an irrigation and may change during the season. During an irrigation, the infiltration rate initially starts out at the highest rate and decreases during the irrigation until a relatively constant basic infiltration rate is reached (fig. 16.1). This basic infiltration rate should be equal to or greater than the sprinkler application rate so that no runoff occurs. Often the best guidance for choosing a sprinkler application rate is determining what has worked on other local orchards with similar soil conditions, slope, and management.

When designing a sprinkler irrigation system, it is preferable to have the application rate chosen correctly at the design stage, but nozzle sizes can be retrofitted if needed. If larger sprinkler nozzles are retrofitted to the system but the pipelines and pump are too small, the sprinkler system will have inadequate pressure for proper operation.

Determining Irrigation Set Time

The irrigation set time is based on the sprinkler system application rate and the soil water depletion since the last irrigation, along with some additional water for

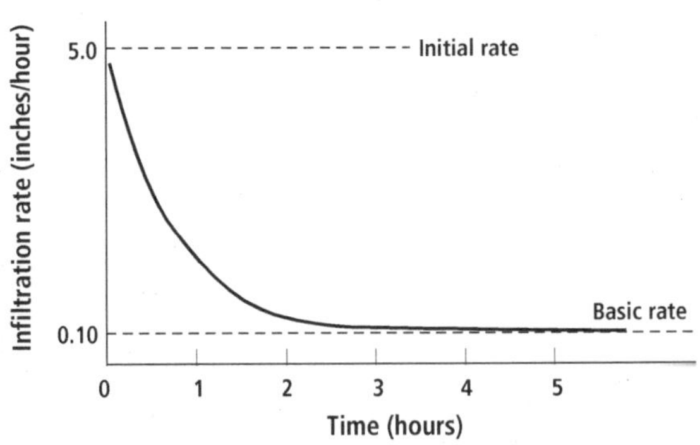

Figure 16.1. Typical infiltration rate curve showing high infiltration rate at the beginning of the irrigation decreasing to a lower, constant final infiltration rate. *Source:* Oster et al. 1992.

irrigation inefficiencies. The soil water depletion can be estimated by evapotranspiration (ET) techniques or by soil moisture monitoring (see chapter 15, "Irrigation Scheduling"). Since no irrigation is 100 percent efficient, additional irrigation water must be applied to ensure adequate irrigation. Efficiencies of 75 to 85 percent are desirable for solid-set sprinkler systems.

The irrigation amount and set time can be obtained by the following equations:

$$\left[\begin{array}{c}\text{desired}\\\text{irrigation}\\\text{amount}\\\text{(in)}\end{array}\right] = \left[\begin{array}{c}\text{soil}\\\text{moisture}\\\text{depletion}\\\text{(in)}\end{array}\right] \div \left[\begin{array}{c}\text{irrigation}\\\text{efficiency}\\\text{(\%)}\end{array}\right] \times \left[\,100\,\right]$$

and

$$\left[\begin{array}{c}\text{irrigation set}\\\text{time (hr)}\end{array}\right] = \left[\begin{array}{c}\text{desired}\\\text{irrigation}\\\text{amount (in)}\end{array}\right] \div \left[\begin{array}{c}\text{sprinkler}\\\text{application}\\\text{rate (in/hr)}\end{array}\right]$$

Example

Assume that the interval since the last irrigation is 10 days; the estimated soil moisture depletion is 2.0 inches (from ET information); the sprinkler application rate is 0.1 in/hr; and the irrigation efficiency is 80%.

$$\text{desired irrigation amount} = 2.0\ \text{in} \div 80 \times 100 = 2.5\ \text{in}$$

and

$$\text{irrigation set time} = 2.5\ \text{in} \div 0.1\ \text{in/hr} = 25\ \text{hr}$$

Irrigation Uniformity

How uniformly or evenly water is applied is also important to good irrigation water management. A uniform sprinkler irrigation system applies water so that each tree receives the same amount of water during an irrigation: no tree will get too much or too little water.

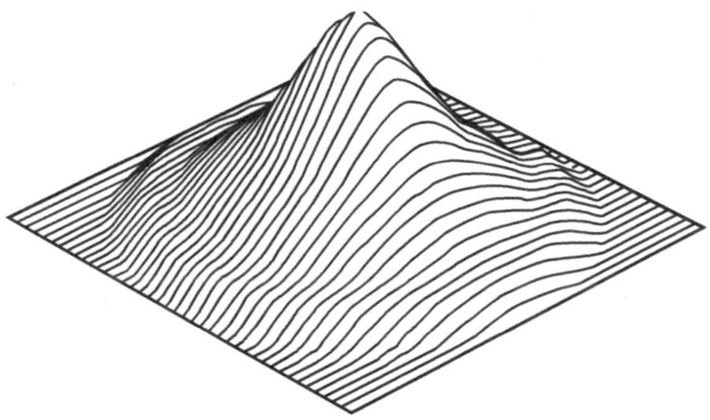

Figure 16.2. Typical distribution pattern of an impact sprinkler. The sprinkler head is located in the center of the plot.

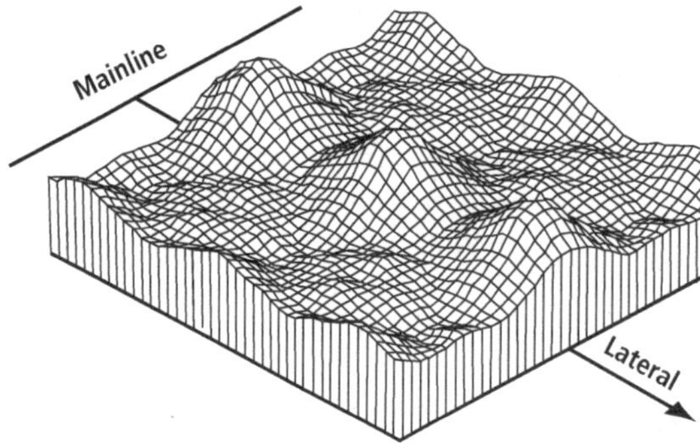

Figure 16.3. Water distribution pattern for overlapped impact sprinklers. The sprinkler heads are located along a lateral line.

A single sprinkler head does not apply water evenly over the area it wets. Figure 16.2 shows the pattern of water distribution for a typical impact sprinkler head. Note how some areas have more water applied to them than do others. In order to make the water application more uniform, the patterns from adjacent sprinklers must be overlapped (fig. 16.3). This sprinkler pattern overlap is advantageous because water application is more uniform, but it has the disadvantage of increasing the application rate. Thus, reducing the sprinkler spacing may increase irrigation uniformity, but it can also increase the application rate to the point where runoff can occur. A second disadvantage of reducing sprinkler spacing is the added cost of additional sprinkler heads and the increased size of pipelines supplying the sprinklers. Designing a good sprinkler system is therefore a balancing act of choosing an acceptable application rate and achieving good irrigation uniformity.

SURFACE IRRIGATION

Surface, or flood, irrigation systems are the least hardware-oriented irrigation systems, but they are the most complicated to manage properly. The objective of a

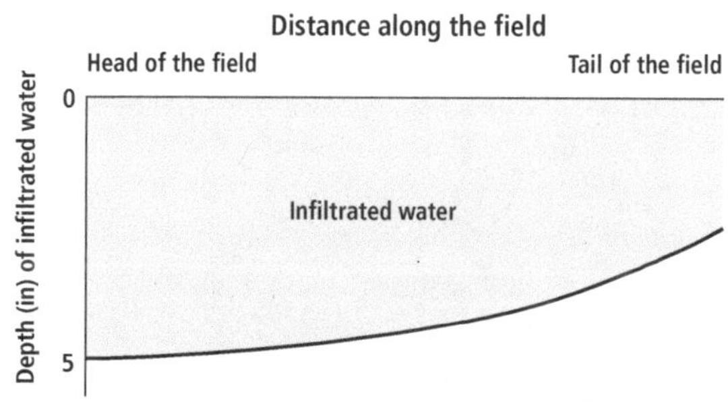

Figure 16.4. Typical infiltrated water distribution for a furrow-irrigated field that has recently been irrigated.

well-managed surface irrigation system is to have water infiltrating for the same length of time at all locations of the field or orchard. This is difficult to accomplish, since water is introduced at the head (highest elevation) of the field and advances down the field. Water infiltrates at the head of the field during the entire irrigation, but other field locations have water on them for a lesser period of time. The less time water is in contact with the soil, the less water infiltrates into the soil.

For furrow irrigation, the head of the field nearly always has more water applied to it than the tail end of the field (fig. 16.4). The exception is if water is allowed to pond at the end of the field. The field location that receives the least applied water is often at approximately two-thirds to three-quarters of the distance down the field when water is allowed to pond at the tail end of the field.

Many border system management variations exist depending on the border length, slope, and operation. In general, it is advantageous to keep field borders as short as practical to keep irrigation uniformity high. The disadvantages of short border lengths are increased labor and pipeline costs.

An additional difficulty in managing surface irrigation systems is measuring the amount of water applied to the orchard. A flow meter, such as a propeller meter, can be installed in a pump discharge pipe. Follow the manufacturer's recommended installation criteria to obtain accurate measurements. Open-ditch or canal water flow rates are difficult to measure. Consulting the irrigation district may help in getting a good estimate of the flow rate to the orchard.

The following formula may be used to determine the average amount of water applied to an orchard:

$$\left[\begin{array}{c}\text{applied}\\ \text{water}\\ \text{flow rate}\\ \text{(cfs)}\end{array}\right] \times \left[\begin{array}{c}\text{irrigation}\\ \text{time (hr)}\end{array}\right] = \left[\begin{array}{c}\text{area}\\ \text{irrigated}\\ \text{(acres)}\end{array}\right] \times \left[\begin{array}{c}\text{average}\\ \text{depth of}\\ \text{applied}\\ \text{water (in)}\end{array}\right]$$

Note: The units in the above equation must be used. If the flow meter display is in gallons or gallons per minute (gpm) rather than in cubic feet or cubic feet per second (efs), the conversions are: 1 cubic foot = 7.48 gallons, 1 cfs = 449 gpm.

The average depth of applied water in the above equation should be the orchard water use since the last irrigation, and it can be estimated by the orchard evapotranspiration (see chapter 15, "Irrigation Scheduling"). Additional water should be applied since no irrigation system is 100 percent efficient. Flood-irrigated orchards are generally less efficient than sprinkler-irrigated orchards. If no irrigation evaluation information is available on the orchard, an irrigation efficiency estimate of 60 to 70 percent is realistic.

REFERENCES

Oster, J. D., M. J. Singer, A. Fulton, W. Richardson, and T. Prichard. 1992. Water penetration problems in California soils: Prevention, diagnoses, and solutions. Kearney Foundation of Soil Science Research Report. Oakland: University of California Agriculture and Natural Resources Publication 21553.

17
Frost Protection

RICHARD L. SNYDER

FROST SENSITIVITY

Pears are relatively insensitive to freezing temperatures before bloom. As the buds break, flowers emerge and small fruit develop, and susceptibility to frost injury increases. Damage is more likely at lower temperatures, and the amount of damage increases with the duration at the minimum temperature. The temperatures at which 10, 50, or 90 percent damage can be expected are given in table 17.1 for pears at various growth stages. The pears were subjected to 30 minutes at the indicated minimum temperatures. These damage figures provide a guideline for expected damage on unprotected pears. However, there may be differences in frost damage within an orchard due to spatial temperature variations and cultural practices.

TYPES OF FROST EVENTS

Advection Frost

An advection frost occurs when cold air blows into an area to replace warmer air that was present before the weather change. It is associated with moderate to strong winds, no temperature inversion, and low humidity. Often temperatures will drop below 32°F (0°C) and stay there all day. Advection frosts are difficult to protect against, but fortunately they are rare in California's fruit-growing regions.

Radiation Frost

Radiation frosts are common occurrences in California. They are characterized by clear skies, calm winds, and temperature inversions. Radiation frosts occur when heat is lost in the form of radiant energy. Under clear nighttime skies, more heat is radiated away from an orchard than it receives, so the temperature drops. The temperature falls faster near the radiating surface, causing a temperature inversion to form (a situation in which the temperature increases rapidly with height above the ground). In fact, the temperature of pear trees during a radiation frost can be considerably colder than the air temperature measured in a fruit frost shelter.

If you measure high enough, the temperature reaches the point where it begins to decrease with altitude (a lapse condition). The level where the temperature profile changes from an inversion to a lapse condition is called the ceiling. A weak inversion (high ceiling) occurs when the air aloft is only slightly warmer than near the surface. When there is a strong inversion (low ceiling), temperature increases rapidly with height. Most frost protection methods are more effective during low-ceiling, strong inversion conditions.

Predicting Minimum Temperatures

Fruit-frost meteorologists have considerable information and experience at forecasting minimum temperatures. Whenever possible, use their forecasts to plan whether protection will be needed. In some cases, you may want to forecast your own minimum temperatures. If so, use the equation below and a psychrometer to estimate the next morning's minimum temperature:

$$\left[\begin{array}{c}\text{Forecast}\\\text{minimum}\\\text{temperature}\\\text{at sunrise}\\(^{\circ}\text{F})\end{array}\right] = \left[\begin{array}{c}\text{Wet-bulb}\\\text{temperature}\\\text{at sunset}\\(^{\circ}\text{F})\end{array}\right] - \left[\begin{array}{c}\text{Temperature}\\\text{at sunset}\\\text{divided by 4}\\(^{\circ}\text{F})\end{array}\right] - \left[\,4\ (^{\circ}\text{F})\,\right]$$

This equation was developed by C. C. Allen, a former National Weather Service (NWS) fruit-frost

Table 17.1. Temperatures at which 10, 50, and 90 percent damage is expected in pears at selected developmental stages after 30 minutes of exposure.

	Percent damage		
	10	50	90
Development stage	°F	°F	°F
Scales separating	17	6	0
Blossom buds exposed	19	10	4
Tight cluster	23	15	9
First white	26	24	15
First bloom	26	23	20
Full bloom	27	26	23
Postbloom	27	26	25

Source: Adapted from Proebsting and Mills 1978, p. 194.

meteorologist, for winter months in California's Central Valley. You should test its predictions against measured minimum temperatures during the winter prior to frost season. This equation should not be used during cloudy, foggy, or windy weather. It may also be inaccurate where cold air drains into a low-lying orchard. However, under clear, calm conditions without cold air drainage, the equation works well.

In addition to predicting the minimum temperature, it is useful to know the temperature drop rate. Assuming clear, calm conditions without cold air drainage, the following equations can be used to estimate temperatures at any time during the night. These predictions are useful to estimate when to employ frost protection methods.

$$\left[\begin{array}{c}\text{Temperature}\\ \text{after a}\\ \text{number of}\\ \text{hours after}\\ \text{sunset (°F)}\end{array}\right] = \left[\begin{array}{c}\text{Sunset}\\ \text{temperature}\\ \text{(°F)}\end{array}\right] - \left[\left[\begin{array}{c}\text{Rate of}\\ \text{change}\\ \text{factor}\end{array}\right] \times \left[\begin{array}{c}\text{Square}\\ \text{root of the}\\ \text{number of}\\ \text{hours after}\\ \text{sunset}\end{array}\right]\right]$$

where:

$$\left[\begin{array}{c}\text{Rate of}\\ \text{change factor}\end{array}\right] = \left[\begin{array}{c}\text{Sunset}\\ \text{– sunrise}\\ \text{temperature}\\ \text{(°F)}\end{array}\right] \div \left[\begin{array}{c}\text{Square root of}\\ \text{the number}\\ \text{of hours from}\\ \text{sunset to}\\ \text{sunrise}\end{array}\right]$$

If the wind speed increases to more than 4 miles per hour, skies get cloudy, or fog forms, all of the above equations give inaccurate predictions.

ENERGY TRANSFER

Energy or heat transfer determines how cold it will get and how effective frost protection will be. The four methods of energy transfer are radiation, conduction, convection, and latent heat.

Radiation

Radiation is electronmagnetic energy transfer. A good example of radiation is sunlight. Because it is very hot, the sun radiates considerable energy to the earth. Although much cooler, objects on earth also radiate energy to their surroundings. Normally, if an object radiates more energy than it receives from other sources, it will cool.

Conduction

Conduction is heat transfer through matter in which the objects do not move. A good example is the transfer of heat through a metal rod if one end is placed in a fire. The heat is transferred by conduction to the other end of the rod. Conduction is important in soil heat transfer, and therefore in frost protection.

Convection

In convection, heated matter physically moves from one place to another and takes heat with it. Air heated by smudge pots is an example of convection because the air, warmed by the heaters, rises and mixes with colder air in the orchard to raise the ambient temperature. Smudge pots also radiate heat to nearby trees but their main protection comes from convection.

Latent Heat

Latent heat is chemical energy stored in the bonds that join water molecules together; sensible heat is the heat you measure with a thermometer. When water condenses, cools, or freezes, latent heat is changed to sensible heat, which causes the temperature of the environment around the water to rise (or fall slower). When ice melts, water warms, or water evaporates, sensible heat is changed to latent heat and the air temperature commonly falls. Table 17.2 shows the amount of heat consumed or released per unit mass for both of these processes.

Frost Night Energy Budgets

During radiation frost conditions, more heat is lost to the sky than is gained from it. Table 17.3 shows a typical nighttime energy balance for citrus; the values are likely to be similar for pear orchards. Energy is lost through upward radiation to the sky. Energy is gained by downward radiation from the sky, by conduction of heat upward through the soil, and by convection of warmer air to the colder trees. Under cloudy or foggy conditions, the downward radiation is increased, depending on the temperature of the clouds or fog, and there is little or no net energy loss. Wind is a factor because it increases the amount of convective energy transfer from the air. During a radiation frost, when wind speeds increase to more than 4 miles per hour, the

Table 17.2 Exchange of latent heat to sensible heat

Process	Heat exchange* (cal/g)
Cooling water from 20°C (68°F) to 0°C (32°F)	+20.0
Freezing water at 0°C (32°F)	+79.7
Cooling ice from 0°C (32°F) to –5°C (23°F)	+2.5
Evaporation (water to water vapor) at 0°C (32°F)	–597.3
Condensation (water vapor to water) at 0°C (32°F)	+597.3
Sublimation (ice to water vapor) at 0°C (32°F)	–677.0
Deposition (water vapor to ice) at 0°C (32°F)	+677.0

Note: *Positive exchange indicates that the air is warming.

Table 17.3. Average energy balance during a radiation frost

Energy transfer	Energy transfer rate per unit area (W/m²)
Downward radiation (from the sky)	+230
Convection (from the air)	+39
Conduction (from the soil)	+28
Upward radiation (from the orchard)	−315
Net energy loss from the crop	−18

Source: Adapted from Mee 1979, p. 337.

increased conductive heat transfer is often great enough to balance the heat loss. Latent heat is a factor only when water is present, so it is generally ignored except when irrigation is used for frost protection.

PASSIVE FROST PROTECTION

Passive frost protection in an orchard, prior to a frost night, helps keep the air temperature near the trees higher. Passive frost protection includes:

- selecting an orchard site that is less prone to freezing temperatures
- choosing less-sensitive varieties or varieties that bloom later in the spring
- maintaining a firm, moist soil with little or no ground vegetation
- removing brush, trees, or structures that block the flow of cold air out of the orchard

Site Selection

Selecting a good orchard site is the most important passive frost protection. The best location is typically on a north-facing slope of a hill. The top and bottom of a hill tend to be colder; north-facing slopes also tend to be cooler during the daylight, which delays bloom. The probability of freezing temperatures drops rapidly during the spring, and a 1- or 2-week bloom delay can reduce frost damage in the long term.

When selecting an orchard site, use climate records or old annual reports from the NWS fruit frost forecaster to determine if the site is prone to freezing (many UC Cooperative Extension county offices maintain these annual reports on file). If records are unavailable, study the topography of the site to evaluate whether it is a sink for cold air drainage. Look for terrain or objects that may block drainage of cold air from the orchard site. Removal of the blocking objects or terrain can sometimes make a site acceptable.

Soil Water Content

Firm, moist soil conducts and stores heat better than does loose or dry soil. Do not cultivate at the beginning of frost season: many growers have cultivated part of an orchard before a frost and lost everything in the cultivated section. If the soil is dry at the beginning of frost season irrigate it to wet the top 1 foot (0.3 m) of soil. It is not necessary to wet the entire soil profile since the temperature below 1 foot deep changes little during the day. Apply from ½ to 1½ inches (1.3 to 4 cm) of water, depending on the type of soil and its dryness. Heavier clay soils require more water than light, sandy soils.

Ground Cover

Ground cover on the orchard floor can reduce the air temperature inside an orchard. For frost protection, an orchard floor with no ground cover is best. If a ground cover is necessary, it should be mowed (chemically or mechanically) as short as possible during the most sensitive frost periods. Research has shown that firm, moist, bare ground is usually warmer than an orchard floor with a ground cover. After periods of strong dry winds, however, the soil surface may dry out, causing the bare ground to be colder. A bare soil surface should be wetted after several days of drying winds.

Ice-Nucleating Bacteria

Reducing the concentration of ice-nucleating bacteria has been shown to lower the freezing temperature on plant surfaces, thus lowering the critical damage temperature. This works because ice melts at 32°F (0°C) but water does not necessarily freeze at that temperature. For water to freeze, the temperature must be below the melting point of ice and something must initiate the freezing process. At very low temperatures physical agitation of "supercooled" water can cause freezing, but at temperatures that damage frost-sensitive plants, ice-nucleating bacteria often initiate freezing on the leaf surfaces.

Although laboratory research using various bacteria-controlling materials to reduce the concentration of ice-nucleating bacteria for frost protection has been positive, field results are still being developed. Copper-containing compounds that kill bacteria and competitive bacteria that do not nucleate ice are most often suggested as protection materials. However, research is still needed to determine what materials work best, when to apply the materials, and how much to apply. Resistance to copper has been documented in populations of ice-nucleating bacteria exposed to repeated field treatments of copper.

Plants are damaged by freezing when ice formation inside the plant, but outside of the cells, dehydrates the cells. Since the freezing is internal, reducing transpiration with antitranspirants is ineffective for frost protection.

There is no known relationship between reducing transpiration and frost damage to pears.

ACTIVE PROTECTION METHODS

Heaters

Heaters have been extensively used in orchards for frost protection, but their use has declined because of rising fuel costs. California regulates the heaters that can be used for frost protection (table 17.4). The list of approved heaters is occasionally updated, so check with your local air resources office for a current list.

Heaters replace the net heat loss from an orchard by radiating heat to the trees immediately around the heaters. They also provide heat by convection in the inversion layer, which can give more protection than from radiation if the heaters are managed properly. The key is to make small fires in many heaters rather than large fires in a few heaters. When many small fires burn, cold air near the ground is drawn to the heater, where it is warmed. The hot air rises until it reaches the height with the same air temperature. It then spreads out, cools slightly, and falls back to the surface. Thus, the many small fires cause a circulation that mixes warmer air aloft with cold air near the ground surface.

Using large fires in a few heaters makes the air too hot, which causes the heated air to rise above the inversion ceiling. Once the hot air has risen above the ceiling, it is warmer than the air around it and continues to rise. In this situation, air does not circulate well within the inversion layer and the heaters are less effective. If there is no inversion, the heaters provide only radiation, affording less protection. Placing heaters on the side of a hill where no inversion occurs is not a useful practice. It is better to concentrate heaters in low areas where cold air settles.

Table 17.4. Orchard heaters approved for use in California

Heater	Maximum primary air orifice (in²)
Coke Heater	—
Exch. Model 7" dia Stack	0.606
Heat-Aid	—
Hy-Lo 230	0.606
Jumbo cone	0.196
Lazy Flame 18"	1.212
Lazy Flame 20"	0.606
Lazy Flame 24"	0.606
Lemora	0.606
National Double Stack	0.802
National Junior	1.212
Pipe System: Kittle	—
Return Stack	—
Return Stack Internal	—
Tree-Heet	—

Wind Machines

Wind machines provide frost protection by increasing convective heat transfer within an orchard. They are most effective during strong inversion conditions, when the wind machines mix warm air aloft with cold air near the surface. The wind machines must be started before the inversion forms. Measure the air temperature upwind of the orchard at 5 and 40 feet (1.5 and 12.2 m) above the surface and start the wind machines before the temperature at 5 feet falls to the temperature at 40 feet. The wind machines should be stopped when the air temperature upwind of the orchard at 5 feet exceeds both 32°F (0°C) and the temperature at 40 feet. The amount of protection afforded by wind machines depends on the strength of the inversion. When wind machines are operated properly, the temperature in the protected orchard will be approximately equal to one-third of the difference in temperature at 40 feet and 5 feet.

Wind machines have been successfully used by citrus growers, but the method is less effective for deciduous trees (see Crawford 1965). The strength of an inversion tends to be less in regions where deciduous fruits and nuts are grown, and the tree branches inhibit air mixing within inversions. Wind machines may be more effective in small valleys, where strong inversions form and the air above the trees can be mixed into the orchard, although verification was not found in the literature for this possibility.

Helicopters

Helicopters have been successfully used for frost protection in many crops. Helicopters work best when a strong inversion exists because the rotating blades force warm air at the top of the inversion downward to the surface. Compressing air also causes it to warm. Barometric pressure is higher near the surface, so air forced downward by a helicopter is compressed and warmed. Consequently, some protection is added even during radiation frost when there is no inversion.

Although research on helicopter use is limited, some pilots claim that best results occur when they fly at the ceiling height where the air temperature is highest. An electronic thermometer mounted on the helicopter can be used to find the correct flight elevation. Pilots often fill their spray tanks with water to add weight to the helicopter and provide more downward thrust of air.

The most common method of determining where protection is needed is to have someone on the ground measure the temperature and direct the pilot to cold areas. Some growers have installed lights with thermostats at the tops of the trees. When the temperature falls to a preselected value the lights come on. The pilot then flies over the areas with lights until the rising temperature causes the lights to go off.

Over-Tree Sprinklers

Over-tree sprinklers have been successfully used for frost protection in pears. The major limitations to sprinkler use are the cost of power to maintain a pressure high enough to spray over the trees, and the breakage of ice-coated branches. Frost protection comes from the release of latent heat as water freezes on the leaves, buds, flowers, or small fruit. Water must be applied uniformly to all the trees and at a rate high enough to replace the net heat loss and the additional heat loss due to evaporation. Approximately 7.5 times as much water must be frozen as evaporates to balance the net heat loss resulting from evaporation. Low humidity and high winds increase evaporation rates, so greater applications are needed under these conditions. Table 17.5 gives baseline over-tree sprinkler precipitation rates. Actual sprinkler precipitation rates should be higher if distribution uniformity is less than 100 percent. Divide estimated rates by the distribution uniformity of your sprinklers to calculate the actual precipitation rate you should apply.

Over-plant sprinklers should be started when the wet-bulb temperature is still above the critical damage temperature. This is necessary because the temperature of wet plant parts can fall as low as the wet-bulb temperature if sprinkling stops. The sprinklers should rotate once every 30 to 60 seconds. Otherwise, the wet plant temperatures can fall to the wet-bulb temperature, which may be below the critical damage temperature.

Wet-bulb temperatures are measured using a wet-bulb thermometer. Wet-bulb thermometers are part of a psychrometer, which can be purchased at most farm supply stores. If a psychrometer is unavailable, wet-bulb temperatures can be estimated from measured dewpoint and air temperatures using table 17.6. Private fruit frost forecasters frequently provide dewpoint temperatures in their forecasts; the dewpoint can also be measured with a dewpoint hygrometer. The conversions in table 17.6 are appropriate for a barometric pressure of 1,000 millibars (29.52 inches of mercury), and they are reasonably accurate from sea level up to about 2,000 feet (610 m) elevation.

Table 17.6 can also be used to determine the air temperature for starting and stopping the sprinklers. Since we want to ensure that the wet-bulb temperature does not fall below the critical damage temperature, find that critical temperature on the top (wet-bulb scale) and the dewpoint temperature on the side scale. Then locate the corresponding air temperature inside the table. The sprinklers should be started and stopped when the measured air temperature is above that selected from table 17.6. Be sure to measure the air temperature and dewpoint upwind of the protected orchard.

Table 17.5. Estimated sprinkler rates for cold protection of an 0.8-inch-thick unprotected pear leaf at selected temperatures (assuming 100% distribution uniformity)

Temperature (°F)	Sprinkler rates (in/hr) at wind speeds					
	1.1 mph	2.2 mph	3.4 mph	4.5 mph	5.6 mph	6.7 mph
28.4	0.06	0.07	0.09	0.11	0.12	0.13
26.6	0.08	0.11	0.14	0.16	0.17	0.19
24.8	0.11	0.15	0.19	0.21	0.23	0.25
23.0	0.14	0.19	0.23	0.26	0.29	0.31
21.2	0.17	0.23	0.28	0.31	0.32	0.38
19.4	0.20	0.27	0.32	0.37	0.41	0.44

Source: Adapted from Gerber and Martsolf 1979, p. 329.

Under-Tree Sprinklers

Under-tree sprinklers provide frost protection because water releases latent heat as it cools and freezes on the soil surface. If there is no freezing on the orchard floor, the heat released from cooling is sufficient to replace radiant heat losses. However, under these conditions, using the sprinklers may be unnecessary, since freezing may not occur. Under more severe freezing conditions, most of the protection from under-tree sprinklers comes from latent heat release as the water freezes on the orchard floor. The latent heat is changed to sensible heat and then radiated or convected into the trees for protection. To be effective, the net heat gain (heat gained from freezing minus heat lost to evaporation) must balance the net heat loss from an unprotected orchard, approximately 18 watts per square meter (22 watts per sq yd).

Without irrigation, the soil surface temperature can drop well below 32°F overnight. Irrigation maintains the soil surface temperature near 32°F as long as adequate water is applied. Based on trials in an orchard with a permanent-set sprinkler system with an application rate of 0.08 inches per hour (0.2 cm/hr), under-tree sprinklers were found to be effective under typical radiation frost conditions. With calm winds, air temperatures at 6.6 feet (2.0 m) above a protected orchard were 4° to 5°F (2.2° to 2.7°C) higher than in an unprotected orchard. Therefore, permanent-set under-tree sprinklers usually provide good frost protection.

Under-tree sprinklers should be started and stopped using the wet-bulb temperature (see discussion in over-tree sprinkling, above). This is necessary because lower parts of the tree may be wetted by the sprinklers. The temperatures of wet buds, flowers, or fruit may initially

Table 17.6. Air temperatures for starting and stopping sprinklers corresponding to dewpoint and wet-bulb temperatures

Dewpoint	Air temperatures (°F) at wet-bulb temperatures (°F)										
(°F)	22	23	24	25	26	27	28	29	30	31	32
32	—	—	—	—	—	—	—	—	—	—	32
31	—	—	—	—	—	—	—	—	—	31	33
30	—	—	—	—	—	—	—	—	30	32	34
29	—	—	—	—	—	—	—	29	31	33	34
28	—	—	—	—	—	—	28	30	32	33	35
27	—	—	—	—	—	27	29	31	32	34	35
26	—	—	—	—	26	28	30	31	33	34	36
25	—	—	—	25	27	29	30	32	33	35	37
24	—	—	24	26	27	29	31	32	34	35	37
23	—	23	25	26	28	29	31	33	34	36	38
22	22	24	25	27	28	30	32	33	35	36	38
21	23	24	26	27	29	30	32	34	35	37	39
20	23	25	26	28	29	31	33	34	36	37	39
19	24	25	27	28	30	31	33	34	36	38	39
18	24	26	27	29	30	32	33	35	37	38	40
17	25	26	28	29	31	32	34	35	37	39	40
16	25	26	28	29	31	33	34	36	37	39	41
15	25	27	28	30	31	33	34	36	38	39	41

drop to the wet-bulb temperature, which is lower than air temperature.

Hand-moved sprinklers can also be used for under-tree frost protection. Once they are started, they should not be stopped and moved during the frost night. Under mild frost conditions with light winds, experiments have shown little difference in air temperatures in orchards with various sprinkler spacings. However, as conditions become more severe, spacing sprinklers closer together gives greater protection. The spacing to use depends on the frost forecast. If the forecast is for extreme cold and windy conditions, it is better to concentrate the sprinklers in one area and wet the entire surface. Select the area most prone to frost damage for your sprinklers or concentrate sprinklers on the upwind part of the orchard. Since wind drift carries heat released by the freezing water, concentrating sprinkler lines upwind adds to protection downwind. If the forecast is for a mild frost, spacing the sprinklers at every second or fourth middle may provide protection for the entire orchard.

Micro-sprinklers

Micro-sprinklers can be used for frost protection in some tree crops. The principle is the same as for conventional under-tree sprinklers: heat released by water freezing on the surface radiates and convects to the air and trees. Application rates for micro-sprinklers are usually less than that of conventional sprinklers, and concern has been expressed about their effectiveness in frost protection. Although the application rate is less, the spray trajectory is low and avoids wetting the canopy and cooling the branches, leaves, or buds to the wet-bulb temperature. Evaporation rates are lower than for conventional sprinklers, and the water coats a smaller surface. Both convection and radiation are enhanced by concentrating the spray under the trees.

In field trials comparing several micro-sprinkler application rates, there was little difference in air temperature effects at various rates. However, there was a 1° to 2°F (0.5° to 1.1°C) difference in exposed thermometer readings between applications of 14.6 and 40 gallons per minute per acre (137 and 374 liters/min/ha). The temperature recorded by an exposed thermometer is similar to that expected for an exposed bud. In the low-rate treatment, the soil surface under the site where the temperatures were measured was dry. The soil was wet in the high-rate treatment. Since air temperatures were the same and the exposed temperatures were different, the difference in protection probably resulted from upward radiation differences between the wet and dry surfaces. The soil surface temperature was approximately 3°F (1.7°C) higher with the 40 gallons per minute per acre rate.

In field trials, the temperature was measured outside of or at the fringe of the wetted surface (a greater benefit was expected closer to the sprinkler heads). Our tests were conducted only during mild radiation frosts; the protection may be less under more severe conditions. Some problems with heads freezing and blowing off were observed in the low-rate treatment. This problem may be worse under dry, windy conditions. Based on our experience, micro-sprinklers help provide frost protection during mild frost but the spray should be concentrated under the trees. Higher application rates afford more protection.

Surface Irrigation

If surface water is available it can be used for frost protection. The protection comes from latent heat that is released as the water cools. The water should not be allowed to freeze, since freezing water creates an insulating layer of air between the water and the ice above it, inhibiting the heat transfer from the water to the air. Surface water should flow to the lower end of the field before the critical damage temperature is reached. Tailwater return systems should not be used because warm water provides more protection than the cold water returned from the bottom of the field. It is usually not cost-effective to heat the water unless the energy is going to be wasted anyway. Water should be concentrated under the trees rather than out in the middles, where much of the energy is radiated directly to the sky with less frost protection benefit. Make furrows as wide as possible to maximize the release of energy per unit area.

Active Protection Combinations

Wind machines and heaters

Combining heaters with wind machines has been shown to improve frost protection more than using wind machines or heaters alone. One wind machine with 3.5 heaters per acre (8.6 per ha) provided better protection than 1 wind machine or 17 heaters per acre (2.5 wind machines or 42 heaters per ha) (Brooks and Schultz 1958). Heaters should be distributed throughout the orchard and on the perimeter rather than concentrated near the wind machines (Crawford 1965). Heaters around the perimeter improve edge protection because they warm the cold air drawn into the orchard by the wind machines.

Wind machines and under-tree sprinklers

Using under-tree sprinklers in combination with wind machines has been proposed for frost protection but information on its effectiveness is limited. Researchers in Washington State found that using wind machines in addition to sprinklers provided an additional protection of 1° to 2°F (0.5 to 1.1°C) over sprinklers alone (Evans 1999). This was observed only under strong-inversion, calm conditions. They noted that there is no advantage to adding wind machines under higher evaporation conditions without an inversion. A similar positive effect is likely when wind machines and micro-sprinklers are used in combination. Wind machines should not be used with over-tree sprinklers. Evaporative cooling caused by the wind machines may lead to damage.

Wind machines and surface water

Running surface water in combination with wind machines has improved protection over using wind machines alone. The extra heat added by the cooling water improves mixing (convection heat transfer) and increases radiation to the trees.

Helicopters and irrigation

Using helicopters with irrigation on the orchard floor should provide more protection than using helicopters or irrigation alone. Mixing air within an inversion and releasing latent heat on an orchard floor can only have a positive effect during radiation frosts. However, using over-tree irrigation with helicopters may be detrimental. Increasing wind flow over the wet buds, flowers, or small fruit could increase evaporation and cool the wet parts as low as the wet-bulb temperature, which is usually lower than air temperature.

REFERENCES

Ballard, J. K., and E. L. Proebsting Jr. 1981. Frost and frost control in Washington orchards. Pullman: Washington State University Extension Bulletin 0634.

Brooks, F. A., and H. B. Schultz 1958. Orchard heating effectiveness. California Agriculture 12:4–5, 13.

Crawford, T. V. 1965. Frost protection with wind machines and heaters. Meteorological Monographs 6(28): 81–87.

Evans, R. G. 1999. Frost protection in orchards and vineyards. Avocadosource.com Web site, http://www.sidney.ars.usda.gov/personnel/pdfs/Frost%20Protection%20in%20Orchards%20and%20Vineyards.pdf

Gerber, J. F., and J. D. Martsolf. 1979. Sprinkling for frost and cold propection. In B. J. Barfield and J. F. Gerber, eds., Modification of the aerial environment of plants. St. Joseph, MI: American Society of Agricultural Engineers. 327–333.

Mee, T. R. 1979. Man-made fog. In In B. J. Barfield and J. F. Gerber, eds., Modification of the aerial environment of plants. St. Joseph, MI: American Society of Agricultural Engineers. 334–352.

Proebsting, E. L. Jr., and H. H. Mills. 1978. Low temperature resistance of developing flower buds of six deciduous fruit species. Journal of the American Society for Horticultural Science 103:192–198.

18

Nutrition

PATRICK H. BROWN AND FRANZ J. A. NIEDERHOLZER

Pear trees require 14 elements for normal growth and reproduction. These essential elements vary greatly in the concentrations normally found in plants and are classified as either macronutrients (relatively large concentrations) or micronutrients (see table 18.1).

Each of these elements is essential for particular functions in the plant. Nutrients are also important in promoting disease resistance and improving fruit quality, and the balance between the various elements can affect plant health and productivity. Several elements (chlorine, boron, sodium) may be toxic to the tree if present at excessive levels in the soil or irrigation water. Optimizing pear productivity and quality requires an understanding of the nutrient requirements of the tree, the factors that influence nutrient availability, and the methods used to diagnose and correct deficiencies. This chapter briefly discusses important principles of plant nutrition and gives specific recommendations for diagnosing and correcting deficiencies.

THE SUPPLY OF NUTRIENTS TO THE PLANT

Plants obtain nutrients from the soil as ions dissolved in the water the plant takes up or by diffusion from the soil to the root surfaces. In both cases water must be present for a nutrient to be available. Plant species also differ in the effectiveness with which they obtain nutrients from the soil and in the responses (symptoms) they show to a deficiency or excess of a given nutrient. Thus the nutrient requirements of pears are different than the nutrient requirements of other tree crops, such as walnut. Many factors affect plant nutrition by influencing the availability of nutrients to the root or the effectiveness of root uptake of these elements. For example, soil factors such as type and texture, moisture, pH, and depth as well as plant factors including root distribution, rootstock, fruit load, and competition all influence the nutrition of the tree. Environmental factors such as temperature, disease, salinity, and the presence of high levels of other ions may also influence plant nutrition.

Plants may be deficient in one or several nutrients as a result of the following conditions:

- An absolute deficiency of the element in the soil. This may occur in extremely sandy soils or isolated areas in which there are insufficient levels of one or more nutrients in the soil. Nutrients that are used in large amounts, such as nitrogen or potassium, may become deficient as a result of crop uptake. This condition can be remedied only by the addition of fertilizers to the soil.
- Unfavorable soil conditions for nutrient uptake. The pH of the soil has a significant effect on nutrient availability. High soil pH (>7.5) greatly limits the solubility of many elements in the soil (i.e., zinc, copper, manganese, iron), while low soil pH can lead to deficiencies (phosphorus, calcium) and toxicities (aluminum, iron, manganese). Similarly, low soil temperature, poor aeration, low water availability, or the presence of a hardpan can limit the ability of the plant to obtain nutrients by limiting root growth. Under these circumstances, addition of more nutrients may not alleviate the deficiency. The solution lies instead in correcting the soil conditions that limit nutrient availability. Adding lime to raise or sulfuric acid to lower the pH, using gypsum or organic matter as well as other soil amendments to improve soil structure, and maintaining optimal water and soil conditions can influence nutrient availability to the plant. In some instances adding fertilizers may be effective, though only a small percent of the supplied nutrient will be available to the plant.

Table 18.1. Nutrients required by the pear tree, in order of relative amount

Macronutrients	Micronutrients
nitrogen (N)	iron (Fe)
phosphorus (P)	manganese (Mn)
potassium (K)	chlorine (Cl)
calcium (Ca)	boron (B)
magnesium (Mg)	copper (Cu)
sulfur (S)	zinc (Zn)
	nickel (Ni)
	molybdenum (Mo)

Because nutrient deficiencies have distinct causes, it is essential that you consider all aspects of the orchard and the production system before deciding on a course of action.

DIAGNOSING ORCHARD NUTRIENT STATUS

Soil Analysis

Measurements of nutrient concentrations in the soil or in the tree can be used to determine whether nutrient deficiencies are limiting crop production. However, because of the large soil volume occupied by pear roots and the long lifetime of pear trees, it is often difficult to obtain a soil sample that is representative. This is the primary limitation to soil analysis for pears, as well as for other tree crops. Nevertheless, soil analysis provides information on nutrient levels and soil conditions (pH, lime), which may be important in predicting the occurrence and determining the cause of a deficiency.

Pear tree roots spread through a large volume of soil, and the variability of the soils can be quite high. Most soils differ in composition with depth: surface soil can be quite different from the soil a foot down, and the soil further down could be different again. Samples should be taken from the site of most-active root growth. This is usually the surface soil to a depth of 24 inches (0.6 m), but it will vary depending on the irrigation method used, the texture of the soil, and the tree's rootstock. Deeper soil sampling may be necessary to identify textural changes with depth that might influence tree growth or irrigation management. For a thorough analysis, soil samples should be taken from three to ten different spots around the tree and composited. Replicate samples should then be taken across the orchard. It is quite common for different soil types to occur within a single orchard. Nutrient deficiencies can be linked to localized soil differences, such as those associated with old riverbeds, differences in topography, sand deposits, cuts or fills, or old corral or pasture sites. In these cases it is necessary to sample from within each soil type and relate the findings only to trees growing in that soil type. This difference in soil type can be very localized and may only impact a few trees in an orchard.

Interpretive guides

It is more difficult to interpret soil analysis data than it is to interpret leaf analysis data. A soil can contain considerable amounts of a nutrient, and yet trees grown on this soil may suffer from deficiency. This shows that the nutrient is in a form that is unavailable to the tree or that the roots are not healthy enough for nutrient uptake. To correct this, various methods of extracting nutrients from soils have been used that better reflect the "plant available" nutrients in a soil. The methodology used and the principles behind the choice of extractant will not be discussed further here. In general, soil testing laboratories interpret the soil analysis based on the procedures utilized in their laboratory.

Determining the soil pH can be an excellent guide to the diagnosis of nutrient deficiencies. Soil pH affects the availability of mineral nutrients. Low pH (<5.5) may result in deficiencies of calcium, magnesium, phosphorous, or molybdenum and perhaps excesses of manganese, iron, or aluminum. High pH (>7.5), on the other hand, may immobilize manganese, zinc, iron, or copper, making them unavailable to the plant. In the case of excess salts in soil, the approximate conductivity of the saturation extract (EC_e) at which yields have been reduced is about 2.5 mmhos/cm (at 25°C) for pear. Soil chloride (Cl^-) in excess of 10 meq/liter and boron in excess of 1.5 ppm (in saturation extracts) can also reduce yield and long-term productivity. High levels of calcium carbonate ($CaCO_3$, lime) in the soil can induce deficiencies of iron, manganese, or zinc and may also make acidification of the soil difficult. The presence of any soil physical characteristic that is likely to limit root growth or water penetration is also likely to affect nutrient uptake.

Plant Analysis

For pear, leaf analysis is more useful in diagnosing mineral deficiencies and toxicities than is soil analysis. The mineral composition of a leaf depends on many factors, such as its stage of development, climatic conditions, availability of mineral elements in the soil, root distribution and activity, irrigation, and so on. The beauty of using leaf samples is that the tree integrates all these factors so that leaf analysis provides an estimate of which elements are available to the roots of the plant. The main problem with leaf analysis is that it does not tell us why the nutrient is deficient (see "Supply of Nutrients to the Plant," above).

Sampling procedure

Concentrations of nutrients in leaves vary with time, leaf age, position in the canopy, and the presence or absence of fruit. Trees within an orchard may also vary in their nutrient status as a result of differences in soils, water, or sunlight exposure. Biological factors that influence nutrient status include rootstock, crop load, and variety. Because of this variability, sampling techniques must be standardized if valid comparisons are to be made. The choice of sampling method also depends on the purpose of the survey. If the aim is to identify the problem only in an isolated tree or area, sampling of just a few poor and some good trees should suffice. If a determination of overall nutrient status in a large orchard is required, a more extensive survey containing trees from many sites within the orchard will be needed.

The accepted leaf sampling procedure is as follows. Collect fully expanded nonfruiting spur leaves from spurs at about 6 feet (1.8 m) from the ground picked at random around the tree. Collect 4 to 10 leaves from each tree and sample 10 to 20 trees in each orchard block. Leaves that have been sprayed with micronutrients should not be analyzed for that nutrient since all the surface contamination cannot be removed. Do not sample any leaves that have received nutrient sprays in the current year. Keep the leaves in labeled paper or plastic bags and submit them to the analytical service within 24 hours of collection.

Time of sampling

Collect leaf samples from late June through July. This time is best because the critical nutrient levels established through experimentation and observations (see table 18.2) are based on samples taken at that time. However, if you want to compare a good tree against a poor one, you can sample leaves at any time and compare the results between trees. Leaves sampled at times other than from late June through July may have nutrient concentrations different than those recommended in table 18.2, and results for them must be interpreted with care.

Interpreting leaf analyses

Results of tissue analysis are reported as the concentration of a nutrient on a dry weight basis. For macronutrients, concentrations are reported on a percent (%) basis (grams of nutrient per 100 grams dry weight), while micronutrients are reported in parts per million (ppm, micrograms of nutrient per gram dry weight). For each element, the laboratory will usually identify the critical value or the adequate range to aid in interpretation of the results. The critical value (CV), or critical level, refers to the nutrient concentration at which plant yield is 95 percent of maximum or at which distinct symptoms of deficiency are present. Tissue nutrient concentrations below this level cause poor plant growth and reduce yields. The adequate range refers to the nutrient concentration range at which growth is optimal. Above this nutrient concentration, plant growth may be inhibited by nutrient toxicity, or excessive leaching of nitrogen may occur. These values are specific for each nutrient and differ between species and to a lesser extent between cultivars. The nutrient recommendations supplied by the testing laboratory must be based on the crop being tested (in this case, pear). On occasion, some testing laboratories may use outdated or nonspecific recommendations. Reputable testing laboratories base their recommendations on published values that may have been adapted for local conditions and specific cultivars. Table 18.2 presents the currently accepted critical values and ranges for pears in California.

Although critical values are very useful in determining the nutritional status of an orchard it is important to realize that this technique is far from perfect. For instance, critical values are based on the health of the tree and not on the yield or the quality of the crop. In addition, the expression of a nutrient deficiency symptom occurs only after considerable yield reduction has occurred (usually greater than 10 percent), and in some instances an incipient nutrient deficiency (no leaf symptoms) may impair growth, impact fruit yield and quality, and enhance susceptibility to disease. Critical values based on nutrient deficiency symptoms should be considered the minimum nutrient concentration necessary for optimal growth. For several elements (nitrogen and boron, and the toxic elements sodium and chlorine) nutrient concentrations in excess of the optimum may impair plant growth.

COMMON NUTRIENT DEFICIENCIES IN CALIFORNIA PEARS: IDENTIFICATION AND CORRECTION

This section describes the symptoms associated with the most common nutrient deficiencies and toxicities occurring in pears in California. Often the pattern of symptom development (e.g., the site of deficiency with-

Table 18.2. Summer leaf nutrient concentrations and recommendations for pear in California

Nutrient	Critical level	Adequate range	Excessive
zinc	17 ppm	20–50 ppm	—
manganese	16 ppm	20–50 ppm	> 200 ppm
boron	22 ppm	25–45 ppm	> 100 ppm
copper	4 ppm	6–20 ppm	> 50 ppm
sulfur	0.15%	0.16–0.3%	—
phosphorous	0.13%	0.15–0.35%	—
magnesium	0.25%	0.25–0.5%	> 0.9%
potassium	1.0%	1.2–2.0%	—
nitrogen	2.2%	2.3–2.6%	> 2.6%

in the tree) can be used to identify a nutrient deficiency. Because of differences in mobility of the various elements within the tree, symptoms of nutrient deficiency appear first in different portions of the canopy. For example, deficiency of nitrogen or potassium appears first in the oldest tissues, whereas a deficiency of iron, manganese, boron, zinc, or copper appears first in the youngest leaves of the tree.

Nitrogen

Nitrogen deficiency is the most common nutrient deficiency in California pears. For a thorough discussion, see chapter 19, "Nitrogen Fertilization."

Potassium

Potassium deficiency symptoms include upward rolling of the leaf edges (boat shaped) with silvering or burning of the edges as symptoms become more severe. Leaves may look as if they are drought stressed (color plate. 18.1). Potassium deficiency occurs in all Californian pear districts, and evidence suggests that pears are particularly susceptible to potassium deficiency. Soils with a high clay content and sandy soils are most prone to potassium deficiency. In orchards that have been leveled the exposure of clay subsoils in "cut" portions can encourage potassium deficiency. Potassium is most available in the surface soil, and if the surface soil is dry potassium availability can be reduced.

Adequate potassium nutrition is needed to attain good fruit size. Monitoring leaf potassium levels and maintaining good fertilization practices is essential to orchard profitability.

Since potassium is rapidly fixed (bound) to clay particles in the soil, it is often difficult to correct a potassium deficiency in soils with a high clay content. In these soils a massive soil potassium application is the best way to treat potassium deficiency. Postharvest application rates of 15 to 25 pounds (6.8 to 11.4 kg) of potassium sulfate per tree shanked or trenched to around 48 inches deep (1.2 m) is the recommended treatment. Such an application should last 3 to 5 years. Potassium chloride is less expensive than potassium sulfate, but chloride may cause root burn. Do not broadcast potassium fertilizer in clay soils; place it where it can be used by roots, since potassium does not move in clay soils. In sandy soils, potassium fixation is not a problem, and trees respond well to soil potassium applications. In sandy soils, split applications into as many separate applications as is economically feasible.

Foliar potassium nitrate applications do not effectively correct orchard potassium deficiency unless numerous and costly applications are made. Foliar sprays get some potassium into trees quickly and may be effective as a stopgap measure until a soil application is possible.

In both heavy and light-textured soils applying potassium through the drip irrigation system (fertigation) or placing the fertilizer directly in the drip or microspray zone is likely to be effective. This has been demonstrated in prunes and is likely to be effective in pears. Potassium levels in prune leaves respond more rapidly to fertigation than to massive soil applications.

Phosphorous

Phosphorous deficiency can be a problem in pear orchards on volcanic soil in the Sierra Foothills of California or the Mid-Columbia region of Oregon and Washington. Phosphorus deficiency may also appear in pear orchards in heavy crop years. Phosphorous-deficient pear trees have little vigor, produce light crops, and have pale leaves with wavy edges (color plate 18.2).

Phosphorous is less mobile in all soils than potassium, so locally concentrated fertilizer applications are required. Trench or shank 15 to 25 pounds (6.8 to 11.4 kg) of treble superphosphate per tree under the tree canopy. High-grade chicken manure at a rate of 6 tons per acre (2.2 t/ha), or 125 pounds (56.8 kg) per tree, is also reportedly successful in treating phosphorous deficiency.

Zinc

Zinc is often deficient in California pear orchards. Annual zinc applications are often necessary to maintain growth, crop load, and fruit size. Zinc deficiency symptoms include small, narrow, wavy-edged leaves with yellow spots between the veins. Symptoms appear on the youngest leaves first. Severe zinc deficiency causes delayed bud break and the development of small, closely spaced leaves, referred to as "little leaf" (color plate 18.3).

Foliar applications are the best way to get zinc into pear trees. Zinc applied in many soils may not be available to trees. Spring or fall are the best times for foliar zinc sprays. Current recommendations for spring sprays are one of the following:

- basic zinc sulfate: 4 pounds per 100 gallons of water (1.9 kg/400 l)
- zinc sulfate: ½ to 1 pound per 100 gallons of water (240 to 480 g/400 l)
- zinc oxide: 2 pounds per 100 gallons of water (960 g/400 l)
- zinc chelate: 2 pounds per 100 gallons of water (960 g/400 l)

Spray in April or early May. Later applications are not as effective unless new growth is present. In the fall (November), zinc sulfate (36%) can be used in the dormant and crack bud or leaf drop stage. Separate zinc sulfate and oil applications by 30 days to avoid injuring plants. Recent work in walnut demonstrates the importance of applying zinc to leaves before they mature. In walnut the use of zinc sulfate with the addition of a spreader to the spray mix, buffered to pH 5.0, gave the best results. This formulation should be tested on pears on a small scale prior to its acceptance. Since foliar application of zinc contaminates leaf surfaces do not analyze leaves for zinc status after they have been sprayed.

Iron

Pear tree iron deficiency produces iron chlorosis in many California orchards. Pale yellow young leaves are the first sign of this condition (color plate 18.4). In almost all cases iron deficiency is caused by high soil pH and is not due to a lack of iron in the soil. Consequently, decreasing soil pH to increase iron availability is the best long-term solution to iron chlorosis. Materials such as sulfur, sulfuric acid, or N-phuric (urea + sulfuric acid) are effective in acidifying soil. N-phuric has recently produced excellent results in trials in several Lake County, California, orchards at the rate of 50 gallons per acre (77 l/ha) N-phuric (apply half in spring, half in fall). For problem soils, applications may be required for 4 years before results become evident.

Soil pH is also increased when irrigation water with more than 200 ppm of bicarbonate is used. If iron deficiency is encountered, test soil pH and irrigation water quality (salinity and bicarbonate level). These data can be used to estimate the rate of acidifying material required.

Soil-applied chelated iron has been shown to relieve iron chlorosis. Apply Sequestrene 138 at an annual rate of 0.2 pounds (91 g) per tree in equal amounts between April and harvest. This material must be applied locally under trees to ensure uptake.

Boron

Severe boron deficiency can cause blossom blast, a condition that causes blooms to die without setting fruit. Less-severe boron deficiency causes poor fruit set with no other visual signs of deficiency. Severe boron deficiency is characterized by reduced shoot growth, deformed young leaves, and death of the growing tip (color plate 18.5). In many regions of California boron toxicity, usually associated with irrigation water high in boron, also severely limits crop yield. Because both boron deficiency and boron toxicity can be difficult to recognize visually, the regular analysis of tissue boron levels is critical. In orchards where boron is needed, adequate blossom boron content is provided by summer or postharvest foliar applications of soluble boron material (e.g., Solubor, etc.) at 1 to 2 pounds per 100 gallons of water (480 to 960 g/400 l). This application is specifically designed to ensure adequate boron in the flower buds and has been routinely recommended in large areas of Oregon and Washington. This application of boron, however, may not be adequate to correct a severe boron deficiency. Since excessive boron application can reduce fruit set and cause toxicity problems, careful supervision of spray applications is critical.

In sandy soils where high-purity irrigation water is used, boron deficiency can be severe enough to limit tree growth and productivity. Typical symptoms, include delayed and poor bud break, extremely low fruit set, small and misshapen leaves, and death of the growing points. In these cases soil application of 1 to 2 ounces (28 to 56 g) of Solubor per tree may be required annually as guided by tissue analysis results.

Tissue analysis of boron concentrations in leaves can accurately determine whether boron is deficient (see table 18.2). Determination of boron toxicity, however, cannot be reliably made using leaf analysis alone. Experimentation is currently underway to develop improved methods of analyzing tree boron status.

Calcium

Calcium deficiency has not been observed in California pear orchards. However, low levels of fruit calcium are a major problem for Anjou pear and several apple varieties. Low calcium causes storage problems for these fruits. In California, Bartletts have not responded to calcium sprays. For Anjou pears, recent research from Washington state recommends five calcium chloride sprays of 1.5 pounds of material per 100 gallons water (720 g/400 l) between June and August for control of cork spot, alfalfa greening, and black end.

Sulfur

Sulfur deficiency has been reported in Oregon and Washington but not in California. Orchards fertilized with ammonium sulfate or potassium sulfate or acidified with sulfur-containing materials should not require additional sulfur application. Sulfur deficiency may result in the appearance of pale yellow leaves (similar to nitrogen deficiency symptoms) in orchards not treated with sulfur-containing materials. For example, pale yellow leaves were observed in a Medford, Oregon, orchard fertilized with high rates of urea. Using ammonium sulfate instead of urea solved the problem.

Magnesium

Magnesium deficiency has been reported in young pear orchards in California. The deficiency symptoms (leaves with wide yellow edges and a green Christmas tree pattern in the center of the leaf) were noticed on lower leaves of vigorous shoots of noncropping trees (color plate 18.6). These deficiency symptoms did not appear once shoot growth vigor decreased as the trees began to crop. Since early tree vigor is essential to orchard establishment the application of magnesium sulfate according to the manufacturer's specifications may be beneficial.

Copper

Copper deficiency is rare in California pear orchards, given the common use of copper sprays to control fire blight. When present, copper deficiency can be effectively treated with spring sprays of copper chelate at less than 1 pound of material per 100 gallons (480 g/400 l) of water.

Manganese

Manganese deficiency appears as large yellow green spots between the veins of mature pear leaves. This deficiency can be corrected by foliar applications of manganese sulfate at a rate of 1 to 2 pounds per 100 gallons of water (480 g per 400 l) applied in dilute applications of 400 gallons per acre (3,740 l/ha), for a total of 4 to 8 pounds of manganese sulfate per acre (4.48 to 8.96 kg/ha). In the Pacific Northwest, low leaf manganese concentrations are associated only with high-pH soils. Acidification of high-pH soils should increase manganese availability and may make sufficient manganese available to maintain plant growth.

REFERENCES

Ohlendorf, B. 1999. Integrated pest management for apples and pears. 2nd ed. Oakland: University of California Agriculture and Natural Resources Publication 3340.

19

Nitrogen Fertilization

STEVEN A. WEINBAUM

The potential for pear tree growth and productivity depends in part upon the availability and uptake of nitrogen (N), as well as other elements essential for plant growth (table 19.1).

THE NITROGEN CYCLE IN ORCHARDS

Making up 78 percent of the atmosphere by volume, gaseous nitrogen is not immediately available for use by pear trees unless the nitrogen gas is reduced to available ammonium. Ammonium becomes available through nitrogen fixation by bacterial microorganisms or by the synthesis and decay of organic matter. Commonly used forms of fertilizer nitrogen are ammonium (NH_4^+), nitrate (NO_3^-), or urea (NH_2–CO–NH_2)

The nitrogen in soils is derived from biological and industrial processes that take place in the nitrogen cycle in nature (fig. 19.1). Generally, soils contain about 0.05 to 0.1 percent nitrogen, of which only 1 to 3 percent becomes available to plants on an annual basis in the forms of ammonium or nitrate. Organic nitrogen is usually produced by decomposing leaves or released by soil microflora. In the upper humus layer of soil, the remaining nitrogen occurs mainly as proteinaceous material or as conjugates with other compounds in the dead remains of plants and animals.

Various bacteria and fungi are involved in breaking down nitrogen-containing organic compounds into simpler molecules. A soil bacterium, *Nitrosomonas,* oxidizes ammonium ions to nitrite, and another bacterium, *Nitrobacter,* oxidizes the nitrite to nitrate. As a result of the activity of these and other soil microorganisms, the complex nitrogen-containing substances originally synthesized by plants and animals are broken down and converted to the ammonium and nitrate ions absorbed by roots.

Nitrogen circulates continuously through the nitrogen cycle. Nitrogen in soil solution, mainly in the forms of nitrate and ammonium, is readily taken up by growing trees through their roots. The tree combines this nitrogen with the products of photosynthesis to produce free amino acids that are used as biochemical building blocks for such life-supporting substances as the nucleic acids, proteins, and a wide range of secondary nitrogenous compounds used in plant growth. Approximately 50 percent of the leaf nitrogen is translocated back into the tree before natural leaf fall, and some portion of the nitrogen carried to the orchard floor in leaf litter is recycled within the soil-plant system depending on external conditions such as moisture and temperature.

LOSS OF NITROGEN FROM ORCHARDS

Nitrogen is lost from trees when the harvested crop and prunings are removed from the orchard. The crop is the predominant avenue of tree nitrogen loss, and the magnitude of nitrogen removal from the orchard in the harvested crop varies with yield. Inappropriate fertilization practices accentuate soil nitrogen losses. Nitrogen applied in excess of that needed to support nearly maximum yields is less efficiently recovered by the tree. Fertilizer nitrogen applied during the dormant period is more likely to be lost from the orchard (via volatilization, denitrification, and leaching) than nitrogen

Table 19.1. Essential plant nutrients for pear production

Element	Symbol	Source
Macronutrients		
Carbon	C	air (carbon dioxide gas)
Hydrogen	H	soil (water)
Oxygen	0	air, soil
Nitrogen	N	soil
Phosphorus	P	soil
Potassium	K	soil
Calcium	Ca	soil
Magnesium	Mg	soil
Micronutrients		
Zinc	Zn	
Iron	Fe	soil
Manganese	Mn	soil
Copper	Cu	soil
Boron	B	soil
Zinc	Zn	soil
Molybdenum	Mo	soil
Chloride	Cl	soil
Nickel	Ni	soil

applied between March and September when the trees are actively growing.

Nitrogen can be lost from the soil to the atmosphere by microbial reactions that convert nitrate to gaseous nitrogenous compounds. These denitrification reactions are accentuated under anaerobic conditions caused by waterlogging or soil compaction. This loss mechanism may be much more significant than leaching, runoff, and erosion on heavy-textured, high-clay soils on which many pears are grown in California.

Irrigation management is intimately related with the efficiency of fertilizer nitrogen usage. Overirrigation enhances nitrate leaching. The volume of water drained below the root zone and the rate of fertilizer nitrogen applied are the principal factors that influence the magnitude of nitrate leaching. Nitrogen should not be introduced through the irrigation system (sprinklers, drip, etc.) at the initiation of the irrigation set, as excessive nitrate leaching beyond the root zone may occur. The best results are obtained when the fertilizer is injected into the system toward the middle of the set and is terminated before the set is completed to allow for flushing of the system and sufficient movement of nitrogen into the soil. Leaching of nitrate may occur irrespective of whether the source of the nitrogen is organic (animal manures, compost, etc.) or inorganic (ammonium, nitrate, urea, etc.). Broadcast applications of ammoniacal forms of nitrogen may permit more ammonium adsorption to clay particles and reduce the nitrogen leaching potential of the fertilizer applied, but they are more vulnerable to possible losses through volatilization, particularly as the soils dry.

Orchard studies suggest that fertilizer injection through low-volume irrigation systems (such as drip, micro-sprinklers, etc.) results in a higher percentage recovery of fertilizer nitrogen than occurs in more conventional systems. Some researchers indicate that they could maintain tree productivity under drip irrigation even though they reduced somewhat the amount of fertilizer nitrogen applied under the more conventional, higher-volume systems. Presumably, low-volume irrigation stimulates a higher density of fibrous roots in a restricted volume of soil, and this rooting pattern more effectively intercepts the fertilizer nitrogen. Fertilizer nitrogen is absorbed more efficiently when applied frequently and in small doses over the growing season as compared with large doses of fertilizer applied infrequently.

Volatilization of ammonia (NH_3) could be responsible for significant losses of soil nitrogen or ammonium-containing and ammonia-forming (urea) fertilizers (table 19.2). Nitrogen in the form of nitrate (NO_3^-) is not subject to loss by volatilization. Volatilization of ammonia is markedly reduced by soil incorporation as opposed to application on the soil surface. Losses are accentuated at soil pH values above 7.0, high temperatures, soil moisture loss through drying, and high concentrations (such as occurs in banding) of ammonia-forming fertilizers. Lower volatilization losses occur as the cation exchange capacity (CEC) of the soils increase. Thus, some of the greatest losses (>50%) can occur on high pH (>7.0) sandy soils, or during warm temperatures when urea, particularly, is applied and left on the soil surface for longer than 1 to 2 days.

NITROGEN FERTILIZATION

The timing and rates of fertilizer applications may greatly influence the recovery of fertilizer nitrogen by the tree. Failure of the tree to recover the fertilizer-applied nitrogen represents an economic loss and contributes to the leaching of nitrate (NO_3^-) below the root zone and ultimately to pollution of the groundwater.

Table 19.2. Nitrogen-containing fertilizers

			Equivalent acidity and basicity (in lb $CaCO_3$)	
Fertilizer	**Formulation**	**Nitrogen (%)**	**Acid**	**Base**
Ammonium nitrate	NH_4NO_3	33.5–34	62	—
Ammonium sulfate*	$(NH_4)_2SO_4$	21	110	—
Anhydrous ammonia	NH_3	82	148	—
Aqua ammonia	NH_4OH	20	36	—
Calcium ammonium nitrate solution†	$Ca(NO_2)_2 \cdot NH_4NO_3$	17	9	—
Calcium nitrate	$Ca(NO_3)_2$	15.5	—	20
Manure (dry)‡	—	1–3	—	—
Sodium nitrate§	$NaNO_3$	16	—	29
Urea	$CO(NH_2)_2$	45–46	71	—
Urea ammonium nitrate solution†	$NH_4NO_3 \cdot CO(NH_2)_2$	32	57	—

Notes: *Will acidify the soil when used exclusively over a period of time.
†Often injected through low volume irrigation systems because of its liquid form.
‡Also improves soil structure but can contain high salt concentrations.
§Not recommended in most soils because of the deflocculating effect on soil structure, which decreases water movement in soils.

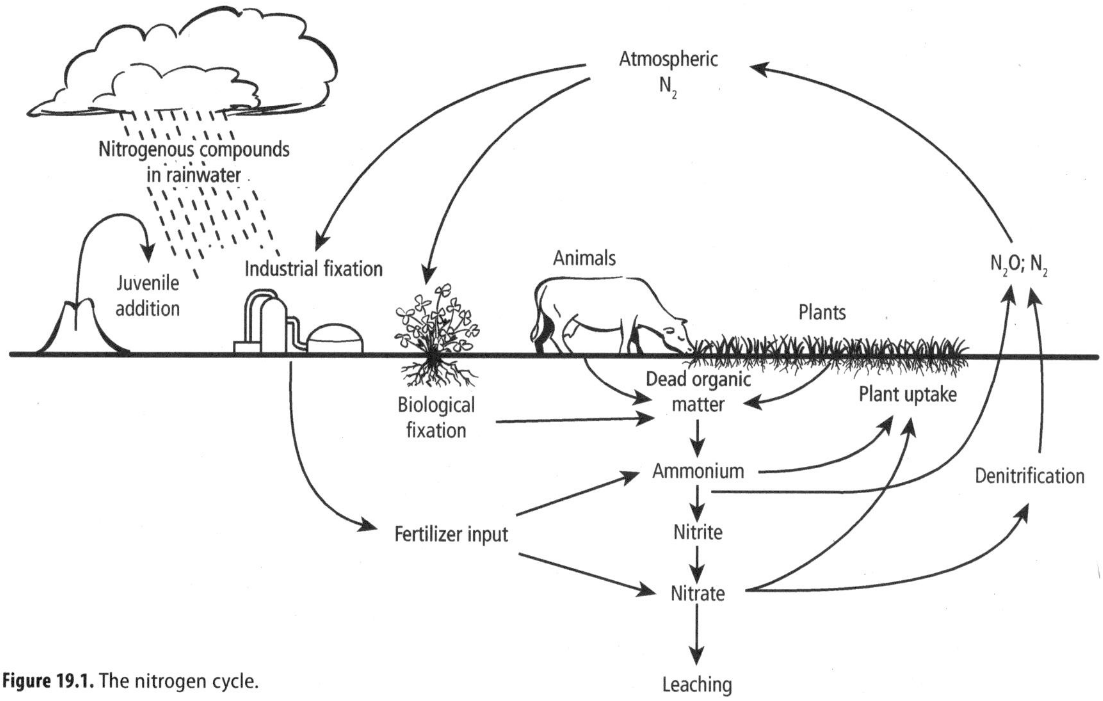

Figure 19.1. The nitrogen cycle.

Timing

Deciduous fruit trees absorb and assimilate nitrogen from the soil much more efficiently when trees are actively growing (spring and summer) than during the dormant season. This period of rapid soil and fertilizer nitrogen uptake occurs between the onset of vegetative growth in spring and the decline of leaf function in late summer. Pear trees, like other fruit tree species, appear to exhibit two periods of comparatively intensive root growth, occurring during active growth resumption in early spring and following fruit maturation and harvest. Nevertheless, the periodicity of root growth does not appear to influence nitrogen uptake. The capacity for nitrogen uptake declines as the tree approaches leaf fall in autumn, and uptake is minimal when the tree is leafless and dormant. Consequently, it is not advisable to apply fertilizer nitrogen between October and March. The nitrogen used by the tree in early spring for bloom, fruit set, and the early stages of fruit growth comes from the supply absorbed from the soil the previous season and from leaf nitrogen translocated back into woody tree parts before leaf fall. Trees require healthy foliage to absorb nitrogen effectively. Thus, late-fall and winter applications of fertilizer nitrogen are not effectively absorbed by the tree at that time. At best, the nitrogen applied in late fall and winter remains in the soil while the trees are dormant. However, significant amounts of this nitrogen can be denitrified or lost to leaching from winter rainfall. Fertilizer nitrogen not lost during the winter may be taken up in spring and used to support shoot growth, not for flower development, early fruit set, or early fruit growth. Although fertilizer nitrogen applied in the spring is absorbed effectively by the tree, it may stimulate undesirable vegetative growth if applied in excess. Smaller, more-frequent applications of fertilizer nitrogen during the growing season may be used more efficiently by the trees. This is especially true in coarse-textured (sandy) soils.

Rates

The appropriate rate to apply fertilizer nitrogen varies both from orchard to orchard and from year to year within any given orchard. Tree nitrogen status, yield, soil depth and texture as well as organic matter content, irrigation method, fertilizer application method, and nitrate content of the irrigation water influence the need for fertilizer nitrogen, the availability of nonfertilizer sources of nitrogen, and the efficiency of fertilizer nitrogen recovery.

Since the climate in California's Central Valley is mostly semiarid, the average organic matter content of the soil typically does not exceed 1 to 2 percent. Of this organic matter, 5 to 6 percent is nitrogen, but only about 1 to 3 percent of the nitrogen in soil organic matter becomes available for plant uptake each year (as a result of soil microbial activity). This amounts to about 20 to 60 pounds per acre (22.4 to 67.2 kg/ha) of nitrogen per year. Sources of nitrogen for plant use other than currently applied synthetic fertilizers may include residual fertilizer nitrogen, nitrogen "fixed" biologically by leguminous cover crops, and the application of high-nitrate irrigation water. In some San Joaquin and Sacramento Valley locations, wells are exhibiting increasing nitrate (NO_3^-) concentrations. As a result, many wells have

Table 19.3. Correlation between leaf analysis and tree performance in pear

Tree N status	Leaf N (% dry wt)	Vegetative appearance	Relative yield
Deficient	< 1.9	weak growth, pale leaves	< 75%
Low	2.0–2.1	—	90–100%
Adequate	2.1–2.4	—	100%
High	2.4–2.6	—	100%
Excessive	> 2.6	excessive growth, watersprouts, shade-reduced fruiting	80–100%

become significant sources of "free" fertilizer nitrogen. The nitrogen in irrigation water should be considered a part of the total fertilizer budget. As trees approach maximum yields, the capacity of the trees to absorb additional nitrogen from the soil is reduced. The residual fertilizer nitrogen in the soil is increasingly vulnerable to loss. Thus, overfertilization is not cost-effective and may contribute to nitrate pollution.

The demand for nitrogen varies with tree age and yield. Less nitrogen will be removed in the crop of immature (i.e., lower-yielding) trees as compared with mature trees. More nitrogen will be utilized for vegetative growth in the young trees, however. A general rule of thumb is to apply 2 pounds of nitrogen per ton of fruit harvested per acre (2.24 kg nitrogen per metric ton of harvest per ha). The crop removes about 1.3 pounds of nitrogen per acre per ton (590 g/t) of fruit, and the additional amount will supply the nitrogen for vegetative growth and accommodate the inefficiency of fertilizer recovery by the trees. Another recommendation is to apply no nitrogen if the average shoot length is greater than about 12 inches (0.3 m).

If sufficient soil nitrogen is available, pear trees will not respond (in terms of yield, tree growth, and leaf nitrogen concentration) to additional applications of fertilizer. Available data indicate that leaf nitrogen concentrations between 2.1 percent and 2.4 percent in July are adequate (table 19.3), and trees may not respond to additional nitrogen. If a sod cover is particularly heavy, nitrogen rates should be increased by about 40 pounds per acre (44.8 kg/ha). If legumes are grown as cover crops, nitrogen applications can be reduced by approximately 40 pounds per acre.

TREE NITROGEN STATUS

A number of factors indicate the nitrogen status of Bartlett pear trees. The length of annual shoot growth, fruit size, leaf size and color, and leaf nitrogen concentration vary with tree fertility. Fire blight develops most rapidly on succulent new growth, so nitrogen should not be applied in excess. Experimental evidence suggests that besides nitrogen, the status of potassium (K) influences the severity of fire blight, and the number of strikes increases as the ratio of nitrogen to potassium in the leaf increases. Leaf potassium concentrations above 1.4 percent may be more protective. In mature pear trees (over 15 years old), 8 to 10 inches (20.5 to 25.5 cm) of new growth each year indicates adequate nitrogen availability fo support tree performance. Less growth may indicate insufficient nitrogen availability.

Correction of nitrogen deficiency occurs readily following application of any nitrogen-containing fertilizer. The form of fertilizer makes little difference with respect to its ability to supply nitrogen to the tree. The choice of material should be based, therefore, on economics or other considerations. For instance, ammonium fertilizers are usually the least expensive; but they also lower the pH of the soil (see table 19.2) and are subject to volatilization if not incorporated immediately. Nitrate fertilizers offer the advantage of being immediately available for uptake by the plant. Manure and other organic fertilizers may improve soil structure and water penetration; however, some of these materials may release nitrogen for uptake more slowly than commercial fertilizers and may add undesirable salts to soil. With any fertilizer, take into account the potential beneficial and detrimental effects of the other nutrients making up the fertilizer material.

Leaf sampling and analysis should be conducted each summer to monitor the effectiveness of the nitrogen management program and to detect gradual changes in orchard nutrient status (see chapter 18, "Nutrition"). Excess nitrogen availability may result in excessive growth (including production of watersprouts) and reduced yields. Overfertilization that causes excessive vegetative growth not only may increase pruning costs but is responsible for increased shading, which reduces flower formation and fruitfulness in the lower, interior portion of the tree canopy (see chapter 9, "Canopy and Light Management").

20
Insect and Mite Pests

ROBERT A. VAN STEENWYK, LUCIA G. VARELA, AND MARIO S. MORATORIO

The major insect pest of pears is codling moth. Control of codling moth relies on pheromone mating disruption combined with one or two applications of either a selective or broad-spectrum insecticide, depending on the codling moth population density. If a broad-spectrum insecticide is used, secondary pests and occasional pests may increase in number because of the elimination of beneficial insects and mites that maintain the secondary pests under control. The secondary pests of pears are pear psylla and twospotted spider mite. Population increases of these secondary pests require additional insecticide or miticide applications for control. The occasional pests of pears are plant bugs (lygus, stink, and boxelder bugs), obliquebanded and fruittree leafrollers, speckled and humped green fruitworms, European red mite, San Jose scale, pear mealybug, rust mite, blister mite, pear thrips, western flower thrips, and pear slug. Population increases of occasional pests may cause economic damage in some years under favorable environmental conditions and may require insecticide or miticide applications.

A pest control strategy for pears must be based on sound pest sampling plans and treatment levels. Sampling techniques have been developed for many pear pests. For a comprehensive treatment of the use of these sampling plans and treatment levels and related matters, see the University of California Division of Agriculture and Natural Resources publication *Integrated Pest Management for Apples and Pears* (Ohlendorf 1999). For further information on specific insecticides, see the UC IPM *Pest Management Guidelines for Pear* (Elkins et al. 2002,) which is available for free downloading from the UC IPM Web site at http://www.ipm.ucdavis.edu.

MAJOR PEST

Codling Moth
Cydia pomonella

Codling moth was first observed attacking apples in the northeastern United States in the 1700s, and it was found in California in the 1870s. Although it was first observed attacking apples, its preferred host, codling moth will readily attack pears. Codling moth is now the major pest of pears, apples, and walnuts in California. It has also been reported to attack apricots, plums, and peaches, although these reports are infrequent and usually associated with special conditions.

Description

Adults are about ½ inch (12.5 mm) long with wavy gray and brown bands across the forewings (color plate 20.1). Each forewing has a copper-colored spot at the tip. Newly laid eggs are translucent white, flattened, and circular. As the young embryos develop within the eggs, the eggs turn to a reddish orange color (red ring stage). Shortly before hatching, the completely formed embryos with a black head can be seen in the egg (black head stage, color plate 20.2). There are five larval instars. The young first instar larvae are whitish, about 1/16 inch (1.6 mm) long, and have a dark head and thoracic shield. Mature larvae are about ¾ inch (19 mm) long and pinkish, with a mottled brown or black head. The mature larvae leave the fruit and spin silken cocoons, usually under the bark of the tree. The larvae form brown pupae about ½ inch (12.5 mm) long, and adults emerge in about 1 to 2 weeks, depending on the temperature (they emerge earlier in warmer temperatures). In late summer and fall, the mature larvae do not form pupae but enter a diapausing, or overwintering, condition as mature larvae.

Seasonal development

Codling moth overwinters as mature larvae in a diapausing condition inside thick silken cocoons, usually under the loose bark on the tree trunk (color plate 20.3). The larvae change into pupae in early spring (February to early March, depending on the location in California). Adult emergence usually begins from mid-March to early April in the Sacramento Valley and from late March to mid-April in the North Coast region.

A seasonal development model for codling moth has been developed that allows growers and pest control advisors to better time control procedures. The model uses air temperature. Codling moth development stops at temperatures above 88°F (31°C) and below 50°F (10°C). The model uses these upper and lower air temperatures and the current local air temperature to calculate degree-days (DD). One degree-day is 1°F above the minimum threshold (50°F) for 24 hours. The degree-day model is initiated and the accumulation of degree-days begins at codling moth biofix, the date mating takes place and viable eggs are deposited. Adult mating takes place when sunset temperatures exceed 62°F (16.5°C). However, male flight can occur at temperatures lower than 62°F, and males can be captured in pheromone traps at temperatures below 62°F. Thus, the setting of the biofix date for the first codling moth flight requires both moth flight as indicated by pheromone trap catch and sunset temperatures that exceed 62°F. This is because the codling moth degree-day model starts at the beginning of oviposition (egg laying), not at adult male flight. The UC IPM Web site (http://www.ipm.ucdavis.edu/) provides the date in the spring that the sunset temperature exceeds 62°F (16.5°C) for a number of weather stations throughout the state. If sunset temperatures are not available, a maximum air temperature of about 75°F (24°C) is correlated with a sunset temperature of 62°F (15°C).

Codling moth females begin oviposition about 60 degree-days after mating. The eggs are deposited on or near fruit and hatch in about 160 degree-days. The young larvae usually bore into the calyx end of the developing fruit within about 24 hours after hatching and complete the first instar just beneath the fruit surface. The mortality of these larvae during the first generation is extremely high (over 90%) due to the high density of stone cells on the fruit surface. As the fruit grows, the density of stone cells decreases, which allows for higher survival rates of first-instar larvae later in the season. Larvae that successfully enter the fruit and complete the first instar then tunnel to the core of the fruit and complete their larval development on the immature seed. Larval development takes about 475 degree-days. The mature larvae leave the fruit to pupate under loose bark. Pupal development takes about 430 degree-days, and the first generation is completed in about 1,065 degree-days.

The peak activity of the first flight occurs at about 300 degree-days (fig. 20.1). The flight activity gradually declines to a low point at about 500 degree-days. A second peak of flight activity frequently occurs at about 700 degree-days. The first peak is termed the "A" peak of the first flight, while the second peak is termed the "B" peak of the first flight. The cause of this bimodal flight is not understood. However, the B peak is of major concern for pear growers since a much higher percentage of first-instar larvae from this peak are able to infest the pears and complete their larval development. After the completion of the B peak of first flight, the second flight begins about 1,000 to 1,100 degree-days after biofix. The second flight starts about mid-June to early July, depending on the seasonal temperatures and the location in the state. The second flight in the North Coast region is about 1 to 2 weeks later than that in the Sacramento Delta.

Females from this second flight produce the second generation. Starting in mid-July, a small percentage of the larvae from the second generation do not pupate but enter a diapausing stage; by mid-August, nearly all the larvae have entered the diapausing stage. If most of the larvae of the second generation complete their development before mid-August, they pupate and produce a third generation. Depending on the seasonal temperatures and location in the state, there can be two to three generations of codling moth per year in pears. Because of slightly cooler temperatures, the North Coast region produces about one-half of a generation per season less than the Sacramento Delta (see fig. 20.1). The overwintering larvae spin a thickened cocoon, usually under the bark of the tree, where they remain through the fall and winter.

Damage

Codling moth directly damages pear fruit. The damage may be classified either as stings or deep entries. Stings are produced when young larvae chew through the fruit skin or bore into the flesh for a short distance before dying within the fruit or leaving the area to sting another place on the fruit. The affected area heals and leaves a small scar. In deep entries, larvae bore to the core and feed in the seed cavity area. Larvae may enter through the sides, stem, or calyx end of the fruit (color plate 20.4). Calyx-end entries and stings are more often observed in young fruit because young fruit are harder due to the density of stone cells, and young larvae have a very difficult time penetrating them. More side and stem entries are observed as the fruit matures. Only one codling moth larva is usually observed feeding in a single fruit, since the larvae are cannibalistic.

SECONDARY PESTS

Pear Psylla
Cacopsylla pyricola

Pear psylla is a pest of pears throughout the United States. It was introduced from Europe to the United States, probably in shipments of pear nursery stock, and was first found in the Massachusetts-Connecticut area in 1832. It arrived on the West Coast over 100 years later, reaching Spokane, Washington, in 1939

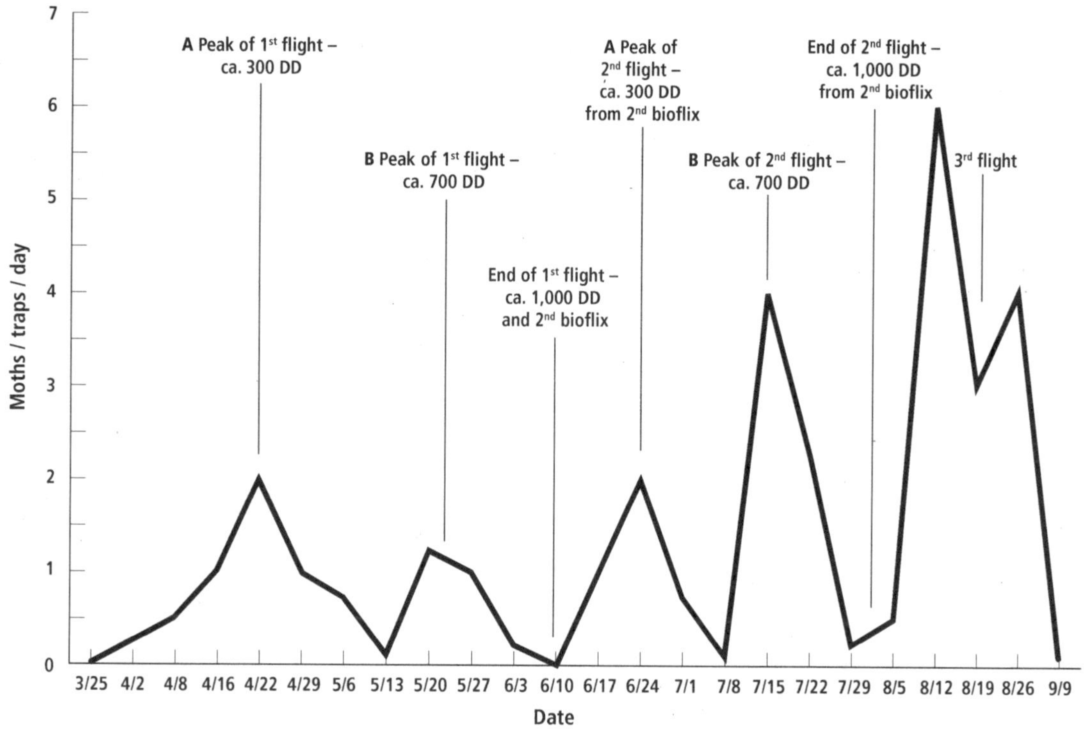

Figure 20.1. Seasonal male codling moth flight pattern showing degree-day accumulation for peak flight activity and generation completion.

and Mendocino County, California, in 1955. Although adults may be found on apple trees and other deciduous fruit trees during winter, pear psylla can complete its development only on pear trees.

Description

Pear psylla has two adult forms: the summer form, which is 1⁄12 inch (2 mm) long, and the larger winter form, which is 1⁄10 inch (2.5 mm) long (color plate 20.5). Pear psylla resembles a minute cicada, with both wings held rooflike over the abdomen. The winter form can be distinguished by its dark, smoky area on the top middle margin of both wings. Recently laid eggs are white and elongated, and they turn yellow as hatching nears. Newly hatched nymphs are tiny and greenish yellow with red eyes and black antennae (color plate 20.6). By the third instar, wing pads appear. The nymphs turn greenish brown in the fourth instar. By the fifth instar, or hard shell stage, they are dark brown; the wing pads are dark and clearly visible (color plate 20.7).

Seasonal development

The adult pear psylla overwinters under bark scales of pear trees or under leaf litter on the ground. Both adult female and male psylla overwinter, but most mating does not occur until January or early February. Occasionally, pear psylla adults can be found sunning themselves on tree branches on clear winter days when temperatures exceed 45°F (7°C). Overwintering females can deposit up to 500 tiny, elongated eggs on or near fruit spurs (color plate 20.8). The eggs are barely visible without the use of a hand lens. The egg-laying period starts at bud swell and finishes 2 to 3 weeks after petal fall. As buds open, females lay eggs along midribs and petioles of developing leaves or on stems and sepals of blossoms. Eggs begin to hatch from 14 to 30 days after being laid. Pear psylla nymphs pass through five instars, feeding on green leaf tissue. Newly hatched nymphs settle down and feed frequently in the shelter of a leaf axis. By the fifth instar, or hard shell stage, psylla nymphs are more active and have moved to the base of petioles. At each new instar, the nymphs increase in size. As the nymphs feed, they secrete copious amounts of a sticky, clear liquid (honeydew). During the first four stages, honeydew encases most of the nymph's body. During the hard shell stage, not as much honeydew is produced and the body is clean.

Pear psylla prefers to lay eggs on young, succulent foliage; nymphs that develop on older leaves may not be able to complete development. Psylla populations tend to concentrate in the upper part of the canopy, possibly because spray coverage of this part of the tree is not as thorough as elsewhere in the tree. Five generations of pear psylla appear in California pear-growing areas. The length of each generation varies, depending on temperatures (shorter in warmer areas). The first generation develops rather uniformly, but succeeding generations overlap considerably.

When populations are low, the adult pear psylla tends to remain in the orchard. However, adults disperse from the orchards if high psylla populations cause the trees to defoliate in late summer. Dispersal normally occurs from harvest to leaf fall when the weather is warm: warm temperatures are needed for any substantial psylla movement.

Damage

When they feed, pear psylla nymphs produce a great amount of honeydew that runs off the leaves and drips onto other tree parts, including fruit. A black fungus, sooty mold, grows on the honeydew, causing the fruit skin to russet (color plate 20.9). Sooty mold on leaves reduces the tree's photosynthetic capacity. Under excessive honeydew production, trees may have black bark due to the growth of sooty mold. Severe pear psylla feeding can cause fruit drop and can stunt, defoliate, and even kill trees. These symptoms, called psylla shock, may reduce fruit set the following year. During feeding, pear psylla injects saliva, which may transmit pear decline, a mycoplasma disease organism. Pear decline damages the phloem, preventing synthesized nutrients from moving down the tree to the roots (color plate 20.10). Diseased trees may decline slowly or collapse quickly (quick decline). Trees suffering from slow decline can recover if psylla density is low. Resistant rootstocks have reduced the severity of pear decline. Cultivars grafted onto *Pyrus communis* are less susceptible than those with *P. pyrifolia* or *P. ussuriensis* roots.

Twospotted Spider Mite
Tetranychus urticae

Twospotted spider mite, which is also called webspinning mite, can be a serious pest of pears. Spider mite populations are usually held under adequate control by predator mites and insects. However, when these beneficials have been eliminated by applications of broad-spectrum insecticides for codling moth and other pests, spider mite populations can rapidly increase to damaging levels.

Description

Twospotted spider mite eggs are large, spherical, and translucent when first laid. As the eggs mature, they become opaque. The eggs are laid on the underside of leaves in colonies of adults and immatures (color plate 20.11). There are three larval, or nymphal, immature stages. The first larval stage has six legs and the two other nymphal stages and adults have eight legs. The adult females are greenish to amber with two large "food spots" in the middle of their bodies (color plate 20.12). The adult male is similar in appearance to the adult female, but males are smaller and have more pointed abdomens.

Seasonal development

Twospotted spider mite overwinters on weeds in the orchard floor or under bark scales on the tree as mature orange females. The males do not overwinter. In the spring, the mites become active and start feeding on weed hosts. As the weather warms, they move up the tree and form colonies on the underside of the leaves. If the mite population is extremely large, they can be found on both the lower and upper leaf surfaces. At first they are located primarily on leaves in the lower inside of the tree canopy. As the season progresses, they move throughout the tree. Twospotted spider mite produces considerable webbing on infested leaves. This webbing is characteristic of these mites and can be used to identify damaging populations. In midsummer, generation time is about 10 days, and there can be 8 to 15 generations per year. In the fall, as the foliage quality declines, the females change into the overwintering form and move to protected places or ground cover to spend the winter.

Damage

Pears are very sensitive to twospotted spider mite feeding. Only two or three mites feeding near the midrib of a leaf produces a black area on the leaf from the midrib to the leaf margin. Higher mite populations may cause severe mite "burn" and defoliation (color plates 20.13 and 20.14). If severe burn or defoliation occurs before harvest, fruit size may be reduced and the tree may bloom again in the fall. Trees that bloom late in the fall are more subject to winter freeze damage. If defoliation occurs after harvest, the next year's crop may be affected.

OCCASIONAL PESTS

Lygus Bugs
Lygus hespersus* or *Lygus elisus

Stink Bugs
***Euschistus conspersus, Thyanta custator, Chlorochroa sayi,* and others**

Boxelder Bug
Bolsea trivittata

Several species of plant bugs may attack pear fruit. Among these are lygus bugs, stink bugs, and boxelder bug. *Lygus hespersus* is the most common of the two lygus bugs, while *Euschistus conspersus* (consperse stink bug) is the most common of the several species of stink bugs. Infestations of these sucking plant bugs are usually local and are associated with ground covers or surrounding vegetation. Lygus bugs and stink bugs are more common than boxelder bug in the Sacramento Delta, while lygus bugs, stink bugs, and boxelder bug are common on the North Coast. Boxelder bug is found in close proximity to ash, maple, or boxelder trees and usually are associated with riparian habitat.

Description

Lygus bug adults are about ¼ inch (6.5 mm) long. Their color varies from green to yellow-brown (color plate 20.15). A distinctive yellow or green V-shaped mark is located on the center of the back. The wingless nymphs are light green or yellow and have four small black dots that are visible in the last three nymphal stages. The eggs are inserted into plant tissue.

Consperse stink bug adults are about ½ inch (12.5 mm) long, shield-shaped, and gray-brown with a green to gray abdomen (color plate 20.16). They have yellow to orange legs and dark-tipped antennae. The body and legs are covered with small black specks. Nymphs vary in color from black and reddish in the early stages to various shades of yellow or brown later. Consperse stink bug eggs are barrel-shaped and are laid in clusters on twigs and leaves.

Boxelder bug adults are about ½ inch (12.5 mm) long, flat, and narrow (color plate 20.17). They are gray to brownish black with red lines on the thorax and wings. The abdomen is reddish. Nymphs are bright red in the early instars but darken as they mature. Eggs are rusty red and capsule-shaped. They are laid on their sides in groups of two or three.

Seasonal development

Adult lygus bugs overwinter in weeds in the orchard, in nearby uncultivated areas, or in alfalfa fields. Their preferred hosts are legumes, particularly alfalfa. Adults become active on warm days in early spring, feed, and lay eggs on a wide range of broadleaf plants. As the groundcover or weeds begins to dry in the spring, adults migrate to green areas such as orchards. If suitable host plants are not present on the orchard floor, adults fly to trees, where they feed on developing fruit. Adults feed on pears but do not reproduce on pears. There are three to four overlapping generations each year.

Adult stink bugs overwinter in weeds and other protected places either inside or outside the orchard. In early spring adults feed on weeds, mate, and deposit eggs on host plants such as curly dock *(Rumex crispus)*, mustard (*Brassica* spp.), common mullein *(Verbascum thaspus)*, and wild blackberry (*Rubus* spp.). Nymphs complete development in May or June. Usually by this time broadleaf weeds have begun to dry, and first-generation adults move into the pear orchards and begin to feed on developing fruit. Often only fruit on border rows are damaged, indicating that the infestation came from outside the orchard. Eggs of the second generation are laid in late June, primarily on broadleaf plants in the cover crop. Second generation adults appear in August and feed until they move to overwintering sites.

Boxelder bug is found only in pear orchards in a close proximity to rivers or steams, which contain boxelder, maple, or ash trees, their preferred hosts. It is one of the most common plant bugs in Mendocino County pear-growing areas along the Russian River. Adult boxelder bugs overwinter in dry, protected places, under bark, in woodpiles, or in any sheltered place. They often enter houses in large numbers in the fall. Boxelder bugs become active in April or May and lay eggs that hatch in 12 to 14 days. Second generation egg laying starts in June or early July. In the fall, adults gather in large numbers on tree trunks and migrate to sheltered places to overwinter. First and second generation nymphs and first generation adults feed on developing fruit and cause the fruit to be deformed.

Damage

Although lygus bugs feed on developing buds, this damage is usually not serious. The most serious damage is caused by fruit feeding. Early-season feeding is usually characterized by raised pustules, which often rupture as the fruit grows, leaving off-white, grainy centers (color plate 20.18). This can cause the fruit to be badly deformed. Late-season feeding causes irregular depressions that resemble stink bug damage (color plate 20.19). Lygus bug injury differs from stink bug injury by the presence of a mass of hard cells on the pear fruit (color plate 20.20).

Stink bug feeding is characterized by shallow dimples or irregularly depressed areas on the fruit surface, with a brown pithy area in the flesh (color plate 20.21). If the damage occurs just before harvest the dimpling may not be present. Feeding damage is generally concentrated near the stem end of the fruit.

Boxelder bug damage is generally limited to areas where host trees are near orchards. Both nymphs and adults feed on developing fruit. Their damage is similar to that caused by stink bugs but it is not restricted to the stem end of the fruit.

Obliquebanded Leafroller
Choristoneura rosaceana

Fruittree Leafroller
Archips argyrosphila

Leafrollers have become a sporadic problem in pear orchards throughout California, but particularly in orchards on the North Coast. One possible explanation for the increase in leafroller damage is that the insecticides used to control codling moth and pear psylla, which previously provided control of leafroller, are no longer as effective for leafroller control. Leafroller populations have also increased in orchards where insecticides used to control codling moth have been reduced or eliminated and replaced with pheromone confusion.

Several species of leafrollers belonging to the family Tortricidae are pests of fruit trees. The obliquebanded leafroller and fruittree leafroller are the two most often found in pear orchards.

Description

Obliquebanded leafroller adults are ¾ to 1 inch (19 to 25 mm) long and are tan with three oblique chocolate-brown bands across their forewings (color plate 20.22). The leading and trailing edges of each band are darker than the rest of the wing. Fruittree leafroller adults, which are smaller than obliquebanded leafroller adults, are ½ to ¾ inch (12.5 to 19 mm) long and have brown wings mottled with gold and white (color plate 20.23). Female moths lay eggs in clusters, overlapping them like fish scales. Obliquebanded leafroller eggs are dull green and are laid on the upper surface of leaves. As eggs mature, they turn tan in color, and the black head capsules become visible before hatching. Fruittree leafroller eggs are laid on twigs and branches and are covered with a secretion. When first laid, they are dull orange, but they turn dark brown in summer and gray in winter.

It is difficult to distinguish between the larvae of the two species. Both have green bodies. Mature obliquebanded leafroller larvae have a brown to black head capsule and thoracic shield (color plate 20.24), while mature fruittree leafroller larvae have a black head and thoracic shield (color plate 20.25). Fruittree leafroller larvae are usually smaller than obliquebanded leafroller larvae. Larvae of both species roll leaves and attach them together with silk to form protective cases, thus their name, and both pupate in the rolled-up leaf lightly webbed with silk.

Seasonal development

Obliquebanded leafroller has at least two generations per year, causing both early- and late-season fruit damage. Fruittree leafroller has only one generation per year, causing only early-season fruit damage.

Obliquebanded leafroller overwinters as second- or third-instar larvae within silken cases covered with fecal pellets in bark crevices or under the bark. In the spring, larvae emerge from the silken cocoons and move to the cluster buds as they begin to open. As the foliage pushes from buds, the larvae tie the terminal leaves together, where they feed. Mature larvae construct tubular shelters from a single large leaf. Pupation occurs in mid-April in the rolled-up leaf and lasts 10 to 12 days. Moths emerge from late April through May, with peak emergence in mid-May. Actual emergence timing depends on spring temperatures.

The timing of the summer-generation egg hatch varies from year to year, but it is generally from early to mid-June. Larvae mature by mid-July or early August. Following pupation, moths begin to emerge in July, with peak emergence in August. Summer-generation adults lay eggs in early August through September. Newly hatched larvae feed for a short time on foliage and develop to the third instar before moving to shelter under the bark and crevices on the limbs where they overwinter.

Fruittree leafroller overwinters in egg masses. Egg hatch begins when the buds begin to swell and continues until bloom. Fruittree leafroller larvae are difficult to distinguish from obliquebanded leafroller, but fruittree leafroller larvae are present for only about 30 days in the spring and are not present during the summer. Larvae feed on leaves, buds, flowers, and young fruit, and mature by late April. Adult activity peaks in late May to early June. In June females deposit eggs in masses on smooth bark of 1- to 3-year-old wood in the upper parts of trees. Eggs hatch the following spring.

Damage

Larvae roll leaves and web them together to form protective shelters. Larvae feed on leaves, buds, and the surface of pear fruit. Fruit with early-season damage may abort or appear at harvest as severely deformed fruit with irregular corky scars. Late-season feeding caused by the summer generation of obliquebanded leafroller is generally shallow on the surface of the skin (color plate 20.26). The damage often takes place where fruit in clusters touch or where a leaf was attached to the surface of the fruit by the worm. This later damage gives the pest the popular name of "skin worm" and is more serious than the damage caused in early spring.

Speckled Green Fruitworm
Orthosia hibisci

Humped Green Fruitworm
Amphipyra pyramidoides

Speckled and humped green fruitworms may cause serious damage to pears, but they do not cause economic damage every year. Speckled green fruitworm is more common than humped green fruitworm. Damage is more common in the Sacramento Delta and Suisun Valley than in the North Coast of California. Both fruitworms feed on a wide variety of deciduous fruit trees, but their preferred hosts are apples, pears, and prunes.

Description

Adults of both species are robust gray to brownish moths with wingspans of 1¼ to 1½ inches (32 to 38 mm). (color plate 20.27). Eggs are laid singly on the leaves. The eggs are spherical and slightly flattened at the top, with vertical ribs and a flat base where they are cemented to the plant. The larvae are green with light cream-colored lines extending along the back and sides, and they have green heads (color plates 20.28 and 20.29). Humped green fruitworm has a prominent hump on the last abdominal segment, bright yellow lines on the sides, and less distinct white lines on the back. Pupae of both species are brown and approximately ¾ inch (19 mm) long.

Seasonal development

Speckled green fruitworm overwinters as a pupa in the soil, while humped green fruitworm overwinters in the egg stage on tree branches and twigs. Speckled green fruitworm emerges as adults from the soil in early spring (February to April) and lays eggs on twigs. Eggs of both species usually hatch at the beginning of the cluster bud stage. Hatching continues beyond petal fall. Larvae feed on leaves, buds, flowers, and young fruit. Fruit feeding usually takes place after petal fall. Larvae of both species mature in May, drop to the ground, and pupate. Humped green fruitworm emerges as adults during the summer and lays overwintering eggs on twigs and branches, while speckled green fruitworm pupae remain in the soil until the next spring. Both species have one generation per year.

Damage

Larvae feed on leaves, buds, flowers, and fruit. Fruit feeding usually begins shortly after petal fall. Green fruitworms feed inside the fruit cluster, rarely consuming the entire fruit. Fruit may abort, or as the fruit matures, damaged areas scab over and the fruit is misshapen (color plate 20.30).

European Red Mite
Panonychus ulmi

European red mite is an occasional pest of pear. However, European red mite is also a secondary host for the beneficial insects and mites that feed on twospotted spider mite. Thus, low densities of European red mite do not cause economic damage and at the same time help beneficial insects and mites control twospotted spider mites. In addition, European red mite does not feed on as wide a range of host plants as twospotted spider mite and does not build up in the ground cover.

Description

European red mite adult females are brick red with four rows of long, curved spines arising from white spots on its back (color plate 20.31). Adult males are brownish and smaller than the females. Eggs are red with a single spine on the upper surface (color plate 20.32). Nymphs are bright red.

Seasonal development

European red mite overwinters in the egg stage on twigs and branches. Eggs hatch in early spring when trees begin to leaf out. There are three nymphal instars. The mite population builds up slowly in the spring. The generation time is very short, and there may be 5 to 10 generations per year. The mites feed and are active on the upper surface of the leaf. However, when large populations are present, the mites also infest the lower leaf surface. European red mite produces very little or no webbing, in contrast to twospotted spider mite. European red mite seems to prefer mild temperatures and shady conditions.

Damage

European red mite damage is not as pronounced as twospotted spider mite damage. European red mite feeding causes leaves to lighten in color and become mottled or stippled. Prolonged feeding by high populations (five or more mites per leaf) causes leaf bronzing. Under high temperatures, leaf burn and defoliation may occur, but this damage is rare.

San Jose Scale
Quadraspidiotus perniciosus

Although several scale insects attack pears, the most damaging is San Jose scale. This scale has a wide host range, including many ornamental trees as well as deciduous fruit trees. San Jose scale is mainly controlled by dormant sprays. It often develops as a pest when spray coverage is poor in the top of the tree. San Jose scale is difficult to detect and often causes severe fruit damage before the problem is noticed.

Description

Female scales give birth to living young called crawlers. The crawlers are bright yellow, with well-developed legs, eyes, and antennae (color plate 20.33). The first instar goes through three stages: crawler, white cap, and black cap. After the initial crawler settles, it begins to secrete a white waxy covering (white cap stage). As scales mature, the covering becomes black (black cap stage), which later turns gray and is difficult to distinguish from the bark on which it is located. The adult scale body beneath the cover is yellow (color plate 20.34).

Males and females are similar until after the first molt. The female scale spends its entire life cycle under a waxy covering and molts twice before maturing. The female cover is round, gray-brown, and has a tiny white knob in the center. The male scale's body is elongated, and its scale cover is oblong.

The males molt four times before reaching maturity. Unlike the female, which does not move after losing its legs, males go through prepupal, pupal, and winged adult stages. The mature male is a tiny yellow insect with wings, long antennae, and a dark band on its back.

Seasonal development

The majority of San Jose scales overwinter in the black cap stage, while some overwinter as mated adult females. When the weather warms in February, overwintering scales undergo the first molt. Both males and females complete their immature stages from March through April. The mature females produce a sex pheromone to attract the males, and mating takes place immediately after male emergence. Females start to produce crawlers in late April and early May. A female can produce crawlers for 6 to 8 weeks at a rate of about 10 per day.

After they find a suitable place to settle, crawlers insert their mouthparts in plant tissue and begin feeding, lose their appendages, and begin secreting a white waxy covering. Females of the first generation mature in June, and second-generation crawlers emerge from late June through August. Some crawlers of this generation settle on the fruit. Under optimal conditions (hot, dry weather), a generation can be completed in 7 to 8 weeks. The third generation is produced in August or September. In some warm years, a partial fourth generation is produced in October and November. Crawlers produced in October and November overwinter as black caps and mature the following spring.

Damage

San Jose scale feeds on bark, leaves, and fruit, sucking plant juices and injecting a toxin. Heavy populations on bark can kill twigs and branches and in severe cases can kill entire trees. When San Jose scale attacks fruit, the fruit produces a red halo around the scale's body (color plate 20.35). Fruit with this red halo is usually culled because of its unsightly appearance.

Mealybugs
Pseudococcus maritimus and *Pseudococcus obscurus*

Originally described as a pest of grapes, the grape mealybug (*Pseudococcus maritimus*) can attack most deciduous fruit crops, including pears. The obscure mealybug (*P. obscurus*), occasionally found in pears, closely resembles the grape mealybug. Both mealybugs are most damaging in older, mature trees with rough bark that shelters overwintering nymphs and egg clusters.

Description

Mealybug eggs are yellowish to orange in color and are laid in a white cottony sack. Crawlers have well-developed legs and are yellow to brown. The young crawlers are covered with a light coating of wax, giving the appearance of being dusted with flour. The sedentary nymph has poorly developed legs and a pink to purple body covered by waxy filaments that give the insect a whitish cast. The wingless adult females are about ¼ inch (6.5 mm) long and look similar to a nymph. The adult females are dark purple with flattened bodies and have a well-developed fringe of waxy filaments (color plate 20.36). The filaments are longest toward the end of the body, becoming shorter near the front. The adult male is much smaller than the female and holds its wings flat over its back.

Seasonal development

Mealybugs overwinter as eggs or crawlers under bark scales on the tree. In the spring, the eggs hatch and young crawlers move to the base of shoots or fruit clusters. They complete the first generation in late May or June. The mature females may move back to the bark scales or into the calyx of the fruit to lay eggs. The eggs hatch in June, and the second generation crawlers move from the bark to feed on the fruit. The crawlers from eggs that are deposited in the calyx end of the fruit remain within the calyx to feed. The second generation is completed in August and September, and the mature females move back to bark scales to lay overwintering eggs. Some of these eggs may hatch to produce overwintering crawlers.

Damage

Mealybugs often feed in the calyx end of maturing pear fruit, which may become soft as the fruit ripens. Because mealybugs cannot be removed after harvest with present washing methods, infested lots of fruit are rejected by both cannery and fresh market handlers. Mealybugs also produce honeydew, which may cause russetting and enable a black sooty fungus to grow on the surface, rendering the fruit unfit for fresh shipment (color plate 20.37).

Pear Rust Mite
Epitrimerus pyri

Pear rust mite is an occasional but serious pest of fresh market pears because of russetting of the fruit finish. Pear rust mite is not a pest of processing pears.

Description

Adult rust mites are extremely small and can be seen only under magnification. Adults and immatures are wedge-shaped and white to pale brown. Eggs are spherical and translucent.

Seasonal development

Rust mite overwinters as adults under leaf buds or in other protected areas. As the buds open in the spring, the mites feed on green tissue at the base of the buds. Later they move to foliage or fruit. Because tender foliage is preferred, populations on leaves decrease as leaves mature and harden. On fruit, damaging populations sometimes develop around the calyx soon after petal fall and may remain active throughout the season. There are many generations per year, with each generation taking from 10 to 14 days. After harvest the mites seek overwintering sites.

Damage

Mite feeding on leaves causes leaves to bronze. This injury may stunt the growth of young trees. However, the damage is minor compared to the damage on fruit. Mite feeding on fruit causes russetting of the fruit surface, making the fruit unsuitable for fresh market (color plate 20.38). Damage is more noticeable on Bartlett fruit than on pears with a naturally russeted surface, such as Bosc.

Pearleaf Blister Mite
Phytotus pyri

Pearleaf blister mite is a minor pest of pear. However, it has become more common with the reduction in the number of insecticide applications for codling moth control.

Description

Adults are conical, very small, white mites that cannot be seen without magnification. The immature mites resemble the adults but are smaller.

Seasonal development

Adult mites overwinter in bud scales. When weather warms in the spring, the mites emerge from the bud scales and attack young fruit and leaves. They feed just below the fruit or leaf surface, causing blisters to form. The mites develop in the blisters and produce a number of generations during the spring months. As weather warms and the pear leaves mature and harden off, pearleaf blister mite reproduction slows. In the fall, the blister mites move back to bud scales to overwinter.

Damage

Extremely large populations of pearleaf blister mite may cause bud failure. A more common type of damage on foliage is an oval, brown blister about $\frac{1}{4}$ to $\frac{1}{2}$ inch (6.5 to 12.5 mm) in diameter with a surrounding halo of clear tissue (color plate 20.39). Damage is more prevalent on Bartlett fruit than on pears with naturally russeted surface like Bosc (color plate 20.40).

Pear Thrips
Taeniothrips inconsequens

Western Flower Thrips
Frankliniella occidentalis

Pear thrips was formerly one of the most damaging pests of pear. However, insecticide applications used to control codling moth and other pests have held pear thrips in check. Pear thrips is now rarely seen in California. Western flower thrips can on occasion damage young pears.

Description.

Pear and western flower thrips are about $\frac{1}{25}$ inch (1 mm) long and are slim with four long, narrow wings fringed with hairs. Pear thrips is dark brown to black, while western flower thrips is yellow. Young thrips larvae resemble the adults but are smaller and wingless. The third instar, the prepupa, is inactive, does not feed, and has wing buds. Pupation occurs in the soil. Eggs are embedded within the plant tissue.

Seasonal development

Adult pear thrips may be found in opening buds, where it feeds and may prevent the blossoms from producing fruit. High pear thrips populations also cause blossom drop when large numbers of eggs are laid in the blossom stems. In addition, pear thrips larvae feed on fruit, causing russetting or scabbing of the fruit surface.

Western flower thrips hibernates as adults in weeds, grasses, and other hosts either on the orchard floor or in adjacent orchards. If overwintering sites are disturbed or dry up in early spring, adults migrate to flowering plants and deposit eggs in the tender portions of host plants. Eggs hatch in 5 to 15 days, and nymphs feed on developing fruit, where they complete their development. They feed by rasping the fruit surface with their mouthparts, which causes russetting. Mature nymphs drop to the ground, where they molt twice and emerge as adults. As native vegetation dries up in the spring, populations decrease sharply and adults migrate to cultivated crops in the vicinity.

Damage

Prebloom feeding by thrips deforms blossoms and leaves, reducing fruit set. The primary damage to pears is caused by egg laying punctures at bloom. After hatching, nymphs often feed on very small fruit, causing small russet areas (color plate 20.41). Nymphs often feed under flower petals that are plastered against fruit during rainy periods.

Pear Sawfly
Caliroa cerasi

California Pear Sawfly
Pristophora abbreviata

Pear sawfly, also known as the pear slug, and California pear sawfly are very minor pests of pear that are easily controlled by any organophosphate or carbamate insecticide. Although abandoned and organic orchards and backyard trees can develop high populations, sawflies are rarely a problem for commercial pear growers. The pear sawfly and California pear sawfly are not true flies but belong to the order of insects that also contains the bees and wasps (Hymenoptera).

Description

Pear sawfly larvae are slimy olive-green to black and slug-shaped, hence the name "pear slug." California pear sawfly larvae are light green and maggot-shaped.

Seasonal development

Both sawfly species overwinter as pupae in the soil. In the spring, adults emerge and female sawflies insert eggs into the leaf tissue. After the larvae have completed their development, they fall to the ground to pupate. California pear sawfly has only one generation per year and remains in the soil as a pupa until the following spring. Pear sawfly produces a second generation in July and August before pupating in the soil to overwinter.

Damage

The larvae feed on foliage and do not feed on fruit. California pear sawfly feeds in circular patterns, leaving oval holes in the leaves (color plate 20.42). Pear sawfly skeletonizes the leaf tissue, leaving only a fine network of veins (color plate 20.43). Only extremely large sawfly populations cause economic damage in pears.

REFERENCES

Bethell, R. S., ed. 1978. Pear pest management. Oakland: University of California Agricultural Experiment Station Publication 4086.

Elkins, R. B., et al. 2002. Pest management guidelines for pear. Oakland: University of California Division of Agriculture and Natural Resources Publication 3455. UC IPM Web site, http://www.ipm.ucdavis.edu/PDF/PMG/index.html.

Ohlendorf, B. L. 1999. Integrated pest management for apples and pears. 2nd ed. Oakland: University of California Division of Agriculture and Natural Resources Publication 3340.

21
Diseases

W. DOUGLAS GUBLER, BROC G. ZOLLER, ROGER A. DUNCAN, AND STEVEN E. LINDOW

Diseases are common in all of California's pear-production regions. While their importance varies from area to area, two diseases, scab and fire blight, are most important in California and occur in all production areas. Fire blight tends to be a more persistent disease in the Delta production area, but growers in all areas must assume that the pathogen is present and under the right environmental conditions will cause disease. Scab is more of a problem in North Coast orchards, but growers in other production areas must also be alert to the disease.

Other diseases such as pear decline and blossom blast are mostly sporadic in occurrence and generally are not thought to be significant problems in California pear production, though at one point pear decline was a significant problem.

Root rots are a consistent threat when irrigation programs allow for long-duration soil saturation. They are, however, easily managed in pears by allowing for adequate drainage and proper irrigation for a particular soil type.

FIRE BLIGHT

Fire blight of pear is the most important and destructive pear disease in California. Losses due to fire blight occur in pear orchards in all production areas. Conducive conditions for disease include rainfall, warm dew periods, and irregular or protracted bloom periods.

The Pathogen

The disease is caused by the bacterium *Erwinia amylovora*. The organism survives in overwintering canker on pear, apple, quince, pyracantha, hawthorne, and crab apple. Disease on these hosts occurs in the spring, and visible gray ooze containing masses of bacteria may serve as a source of inoculum for nearby trees and pear orchards.

Symptoms

Erwinia amylovora attacks flowers, which, while still apparently healthy, exhibit small beads of gray bacterial ooze along the stem approximately 7 to 10 days after infection. As the disease progresses, flowers become blighted, and leaves, shoot terminals, and adjacent fruits wilt, blacken, and die, resulting in disease strikes (color plate 21.1). Strikes at first appear only slightly wilted and less green than healthy shoots and may be difficult to spot high in a tree. Later, strikes are easily seen in the otherwise green foliage. A distinguishing characteristic of these strikes, and one that allows differentiation from bacterial blast, is the "shepherd's crook" nature of the shoot terminal (color plate 21.2). In addition, under conditions of high relative humidity, bacterial ooze can be seen along major veins and the stem of affected leaf tissue, as well as on small fruit (color plate 21.3). Though less common in California, direct infection of succulent shoots, leaves, and fruit may occur during periods of high relative humidity late in the summer and fall.

In California the long, irregular bloom period observed in Bartlett pear can extend the fire blight infection period for many weeks. After infection, the pathogen spreads internally from flowers into small branches and later into larger branches. Externally, infected tissue causes the bark to be sunken and discolored as cankers develop in the wood below. Removal of the bark exposes the orange-red discoloration of newly infected tissue and the brown-black discoloration of older infected tissue (color plate 21.4). Left untreated, canker expansion often kills the tree (color plate 21.5).

Disease Cycle

Erwinia amylovora overwinters in 1-year-old and older branches on pear trees or other hosts (color plate 21.6). Cankers become active in the spring when tree growth resumes. Active or "holdover" cankers exude bacteria to the surface, where the ooze is moved into flowers by insects, mites, wind, and splashing rain (color plate 21.7). The bacteria may multiply in flowers, but infection is thought to require wetting provided by rainfall or dew during high relative humidity and warm temperatures. Movement of inoculum from flower to flower by insects, including bees, is common. Symptoms usually become evident 7 to 10 days after infection. After June or July, when temperatures prevail above 95°F (35°C) and dry conditions occur, tree growth slows, and the risk of fire blight is reduced. However, flowers from protracted or "rat-tail" bloom at these times can still be a site for infections. In late summer and fall, rain or dews may result in shoot blight in young, vigorously growing trees, and resulting cankers may serve as sites of bacterial survival if not removed.

Controlling Fire Blight Using Heat Summation

The degree-hour model

With the development of blossom monitoring techniques using selective media (Miller and Schroth 1972), the concept of heat summation was applied to development of *Erwinia amylovora* in blossoms of Bartlett pears. In this model, 1 degree-hour equals 1°F above 65°F for 1 hour. A temperature of 70°F for 2 hours generates 10 degree-hours. Degree-hours are accumulated each day until 3 consecutive days below 66°F occur; the accumulation of degree-hours is then reduced to zero. This empirical system is based on blossom monitoring experiences of epiphytic populations and field observations of the occurrence of new infections. Monitoring showed that continuously cool weather below 66°F eliminated small epiphytic populations and reduced large populations that were already present in blossoms prior to the cool weather. Thus, bacteria initially colonize flowers in warm weather but decline during an intervening cool period of

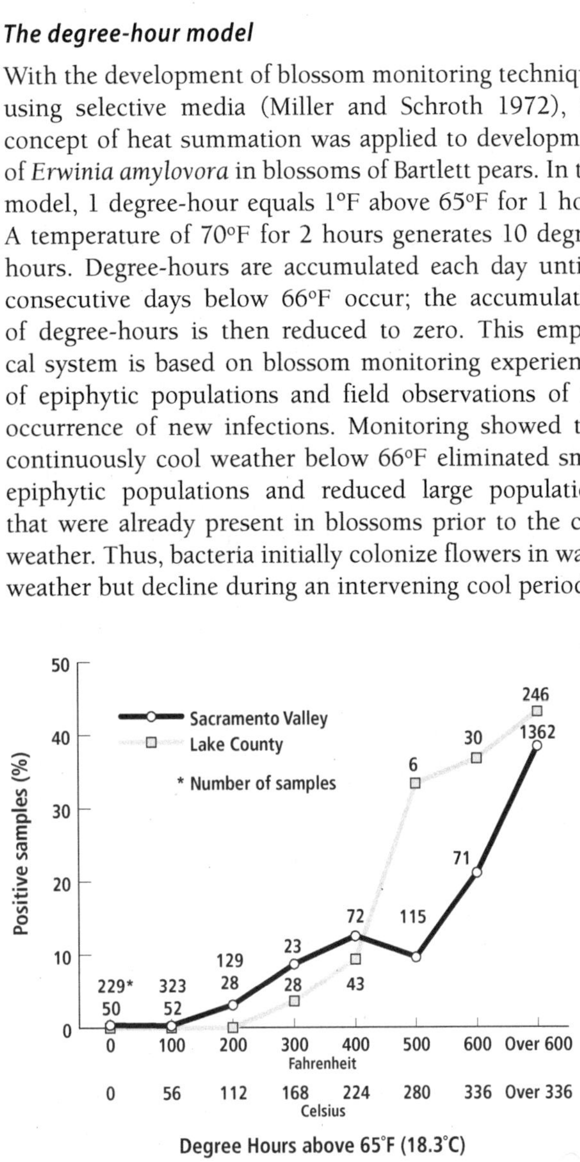

Figure 21.1. Percentage of samples colonized by *Erwinia amylvora* versus degree-hours above 65°F (18.3°C) since the last 3-day period with no temperature above 65°F, 1972–1976. The rate of colonization of pear flowers becomes greater with increases in the accumulated degree-hours prior to sampling. Accumulation of 500 degree-hours is used as a critical threshold to increase treatment frequency if the threshold is reached during major bloom periods with warm night conditions predicted. *Source:* Zoller 2000, p. 104.

sufficient duration (fig. 21.1) (Van der Zwet, Zoller, and Thompson 1988; Zoller 1978; Zoller 2000; Zoller and Sisevich 1979).

In a 5-year period from 1972 to 1976, blight bacteria were detected in only 0.3 percent of blossom samples (100 to 200 blossoms per sample) collected prior to the accumulation of 150 degree-hours; 10 percent of pear blossom samples contained bacteria at 350 degree-hours, and 40 percent of blossom samples contained bacteria at 600 degree-hours (fig. 21.2) (Zoller 2000; Zoller and Sisevich 1979).

Blossom blight infections are randomly initiated when epiphytic inoculations of *E. amylovora* are present during warm weather accompanying precipitation and during warm, humid weather not accompanying precipitation. Although in many years blight outbreaks may be most severe in the "rat-tail" blooms following primary bloom, observations suggest that infections occurring as a result of favorable weather through 15 days past full bloom impact the number of new infections that occur later in the season.

Practical use of predictive systems

Development of the epiphytic populations of *Erwinia amylovora* in pear flowers is a phenomenon that can be used to predict fire blight outbreaks. Monitoring studies in California have shown that a mean temperature threshold near 60°F (15.6°C) has considerable value in predicting the initial presence of epiphytic *E. amylovra* in pear flower (Thompson et al. 1982; Zoller and Sisevich 1973; Zoller 1978). In practice, however, the mean temperature threshold does not attempt to predict infection periods. Moreover, use of the mean temperature threshold system is "all or none," yielding no relative information about the magnitude of the fire blight risk and therefore the frequency of necessary treatments.

Control guidelines using the degree-hour system:

- Applications are not necessary in the Sacramento Valley when weather has been cool enough so that no degree-hours have accumulated over 65°F (18.3°C), even if rain occurs.
- Treatments should be made within 24 hours preceding rain when 1 to 150 degree-hours have accumulated in the Sacramento Valley or 100 to 250 degree-hours have accumulated in the North Coast region. A lack of treatment in this period will result in new infections near inoculum sources.
- Growers must protect orchards within the 24 hours before precipitation or before warm, humid conditions of a least 57°F (13.9°C) and 90 percent relative humid-

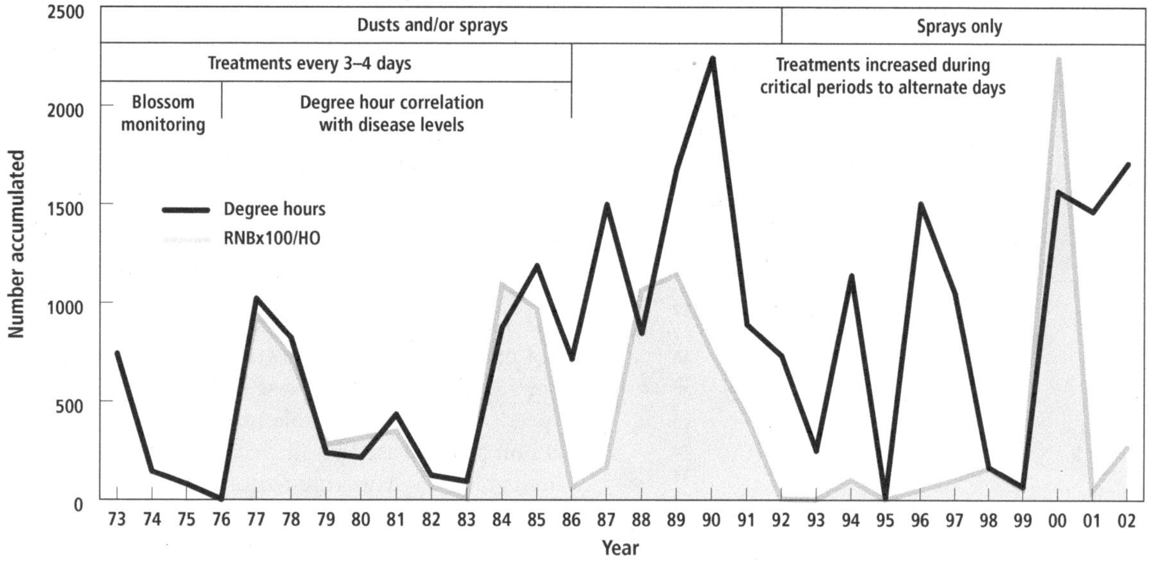

Figure 21.2. Blight history in the mid-Sacramento Valley, 1973–2002. *Note:* RNB × 100/HO = Random new blight infections (times 100) per holdovers observed in weekly inspections through 15 days past full bloom. *Source:* Zoller 2000, p. 104.

ity when more than 150 degree-hours have accumulated in the Sacramento Valley and more than 250 degree-hours have accumulated in the North Coast region. Treatments are suggested every 3 to 4 days when accumulations exceed these levels because the epiphytic population is predicted to be so prevalent that failure to treat within this 24-hour period results in significant numbers of new infections.

- Alternate-day treatments are recommended in the Sacramento Valley if degree-hours surpass 500, and sometimes in the North Coast above 600 degree-hours, in conjunction with major bloom periods.
- If the orchard is being irrigated, the humidity threshold for predicting the initiation of infections is reduced to 57°F with 80 percent relative humidity, as measured outside the orchard. Start each season with a full soil profile and delay the first irrigation of the growing season as long as possible. An orchard that is blooming should not be irrigated.
- The accumulated degree-hour total is not reduced by continuous cool temperatures below 66°F if the total surpasses 400 degree-hours and has coincided with precipitation or warm, humid infection periods of a least 57°F and 90 percent relative humidity. This is because the inoculum potential may have increased due to new infections.
- Eliminating holdover inoculum sources by careful pruning is a valuable management tool that cannot be overlooked. The correlation of infections and weather is based on the assumption that holdover cankers are present, and careful pruning may remove these cankers.

Other models available for fire blight control include the Maryblyt model and the Cougarblite model. Growers should consult UC Cooperative Extension Farm Advisors or pest control advisors for more information regarding use of these models.

Effective fire blight control depends on using bactericides such as coppers, streptomycin sulfate, and terramycin. These products must be applied as preventive sprays on a tight-interval application program. Using these products with the aid of the models for pathogen increase have provided adequate disease control. The biological control agent *Pseudomonas fluorescens* strain A506, commercially available as Blightban A506, usually applied in conjunction with antibiotics, can also control the disease.

Further control of fire blight depends on a program of sanitation that includes removal of the fire blight strikes as soon as they are visible. These strikes should be removed with sharp pruning shears by cutting approximately 6 to 12 inches (15 to 30.5 cm) below the visible margin of the canker. The shears should be dipped in a disinfectant between each cut to prevent spreading the bacterium. Larger cankers should also be removed as soon as possible to prevent further spread in the tree. All infected material should be removed from the orchard and burned.

For many years, breeding pears for resistance to fire blight was a priority in the pear industry. Although one variety did have resistance to flower and leaf infections, it had a severe problem with trunk cankers and was

Table 21.1. The Mills Table: Temperature and moisture requirements for scab infections in spring

Temperature (°F)	Hours of wetting required for infection during spring*			Days until lesions appear
	Light infection	Moderate infection	Heavy infection	
33–41	>48	—	—	—
42	30	40	60	—
43	25	34	51	—
44	22	30	45	—
45	20	27	41	—
46	19	25	38	—
47	17	23	35	—
48	15	20	30	17
49	14.5	20	30	17
50	14	19	29	16
51	13	18	27	16
52	12	18	26	15
53	12	17	25	15
54	11.5	16	24	14
55	11	16	24	14
56	11	15	22	13
57	10	14	22	13
58	10	14	21	12
59	10	13	21	12
60	9.5	13	20	11
61	9	13	20	10
62	9	12	19	10
63	9	12	18	9
64	9	12	18	9
65	9	12	18	9
66–75	9	12	18	8
76	9.5	12	19	9
77	11	14	21	9
78	13	17	26	10

Note: *After infection is established and secondary spores are present in large numbers, hours for reinfection are only two-thirds of figures shown.

therefore not widely utilized. At present, all pear varieties are susceptible and should be grown using control programs.

PEAR SCAB

Pear scab is caused by the fungus *Venturia pirina* (Aderh.). The disease occurs most commonly in the North Coast region of California, but it occasionally occurs in other production areas during moist conditions favorable for disease development. The pathogen can attack all known European pear varieties and causes serious disease in Anjou, Barlett, Comice, Winter Nelis, and Forelle. Pear fruit is susceptible primarily early in the growing season, but pinpoint scab may occur throughout the growing season.

Symptoms

Blossoms, leaves, fruit, and succulent young twigs are susceptible to pear scab infection. Lesions may be irregular to circular. If they contain sporulating colonies of *V. pirina*, they will characteristically be velvety and dark-olive colored (color plates 21.8 and 21.9). As fruit lesions become older, their centers become tan and slightly scabby in appearance as fruit enlarge and lesions crack.

Disease Management

Many foliar pathogens require a minimum duration of leaf wetness for spore germination and infection to occur. Temperature affects the minimum duration of moisture that is required, with shorter durations required at warmer temperatures. In the case of *V. pirina*, infection periods can be effectively assessed by use of modified Mills Tables (table 21.1). Developed at Cornell University for apple scab and slightly modified for pear scab, the Mills Table informs the grower of an infection period based on the fact that a certain number of hours of leaf wetness are required at a given temperature for infection to occur. If rainfall begins when temperatures are averaging 60° to 75°F (15.6° to 23.9°C), foliage and fruit surfaces must remain constantly wet for approximately 9 hours to result in a light scab infection. If the wetness period lasts 12 or 18 hours, the infection can be moderate to heavy. Lesions appear within 8 to 9 days. If infection has already occurred and the conditions above are repeated, the time required for new infections is only approximately two-thirds of that for primary infections.

An infection period is initiated when rainfall begins. If intermittent rainfall with 6 hours of drying occurs prior to reaching the necessary number of hours for infection to occur, the infection period begins anew with additional rainfall. This "zeroing out" of the infection period is particularly true if it is warm and sunny during the drying time. However, should the drying time occur under conditions of high relative humidity and cool temperature, the 6-hour time period for stopping growth in incipient infections should be extended by 3 to 4 hours, especially if the wet period was nearly long enough to constitute an infection period.

In areas where spring rains occur frequently, protective fungicide applications made at an interval of 12 to 14 days (or shorter) may be necessary for scab control. Serious losses have often occurred with rainfall during main bloom to late bloom; fungicide applications are recommended at this time in areas where the disease is prevalent. In areas where rains occur intermittently, growers might choose to use fungicides with a "kick-back" activity that can kill established fungal pathogens. These products can be used up to 96 hours after an infection period has occurred and will kill the pathogen before lesion development. Please consult the *UC IPM Pest Management Guidelines for Pear* (Elkins et al. 2002) for currently registered products and their use. Demethylation inhibiting (DMI) fungicides have shown growth regulating effects on pears and may result in fruit rounding if they are used between white bud and full bloom stage (Sugar et al. 2000). Contact fungicides can be used effectively. Oil sprays made within 10 days of captan and sulfur applications may cause phytotoxicity.

The use of fall eradicants, such as urea, have been shown to significantly reduce the overwintering capabilities of *Venturia pirina*. Urea applied in the fall results in high nitrogen content in leaves, which causes a more rapid decomposition of leaves on the orchard floor and reduces sites where fruiting bodies can be produced. Dormant application of lime sulfur has been shown to effectively "burn out" dormant twig lesions, a major source of inoculum in the North Coast production area.

ROOT AND CROWN ROT DISEASES OF PEAR

Root and crown diseases caused by soilborne fungi inhibit the root's ability to absorb water and minerals from the soil and transport them to the top portion of the tree. Pears are more resistant to root and crown diseases than most other fruit tree species. The most important root and crown diseases of pear are oak root rot (caused by *Armillaria mellea*) and Phytophthora root and crown rot (caused by one of several *Phytophthora* species). Of these two diseases, only oak root rot is of major concern in commercial California pear orchards.

Generally, aboveground symptoms of root rot diseases are a direct result of a compromised root system and can mimic other non-disease-related root problems. Anaerobic soil conditions resulting from too much water, water stress (not enough water), rodent damage, and nutritional deficiencies all may cause symptoms similar to root and crown diseases. Generally, aboveground symptoms include poor shoot growth, sparse or pale foliage, sometimes with marginal necrosis, premature defoliation, dieback of leaves and terminal shoots, decreased productivity and finally death of the tree. To determine the cause of the problem, examination of the root and crown area is essential. Underground symptoms are often striking, making tentative field diagnosis possible. For positive identification, send samples to an appropriate laboratory or consult your local farm advisor.

ARMILLARIA ROOT AND CROWN ROT

Armillaria root and crown rot is caused by the soilborne pathogen *Armillaria mellea*, sometimes called oak root fungus. Other common names for this disease include shoestring and Armillaria root rot. *Armillaria* species are important root rotting pathogens on a wide range of ornamental and agricultural important hosts (Raabe 1979). Cherry, almond, apricot, peach, olive, prune, grape, and many plum species are highly susceptible. In the past, most pear rootstock species were considered immune or highly resistant, including *Pyrus betulaefolia*, *P. calleryana*, French pear, Bartlett-rooted cuttings, and Old Home × Farmingdale. However, many pear orchards on these rootstocks have recently been diagnosed with oak root rot in California. Old Home–rooted cuttings and Winter Nelis are considered to be moderately resistant. Several plant species native to California pear-growing regions are natural hosts of the fungus, such as oaks, willows, cottonwood, and black walnut.

Symptoms

Affected trees usually show a slow, general decline in vigor over many years. Trees that look healthy may suddenly wilt and die in a matter of weeks. Typical symptoms include poor shoot growth; sparse or pale foliage with marginal necrosis; dieback of leaves and terminal shoots; and decreased productivity (color plate 21.10). A critical diagnostic symptom of Armillaria root and crown rot is when trees decline in circular patterns in the orchard and the pattern enlarges each year. These patterns may coalesce into irregularly shaped areas of affected trees.

To confirm the presence of *Armillaria* on a declining tree, remove soil from around the crown, cut away the bark on small sections of the crown and main roots, and look for dense, felty, creamy-white colored, mycelial plaques or fans (color plates 21.11 and 21.12). The mycelium should be visible between the bark and wood. Infected wood feels spongy or "punky" as a result of a white rot and has a strong mushroom-like smell. It may be difficult to locate the mycelium if the fungus is moving up a large root but has not yet reached the crown. Dark brown to blackish rootlike structures called rhizomorphs sometime develop within or on the surface of infected roots (color plate 21.13). These structures may be difficult to find. Large, densely packed, honey-colored mushrooms may form at the base of infected trees in late fall or early winter. Other wood-rotting fungi also have dense white mycelium, so it may be necessary to submit samples to a laboratory for positive diagnosis.

Disease Development

The pathogen survives from one year to the next as mycelium in roots and woody tissue of affected hosts or as rhizomorphs on the surface of the infected material. In the absence of a living host, the mycelium can survive for years on old infected roots or other woody material buried in moist soil.

Armillaria mellea is most active when the soil is cool and moist. It spreads when roots from a healthy tree come in contact with those of an infected tree. When root-to-root contact occurs, rhizomorphs grow into the healthy wood of the new host. No wound is necessary for infection, since rhizomorphs can penetrate directly through the bark of roots. Infection and fungal growth is optimal at soil temperatures of 63° to 75°F (17.2° to 23.9°C) (Bliss 1946). Such temperatures occur during spring, early summer, and early fall in California pear-growing regions.

Once a root is infected, *A. mellea* causes a white rot. In this type of decay, all the major components of the wood cell wall, including cellulose, hemicellulose, and lignin, are degraded. The wood becomes soft, bleached white, and "punky." The mycelium (white, vegetative body of the fungus) grows through the wood from the point of infection up and down the root, eventually reaching the crown, where it girdles the trunk and kills the tree.

The spores produced by *A. millea* mushrooms in late fall or early winter do not contribute in spread of the disease. However, if an orchard was planted in a site where oak, black walnut, or willow trees once grew, the chances are very high that inoculum of *A. mellea* was present at the time of planting and infection of pear trees has undoubtedly occurred. The trees have likely been tolerating the disease for years without noticeable symptoms.

How can pear trees tolerate this disease for many years and then die suddenly? The answer is only speculative at this point, but it may be related to the way in which irrigation methods have changed. Until recently, orchards were flood- or furrow-irrigated. Although the crown area and upper root zone of a tree are saturated for a short time after flood irrigation, they are allowed to partially dry between irrigations. *Armillaria mellea* requires moist conditions to thrive and is very sensitive to drying. This drying period between flood irrigations coupled with the pear tree's natural resistance to the disease may have been enough to keep the pathogen in check. In contrast, most pear orchard are now irrigated with solid-set sprinklers. This enables growers to irrigate more frequently, a practice that allows the production of large pears. Consequently, the orchard floor is often kept fairly moist throughout the entire season, favoring development of the disease. High soil moisture may allow the fungus to overcome the pear tree's natural resistance

In addition to the change in irrigation practices, the move toward denser-planted orchards may lead to a more rapid spread of oak root fungus within an orchard. Closely planted trees means increased root contact and more chances to spread the disease. A recent Armillaria root and crown rot control recommendation is to plant new trees with the root collar at a shallow depth, less than 4 inches (10 cm) deep, rather than with the root collar up to 20 inches (51 cm) below the soil line. Research has shown that pears may be tolerant to infection of roots but succumb quickly when the infection is at the root collar. *Armillaria mellea* rarely colonizes trees above the soil line. Exposure of the root collar region to the air may

- keep the bark dry and offset the influence of excessive moisture
- prevent initial infection of the root collar by rhizomorphs and restrict infection to peripheral parts of the root system
- allow for recovery of infected tissue

Armillaria root and crown rot has been a sporadic problem in pear orchards for many years. In the past, decline in pear trees may have been attributed to other problems such as pear decline. When removing dead or declining trees, be sure to check for signs of oak rot fungus.

Control

Once a tree or an orchard is infected with *A. mellea*, eradication is extremely difficult. Currently, no chemical is registered for controlling oak root fungus on existing, diseased pear trees. Limited success has been reported by removing soil from around the base of an infected tree and exposing the crown for several months. This allows the base to dry out. Diseased trees require diligent care to remain productive. If management practices that encourage disease development are modified and the trees are nurtured with fertilizer and adequate water to reduce stress, infected trees may recover or at least decline more slowly. At some point, declining yields may necessitate removal of infected trees.

Preplant fumigation with methyl bromide is currently the best chemical control strategy available. Proper fumigation kills rhizomorphs and mycelium on small roots, but large infected roots may still harbor viable inoculum of *A. mellea*. When replanting sites with a history of oak root rot, it is advisable to take precautions. Successful, long-term management requires tree removal and soil treatment of large areas. Growers should follow these guidelines before preplant fumigation with methyl bromide:

- Dig out diseased trees. If you want to stop the spread of the disease, remove adjacent, symptomless trees from the perimeter of the infected. Although these trees do not show symptoms, they may have root infections that have not yet moved to the crown. If you do not remove the trees around the perimeter of the oak root fungus area, the disease will not be contained. Roots from one tree can extend far beyond its dripline.

- Remove all woody material greater than ½ inch (12.5 mm) in diameter from the soil as best you can. This is a slow and expensive process but it is very important. Using a screen will help.
- Fumigate the area with methyl bromide before replanting. It is not advisable to rip soil prior to fumigation due to the danger of spreading inoculum to uninfected areas. Apply methyl bromide to large areas at 400 pounds per acre, tarped. Methyl bromide treatment is much more effective if soil is warm (at least 55°F, or 12.8°C) and dry. Dry soil allows the gas to move further in the soil. The drying period also stresses the pathogen. It may be necessary to remove infected trees in the spring or early summer so the soil has time to dry before fumigation. To maximize removal of soil moisture, plant a deep-rooted, nonirrigated cover crop such as mustard, Sudan grass, or safflower. Fumigation should be performed in the fall while the soil is still warm and dry.
- For treating small areas, backhoe an area that is 5 feet by 5 feet by 6 feet deep (1.5 by 1.5 by 1.8 m), remove all large roots as outlined above, and refill the hole. Cover the area with a plastic tarp that is 10 feet by 10 feet (3 m by 3 m) and inject 1.5 pounds of methyl bromide per site. Deep penetration by the gas may be difficult in heavy or stratified soils.

An alternative preplant treatment is metam sodium (Vapam, Soil Prep, and others). Vapam is applied as a drench and therefore moves with the water into the root zone. Thus, it is easier than using methyl bromide for deep treatments. Vapam has the advantage that the soil does not need to be dry prior to application. In fact, maximum depth is achieved when soil moisture is slightly below field capacity. However, Vapam is not as effective as methyl bromide.

Control of oak root fungus through soil fumigation is not completely understood. Fumigation apparently weakens the pathogen but allows an increase in *Trichoderma viride*, a naturally occurring beneficial fungus that is antagonistic to *Armillaria mellea* (Munnecke et al. 1981). Complete eradication of oak root fungus is extremely difficult, and localized follow-up treatments may be necessary.

A permit is required from the county agricultural commissioner to apply methyl bromide or metam sodium. Growers are legally required to read and follow all label directions when using these fungicides. For more information, contact the local county agricultural commissioner; see also *Integrated Pest Management for Apples and Pears* (Ohlendorf 1999) and *The Safe and Effective Use of Pesticides* (O'Connor-Marer 2000).

Studies are currently being conducted by University of California researchers to examine the interactive roles of irrigation strategies, rootstocks, and other root pathogens in controlling Armillaria root and crown rot. Several alternative preplant and postplant management techniques are also being examined. These are long-term projects and definitive answers are still a few years away. Consult *Pest Management Guidelines for Pear* (Elkins et al. 2002) for current information.

PHYTOPHTHORA ROOT ROT

Phytophthora root rot is not usually a major concern on pear trees. Susceptibility and disease severity depends on which species of the pathogen is present, the age and health of the tree, the rootstock, water management, and climatic conditions that occur after the tree is infected.

There are several *Phytophthora* species in California, and some are more virulent than others. Pears are susceptible to *P. cactorum* (Leb. & Cohn) Schroet, *P. cryptogea* Pethybridge & Lafferty, *P. cambivora* (Petri) Buisman, *P. cinnamomi* Rands, and *P. megasperma* Drechsler. *P. cactorum* and *P. cinnamomi* are found on pear trees in Oregon, while *P. cambivora* and *P. megasperma* are more common on pears in California (Ogawa and English 1991). Bartlett and Winter Nelis rootstocks are more susceptible than Old Home, Old Home × Farmingdale, and *Pyrus calleryana* (Cameron 1962).

Symptoms

Symptoms can range from unnoticeable to severe. A small percentage of tiny rootlets are generally being constantly infected by *Phytophthora* and other "root nibbling" organisms with absolutely no visible effect on the aboveground portion of the tree. If the disease progresses and large roots close to the crown are infected, there is a lack of terminal growth, and leaves are small, chlorotic, and drooping (color plate 21.14). Mature trees with partial girdling of the crown may show delayed growth the following season on the affected side of the tree. When conditions are again favorable for disease development, the pathogen may resume growth, eventually girdling the tree. More severe symptoms include dieback of terminal shoots, defoliation, and finally, collapse of the whole tree. These symptoms can mimic oak root fungus or abiotic (non-disease-related) root problems.

To diagnose Phytophthora root rot, remove soil from around the crown and peel back the bark. Phytophthora causes light and dark brown or red lesions on the wood under the bark. A well-defined margin between healthy and diseased wood is usually evident (color plate 21.15). Unlike oak root rot, Phytophthora-infected trees do not develop in circular patches in the orchard, because the disease is not spread by root-to-root contact. Affected trees often are concentrated on the low end of the field and on wetter or heavier soil in the orchard. *Phytophthora* does not produce white mycelial plagues under the bark, rhizomorphs, or mushrooms.

Disease Development

The fungus overwinters as sexually produced spores (oospores), thick walled asexual spores (chlamydospores), or mycelium in soil and infected roots. When temperatures become favorable in the spring, mycelium in infected trees begins growing again. In the presence of free water, motile spores (zoospores) are released from sporangia that are produced from overwintering oospores and chlamydospores. These zoospore have "tails" (flagella) that allow them to swim through water, infecting roots of susceptible hosts.

Zoospores and mycelium continue to be produced during cool, wet soil conditions. Localized spread of Phytophthora root rot takes place largely by flood irrigation and runoff water. Sources of surface irrigation water such as irrigation canals often contain fairly high levels of *Phytophthora* inoculum; therefore, growers must be extremely cautious with water management.

Most *Phytophthora* species can survive for an extended period in moist soil, but they die fairly rapidly in dry soil. The optimal temperature range for disease development is 59° to 73°F (15° to 22.8°C). Although root zones are routinely saturated from rainfall during the winter months, Phytophthora root rot is not a problem in the winter because temperatures are well below those necessary for infection. Also, soil temperatures above the optimum suppress the growth and development of the fungus, and soil temperatures over 90°F (32°F) significantly reduce fungal survival.

Phytophthora root rot damage is most common and most severe in low-lying, poorly drained areas of the orchard. Numerous infections can occur on small feeder roots without obvious effects on the aboveground part of the tree. In many cases, infection occurs right at or below the soil line, where it causes a water-soaked lesion under the bark. If conditions remain favorable, the fungus may continue to grow, and the infection becomes a depressed canker below the bark that may produce a moist, gummy exudate. Later, the affected area becomes shrunken and cracked. Invasion of the roots usually begins at the crown area or near ground level. Cankers may enlarge and spread, girdling large roots, crowns, and trunks, at first causing the tree to grow poorly, produce fewer and smaller fruit, show spare soliage and dieback of twigs, and finally die.

Factors influencing the occurrence, incidence, and severity of the disease include the level and duration of soil moisture, temperature, rootstock, and the vigor of the particular *Phytophthora* species. Certain species primarily rot small feeder and secondary roots, causing a slow decline of the tree. Other species are much more vigorous, rapidly infecting large roots and the crown, causing rapid decline and death of the tree.

Control

Phytophthora root rot is present in most areas and is easily introduced in irrigation water and on farm machinery. Once the pathogen is established in an orchard, it is virtually impossible to eradicate it. However, preplant fumigation with methyl bromide significantly reduces the level of inoculum in an orchard; but fumigation is generally not economically feasible.

Because Phytophthora root and crown rot is directly related to excessive soil moisture, proper water management is the key to controlling this disease. Avoid prolonged soil saturation and do not allow water to stand in contact with crowns of trees. Be sure to provide adequate drainage in low areas of the orchard, in areas where water penetration is poor, or in other areas where rain or irrigation water accumulates. It is advisable to plant trees on mounds or berms 8 to 10 inches (20.5 to 25.5 cm) high to prevent water from accumulating next to the crown. If possible, irrigation should be applied in 12-hour intervals. The graft union must be well above the soil line, because pear scions are usually more susceptible to Phytophthora root and crown rot than are pear rootstocks.

It is best to direct sprinkler irrigation away from tree trunks. If drip or micro-sprinkler irrigation is used, place emitters and sprinklers well away from the trees to avoid flooding the crown. Emitters placed to irrigate very young trees may need to be moved further away from the trunk to accommodate the trees as they get larger.

In general, pear rootstocks are fairly tolerant to Phytophthora root rot, however, none is immune. There is a difference in resistance among rootstock to certain species of *Phytophthora*. In general, *Pyrus betulaefolia* is the most resistant, followed by *P. calleryana, P. communis* (Bartlett), Old Home × Farmingdale, Winter Nelis, and quince.

Attempts in using fungicides on affected trees have produced variable, and usually poor, results.

PEAR DECLINE

Pear decline was first reported in California orchards in 1959. The disease has resulted in loses of over 12,000 acres (4,800 ha) of pears in California and has reduced production in additional pear acreage. The disease was first observed in British Columbia in 1949 and in Washington in 1950. The association between the disease and pear psylla became evident as the disease spread down the West Coast. Pear psylla was demonstrated to transmit the causal organism to healthy pear trees if they first fed on infected trees for short periods; transmission of the causal organism was also demonstrated to occur through budding. Although the pathogen was originally described as a viruslike agent, it was determined to be a mycoparasmalike organism that has recently been renamed as a phytoplasma.

Rootstocks were shown to affect the severity of disease. Pear trees propagated on Asian pear rootstock (*Pyrus serotina* and *P. ussuriensis*) are most sensitive, whereas trees on French-type rootstock (*P. communis*) are generally less effected. Winter Nelis and Bartlett are the most susceptible scion varieties, whereas Flemish Beauty and Comice are somewhat tolerant.

Symptoms

Pear decline symptoms are influenced by tree vigor, environmental conditions, scion susceptibility, and rootstock susceptibility. Generally, trees react to infection by quick decline or collapse of the tree, slow decline, or leaf curl.

Quick decline is most common in vigorous, older trees in the year following infection. Foliage wilts suddenly in the heat of summer and the tree dies within a few days. Quick decline has been most commonly observed on *P. serotina* and *P. ussuriensis* rootstocks. It is rare in California now that these rootstocks are no longer used. Slow decline is characterized by reduced shoot growth, small fruit, and premature reddening of leaves in autumn (color plate 21.16).

Leaf curl occurs most prevalently in young, vigorous trees, especially on decline-resistant rootstocks such as *P. communis, P. calleryana*, and *P. betulaefolia* (color plate 21.17). Leaves thicken, curl, and change to red and purple in late summer or autumn, especially following clear nights (color plate 21.18). Premature leaf drop is common, and yield my be reduced the following year. Bartlett and Old Home exhibit severe purple leaf curl; Comice shows bronzing of leaves with only slight leaf curl; Asian pears rarely show curl symptoms. Slow decline symptoms may be expressed in years following those in which curl symptoms were expressed.

Infected trees planted on *P. serotina* and *P. ussuriensis* rootstocks often exhibit a thin brown line at the graft union, while trees planted on other rootstocks show ridging and grooving at the graft union. Infected trees have reduced numbers of feeder roots, and the feeder roots that do grow have poor vigor and dead tips.

Infected trees are noticeably slow in leafing out in the spring, but by May to June, growth appears to be normal. The titer of phytoplasma in the phloem increases in July, causing the sieve tubes to become necrotic and collapse; new phloem tissue may also collapse due to psylla feeding. The blockage of sieve tubes restricts carbohydrate and nitrogen movement from leaves to the woody plant parts below. Symptoms of decline occur in July and become more severe with time. Red-purple foliage appears as early as mid-August, and leaves drop prematurely, removing vital nutrients from the tree.

Management

Pear decline reduces crop yield. Effective control of decline requires an integrated approach that encompasses pear psylla control, orchard culture, rootstock selection, and chemical treatment of the disease. Controlling pear psylla is valuable because having fewer psylla in an orchard from season to season reduces transmission of the causal organism and damage to the phloem by feeding injury. This increases the trees' vigor, allowing them to produce near-full yields.

Orchard Culture

Nitrogen levels, irrigation practices, crop load, pruning severity, and pest control affect the ability of trees to tolerate pear decline. More frequent irrigation, increased nitrogen, reduced cropping through pruning, and mite control have been proven to help infected trees recover from pear decline and continue to produce for many years. Chemical treatment (injection) of diseased trees using oxytetracycline hydrocloride (terramycin) as a postharvest infection treatment in conjunction with psylla control and good orchard management has also proven valuable in promoting tree recovery.

BLOSSOM BLAST

Blossom blast of pears is caused by the bacterium *Pseudomonas syringae* van Hall. The disease occurs sporadically in all pear-producing regions in California. Though blossom blast can be confused with lack of winter chilling or boron deficiency, it generally results in more severe damage and greater crop loss.

The Pathogen

Pseudomonas syringae is an epiphyte of many crop plants and weeds, and it occurs in high numbers on symptomless pear tissues. The bacterium may cause diseases on numerous crops, including pome fruit and stone fruit. Disease expression always occurs on the aboveground parts of the pear tree. Although the infection process is not well understood, it is well documented that disease incidence and severity increase dramatically after trees have been exposed to predisposition stresses such as cold injury. Epiphytic bacterial populations are known to fluctuate over time and are probably influenced by moisture and temperature conditions.

Symptoms

Symptoms of blossom infections are a blasting of flowers with browning, shriveling, and drying of blossom pears on one or more branches. Cankers may develop on twigs and branches; they are characterized by the uplifting of the outer bark layers, resulting in a paperbark-like appearance. Fruit and leaf infections, though rare, occur as small, sunken, black spots on the surface of the infected tissue. At times, the entire fruit may turn black. A red ring of infected tissue may occur on leaves; after drying and darkening, it falls out, resulting in small holes.

Cold, wet weather favors disease development, and freezing of tissue seems to predispose the tree to blossom infection. Frost protection using overhead sprinklers has been shown to protect against this disease. However, improper or inadequate frost protection may increase freezing injury, significantly increasing disease severity (see chapter 17, "Frost Protection").

False blast caused by boron deficiency looks similar to blast caused by *P. syringae*. However, the latter generally exhibits itself over a relatively wide area of the orchard in a given year, whereas boron deficiency occurs in localized areas. Blossom blast is also more prevalent on the lower branches of trees and in low areas of an orchard and areas blocked to wind or air drainage.

Blasting of flowers as the result of lack of chilling may also occur over a relatively large area. Examine environmental records if blasting or dwarfing occurs predominantly in certain flowers (king flowers) of a cluster.

Management

Blossom blast is difficult to manage because the disease is not well understood and the pathogen appears to be widespread in orchards. Monitoring for the pathogen has no apparent value. In addition, chilling injury and periods of wetness from rainfall, sprinkler irrigation, fog, or dew also may allow populations of the bacteria to increase rapidly, increasing the risk that the disease will develop. If these conditions occur, applications of copper or streptomycin at or after the green tip stage have been shown to reduce the incidence and severity of the disease. The best control is application of copper at green tip followed by streptomycin at full bloom.

Frost control systems should also be used to reduce injury by frost. Although removing orchard floor vegetation by cultivation or herbicides before bloom has not been demonstrated to control blossom blast, it should help avoid the disease by providing warmer nighttime temperatures in the orchards during spring frosts.

REFERENCES

Agrios, G. N. 1978. Plant pathology. 2nd ed. Orlando, FL: Academic Press.

Bliss, D. E., 1946. The relation of soil temperature to the development of Armillaria root rot. Phytopathology 36:302–318.

Cameron, H. R. 1962. Susceptibility of pear roots to *Phytophthora*. Phytopathology 52:1295–1297.

Elkins, R. B., et al. 2002. Pest management guidelines for pear. Oakland: University of California Division of Agriculture and Natural Resources Publication 3455. UC IPM Web site, http://www.ipm.ucdavis.edu/PDF/PMG/index.html.

Jones, A. L. 1992. Evaluation of the computer model Maryblyt for predicting fire blight blossom infection on apple in Michigan. Plant Disease 76:344–347.

Leupschen, N. S., K. G. Parker, and W. D. Mills. 1961. Five-year study of fire blight blossom infection and its control in New York. Ithaca: Cornell University Agricultural Experiment Station Bulletin 963.

Miller, T. D., and M. N. Schroth. 1972. Monitoring the epiphytic population of *Erwinia amylovora* on pear with a selective medium. Phytopathology 62:1175–1182.

Mills, W. D. 1955. Fire blight development on apple in western New York. Plant Disease Report 39:206–207.

Munnecke, D. E., M. J. Kolbezen, W. D. Wilbur, and H. D. Ohr. 1981. Interactions involved in controlling *Armillaria mellea*. Plant Disease 65:384–389.

O'Connor-Marer, P. 2000. The safe and effective use of pesticides. 2nd ed. Oakland: University of California Division of Agriculture and Natural Resources Publication 3324.

Ogawa, J. M., and H. English. 1991. Diseases of temperate zone tree fruit and nut crops. Oakland: University of California Division of Agriculture and Natural Resources Publication 3345.

Ohlendorf, B. 1999. Integrated pest management for apples and pears. 2nd ed. Oakland: University of California Division of Agriculture and Natural Resources Publication 3340.

Powell, D. 1965. Factors influencing the severity of fire blight infections on apple and pear. Michigan State Horticultural Society Annual Meeting 94:1–7.

Raabe, R. D. 1979. Resistance or susceptibility of certain plants to Armillaria root rot. Oakland: University of California Division of Agricultural Sciences Leaflet 2591.

Smith, T. J. 1992. A potential for improved assessment of fire blight risk. Good Fruit Grower (Sep. 1): 17–20.

———. 1994. Learning about fire blight. Good Fruit Grower (March 15): 17–19.

Steiner, P. W. 1990a. Predicting apple blossom infection by *Erwinia amylovora* using the Maryblyt Model. Acta Horticulturae 273:139–148.

———. 1990b. Predicting canker, shoot and trauma blight phases of apple fire blight epidemics using the Maryblyt Model. Acta Horticulturae 173:194–158.

Steiner, P. W., and G. W. Lightner. 1992. MARYBLYT: A predictive program for forecasting fire blight disease in apples and pears. Software, version 4.0. College Park: University of Maryland.

Sugar, D., R. A. Spotts, W. D. Gubler, and G. T. McGourty. 2000. Effect of fenarimol application timing on fruit shape of Bartlett pear. Plant Disease 84(2): 182–184.

Thompson, S. V., M. N. Schroth, W. J. Moller, and W. O. Reil. 1982. A forecasting model for fire blight of pear. Plant Disease 66:576–579.

Van Der Zwet, T., B. G. Zoller, and S. V. Thompson. 1988. Controlling fire blight of pear and apple by accurate prediction of the blossom blight phase. Plant Disease 72:464–472.

Van Der Zwet, T., A. R. Biggs, R. Heflebower, and G. W. Lightner. 1994. Evaluation of the Maryblyt computer model for predicting blossom blight on apple in West Virginia and Maryland. Plant Disease 78:225–230.

Vogel, G., D. Gubler, and J. Ogawa. 1990. Comparison of the standard UC IPM fire blight treatment model with the Maryland blight model. Fresno: California Tree Fruit Agreement.

Zoller, B. G. 1978. Trends in integrated pear insect pest management and *Erwinia amylovora* blight control in California. Oregon Horticultural Society Proceedings 69:57–71.

———. 2000. Fire blight control experiences in the Sacramento Valley of California, 1973–1999. Proceedings of the Western Orchard Pest and Disease Management Conference 74:103–104.

Zoller, B. G., and J. Sisevich. 1973. Seasonal fire blight bacteria population in Upper Lake, California, in relation to spray program and weather in 1972. Abstracts of Reports from 47th Annual Eastern Pest and Disease Management Conference.

———. 1979. Blossom populations of *Erwinia amylovora* in pear orchards vs. accumulated degree hours over 18.3 Celsius, 1972–1976. Phytopathology 69:1050.

22

Postharvest Biology and Technology

ELIZABETH J. MITCHAM AND F. GORDON MITCHELL

European pears, like other fresh produce, are living organisms that must continue their metabolism, even following harvest, in order to remain alive and healthy. To optimize the life of fresh produce after harvest, handling procedures must slow the rate of metabolism without causing irreversible damage to fruit.

CLIMACTERIC FRUIT

Pear fruit are continually respiring, that is, consuming oxygen and producing carbon dioxide in the process of using sugars to produce energy for survival. European pears are classified as climacteric fruit. Climacteric fruit undergo a sudden increase in metabolic activity as ripening begins. This burst in activity is accompanied by an increase in respiration rate or carbon dioxide production and an increased production of ethylene gas. This rise may reach a peak within a few days of its initiation, followed by a continuous and slow decline. Pear fruit reach eating ripeness at or soon after the climacteric peak. Pears must be harvested and stored so as to maintain them at a preclimacteric state until they are to be marketed.

MATURITY AND RIPENING

As pear fruit develop on the tree, they increase in size and accumulate starch reserves. As the fruit ripens, the starch reserves are converted to sugars. Along with the conversion of starch there is often a change in skin color from green to yellow, a softening of the fruit tissues, and the production of flavor volatiles that contribute to typical pear flavor. European pears do not ripen to good eating quality on the tree and may require special treatment to ripen properly following harvest (see chapter 26, "Conditioning and Ripening of Bartlett Pears"). Therefore, fruit are generally harvested mature but firm. Mature fruit are not yet ripe but have developed to a stage where they can be ripened to good quality when harvested. Changes that generally occur in pear fruit as they approach maturity include a change in skin color from green to light green, a decrease in starch and an increase in soluble solids content, and a decrease in flesh firmness.

Each of these changing factors could potentially be used as a guide to fruit maturity and harvest time. In addition, days from full bloom, accumulated heat units (degree-hours above 7°C, or 45°F) and various combinations of these indices have been suggested. However, many of these indices have not been used because they are too time-consuming and costly to measure or the changes are too quick or too subtle to monitor. Many pear-producing regions of the world use flesh firmness guides for pear maturity (see chapter 24, "Postharvest Handling of Winter Pears"). California Bartlett pear maturity standards use a combined index of firmness and soluble solids content that is further modified by fruit diameter (table 22.1). An alternative surface color standard is available. While most legal standards define minimum maturity, California also imposes a maximum maturity standard for Bartlett pears destined for processing. For Red Bartlett pears, skin color at harvest is a greater concern. Unfortunately, in California, red color often fades 1 to 3 weeks before optimal harvest date based on fruit firmness and size. This difficulty has caused the acreage of Red Bartletts to decline in California.

ETHYLENE

Ethylene gas is produced by many fruits and vegetables and is also found in auto exhaust and cigarette smoke.

Table 22.1. California minimum maturity standards for Bartlett pears

Minimum soluble solids (%)	Fruit firmness (lbf)*	
	Diameter 2⅜ to 2½ inches	Diameter 2½ inches and larger
<10	19.0	20.0
10	20.0	21.0
11	20.5	21.5
12	21.0	22.0
13	no maximum	no maximum

Note: *Firmness conversions: 1 lbf = 4.448 N = 0.5 kgf. Diameter conversions: 2⅜ inches = 60.3 mm; 2½ inches = 63.5 mm.

Ethylene is recognized as a growth regulator because it stimulates the ripening of fruit and other plant growth responses. In fact, ethylene gas promotes the production of additional amounts of ethylene by the fruit. For this reason, ripening fruit, decaying fruit, or other sources of ethylene should be excluded from storage facilities used for unripe pears. Only 1 to 10 ppm of ethylene within the fruit is required to induce faster and more uniform ripening. This induction of ripening with ethylene can be a useful tool in postharvest handling, but when ripening is undesirable, ethylene should be avoided. The sensitivity of pears to ethylene is greatly reduced at lower temperatures, and it is negligible at 30°F (–1.1°C). Treatment of pears with SmartFresh (1-methylcyclopropane, 1-MCP) inhibits the effects of ethylene gas during cold storage, but it can also affect the ability to ripen after storage if not properly applied (see chapter 26, "Conditioning and Ripening of Bartlett Pears").

TEMPERATURE AND RESPIRATION

Temperature influences the rate at which metabolism proceeds within the fruit. For every 18°F (10°C) ° increase in temperature, the rate of respiration and other metabolic activities triples. Thus, a pear at 68°F (20°C) respires at 3 times the rate of pears at 50°F (10°C) or at 9 times the rate of pears at 32°F (0°C). Temperature also influences the time required before the fruit commences the climacteric rise in respiration and ripens. At 68° to 77°F (20° to 25°C), ripening commences quickly; at lower temperatures the time before the climacteric rise is extended. As fruit temperature approaches 32°F (0°C), a climacteric rise may not occur regardless of storage time. Above 77° to 86°F (25° to 30°C), the speed of ripening initiation is again slowed, and some ripening changes occur abnormally, resulting in poor texture and flavor.

In addition to inhibiting the initiation of fruit ripening, storage at 32°F (0°C) extends the storage life of the fruit by reducing respiration and other metabolic activities and thereby slowing fruit senescence or death. Many pears can be successfully stored at 30°F (–1.1°C) without freezing as long as the total soluble solids (TSS) content in the core area is at least 11 percent. (TSS is generally lowest in the core and should be checked in this location to determine freezing tolerance.) Low-temperature storage also inhibits the growth of many decay organisms. At 32°F (0°C), growth of many fungi is completely stopped. *Mucor piriformis* and *Botrytis cinerea*, fungi that cause mucor rot and gray mold, respectively, continue to grow slowly at 32°F (0°C), but their growth is greatly inhibited. Bartlett pears can be stored for 2 to 3 months in air and 4 to 6 months in a controlled atmosphere (CA) at 30°F (–1.1°C).

TEMPERATURE MANAGEMENT

The speed with which pear fruit are cooled to storage temperature and the temperature at which they are stored are major factors in determining the storage life of the fruit and the occurrence of breakdown disorders. The lower the fruit temperature during storage, the longer the storage life and the lower the incidence of fruit breakdown, providing that freezing does not occur (see table 22.2).

Delays in cooling also influence the incidence of fruit breakdown in storage. In one study, Bartlett pears were stored for 5 weeks at 33°F (0.5°C) following delays of 0 to 5 days in cooling the fruit (table 22.3). The longer the delay in cooling, the softer the fruit after storage.

Good temperature management includes avoiding fruit heating, cooling the fruit rapidly with forced-air cooling when possible, and maintaining a safe, low temperature in storage. Growers should shade bins of fruit while in the field. Fruit exposed to direct sunlight warms very quickly and is subject to excessive moisture loss. Most importantly, limit the time between harvest and transportation to the packing or storage facility. Bartlett pears should ideally be cooled to 36°F (2.2°C) within 1 day of harvest and to storage temperature within 2 days. At a minimum, cool to 41° to 45°F (5.0° to 7.2°C) within 1 day and to storage temperature within 3 days of harvest.

CONTROLLING WATER LOSS

Water loss can be a serious problem in pear storage because of the storage duration. As with many other fruits, visually noticeable shrivel usually results from water loss of 3 to 4 percent. Water loss is cumulative, however. Although water loss may be occurring, initially no symptoms will be visible. Once the critical threshold of water loss occurs, symptoms become visible. For this reason, it is important to take every precaution to reduce water loss at each handling step even though you may not see symptoms developing.

Rapid cooling of the fruit after harvest minimizes water loss. High relative humidity (90 to 95%) around the fruit is necessary during storage to minimize water loss. This can be accomplished by humidifying the storage room air or by using plastic box liners that maintain high humidity within each box. The disadvantage of plastic liners within boxes or bins is slower fruit cool-

Table 22.2. Fruit breakdown in Bartlett pears cooled within 1 day of harvest and stored for 5 months at selected temperatures

Storage temperature	Percentage of fruit with breakdown disorders
30°F (−1.1°C)	0
33°F (0.6°C)	2
36°F (2.2°C)	3.5

Table 22.3. Watery breakdown in Bartlett pears stored for 5 weeks at 33°F (0.6°C) after delays in postharvest cooling

Delay before cooling (days)	Firmness (lbf)*	Watery breakdown (%)
0	16	0
1	15	0
2	12	5
3	9	>10
4	6	NA
5	4	NA

Note: *Firmness conversions: 1 lbf = 4.448 N = 0.5 kgf.

ing. The plastic liner blocks air flow through the package, but this effect is reduced if the liner is perforated. For this system to be effective, the fruit must be completely cooled before being placed into the liner. This system is not used in California for this reason, but it is widely used in the Pacific Northwest.

Fruit waxes have also been suggested as a means to reduce water loss by reducing surface openings on the pear fruit. A wax that reduces the rate of water loss by more than 40 percent can interfere with the exchange of carbon dioxide and oxygen, causing fruit injury. While waxes can be beneficial for water loss reduction, they are often insufficient by themselves.

High relative humidity in storage air can be obtained by maintaining a very large cooling surface (evaporator coils) that can provide adequate heat removal while the coils are being maintained at a temperature only slightly lower than that of the storage room air. Less than a 2°F (1.1°C) temperature difference between room air and evaporator coils is needed to maintain 95 percent relative humidity. Maintenance of high relative humidity is aided by minimizing air warming as it moves through the room. Placing only thoroughly cooled pears in the storage room, installing heavy insulation, and avoiding heat sources such as equipment operation and door openings reduce air warming. Humidifiers can be used to add moisture to the system and reduce water loss from the pears. Addition of humidity is especially important for pears stored longer than 1 month. The added moisture is needed to supply water to packaging materials and wooden bins, which naturally gain moisture in high-humidity situations. While moisture can be added by wetting the floors, this method is not very efficient, and constant wetting of floors can cause structural problems. Mist or fog systems work best. The finer the mist, the greater the amount of water that will evaporate into the air. The excess water will, however, require extra defrost cycles.

CONTROLLED ATMOSPHERE STORAGE

In combination with low temperatures, controlled atmospheres (CA) that have elevated carbon dioxide and reduced oxygen concentrations can further extend the storage life of pears. These atmospheric conditions reduce fruit respiration, ethylene production, and metabolic activity more than cold storage alone and also reduce the fruit's response to ethylene gas. Combined, these effects slow fruit senescence and extend its storage life. Controlled atmospheres also reduce the growth of decay organisms both directly by inhibiting fungal growth and indirectly by keeping the fruit in a healthy, unripe condition to better resist fungal attack. Fruit harvested early in the season is the best candidate for CA storage. Late-season fruit may be damaged by CA storage because of its advanced maturity. For further information on CA storage, see chapter 25, "Controlled Atmospheres."

PHYSIOLOGICAL DISORDERS

Storage Scald

D'Anjou, Packham, and Bartlett pears are susceptible to storage scald. Storage scald develops during cold storage but may not become visible until after fruit are transferred to warm temperatures for ripening. It is a surface disorder that causes browning of the skin (color plate 22.1). Treating freshly harvested fruit with an antioxidant such as ethoxyquin can greatly reduce the incidence of superficial scald. Ethoxyquin is not registered for use on pears in California but is used in the Pacific Northwest. Controlled atmosphere storage can reduce the incidence of storage scald, especially when low oxygen (1%) is used. More research is needed to address fruit tolerance issues as well as the effectiveness of low-oxygen CA for scald control.

Senescent Scald

Bartlett, Bosc, and Comice pears lose their ability to ripen when stored for extended periods of time. The fruit turn yellow and eventually develop a dark brown discoloration of the skin called senescent scald (color plate 22.2). The browning may develop during storage or after the fruit is removed from storage and subjected to warm ripening temperatures. The fruit is inedible and remains firm, and the skin sloughs off easily. Reducing the storage period is the only control method; however, proper maturity at harvest, rapid cooling after harvest, and low storage temperatures extend the storage life of the fruit and extend the time before senescent scald develops. Controlled atmosphere storage or treatment with SmartFresh can greatly extend (sometimes double) the storage life of pears relative to air storage.

Breakdown Disorders

Core breakdown

Most pear cultivars can suffer losses to core breakdown, also known as senescent breakdown. Symptoms of browning and softening of the tissue in and around the core can develop in storage or after transfer to ripening temperatures (color plate 22.3). Late-harvested fruit are most susceptible. Prompt cooling and good temperature management can reduce the incidence of this disorder.

In general, Bartlett pears grown in cool climates tend to soften more rapidly and be more susceptible to core breakdown than those grown in warmer climates. By contrast, D'Anjou pears grown in warm climates can have a high incidence of mealy breakdown of the flesh.

Temperature management is the most important tool that can be used to maintain the quality and extend storage life of European pears. Controlled atmosphere storage can provide additional storage life and quality maintenance but only as a supplement to good temperature management. Good temperature management is critical following harvest, during storage, prior to and during transportation, and during wholesale and retail handling.

Pink calyx

Cool growing temperatures in Bartlett pears can cause pink calyx, or pink end, a premature ripening condition that is so named because of the pinkish tinge to the calyx area of affected fruit. Early yellowing of the fruit at the calyx basin and accelerated softening indicate premature ripening. Severe losses can occur unless the fruit is harvested at the first signs of ripening, cooled quickly, and sold promptly. Breakdown associated with premature ripening, such as pink calyx, is somewhat different from core breakdown: it occurs at the calyx end of the fruit, and soft, light-brown tissue is clearly visible through the skin (color plate 22.4). Core breakdown is not visible from the outside. This disorder has been related to cool temperatures in the latter part of the growing season. In Oregon, this disorder has been associated with daytime high temperatures below 70°F (21.1°C) if night temperatures are no higher than 45°F (7.2°C) for 2 days, 50°F (10°C) for 9 days, or 55°F (12.8°C) for 21 days. Night temperatures above 55°F (12.8°C) or day temperatures above 90°F (32.2°C) appear to prevent pink calyx.

Watery breakdown

Watery breakdown (color plate 22.5) has sometimes caused severe losses of Bartlett pears in California. In some seasons, this disorder has been responsible for as much as a 10 percent loss of fruit destined for processing outlets. It is caused by a rapid enzymatic flesh breakdown, sometimes during storage but mostly during fruit ripening following storage. Losses from this disorder can also occur in fruit being stored for fresh market. Prompt and thorough cooling and low storage temperatures minimize losses.

PATHOLOGICAL DISORDERS

Blue mold rot caused by *Penicillium expansum* and gray mold caused by *Botrytis cinerea* are the two most common postharvest diseases found on pears in California. Blue mold is a wound-infecting pathogen. The fungus growth is white at first, then becomes pale blue as sporulation occurs (color plate 22.6). For diagnostic purposes, the decayed tissue can be easily separated from healthy tissue in the fruit. The spores spread easily in the air.

Gray mold is also generally a wound pathogen; however, it can infect blossoms during bloom, resulting in calyx-end rots, and once established it can spread from fruit to fruit (nesting). Stem infections are also common, and they may grow into the flesh of the fruit during storage. The gray mold lesion is brown in color and develops gray spores if humidity is adequate (color plate 22.7). Both blue and gray molds have a visually distinct margin. In blue mold, there is a sharp edge to the lesion, while in gray mold there is no clear textural distinction between the lesion and the surrounding healthy tissue. For both blue and gray mold, careful handling to avoid wounding of fruit and sanitation of water systems are important for disease prevention. Benzimidazole fungicides are often used; however, resistance has been noted in both *Penicillium expansum* and *Botrytis cinerea* populations. New fungicides with registrations include fenhexamide (hydroxyanilides), fludioxonil (phenylpyrroles), and pyrimethanil (anilinopyrimidines) for gray mold and the latter two also for blue mold control. Biological controls (e.g., fermentation preparations of *Pseudomonas syringae*) are also available as postharvest treatments for managing gray mold and Penicillium decays.

Bull's eye rot, caused by *Neofabraea (Pezicula) malicorticis* (asexual state *Cryptosporiopsis curvispora*), causes serious losses in the Pacific Northwest, British Columbia, and certain areas of northern Europe. While this disease is present in California, it is not found frequently. Visible lesions are found only after a long time in storage. Lesions are usually round and have a light-colored center. Concentric circular bands of conidia may be evident, forming a target-shaped pattern (color plate 22.8). Field control of cankers is important for disease control.

Mucor rot, caused by *Mucor piriformis*, can also cause severe losses to pear fruit in the Pacific Northwest, but it is not a concern in California pears, although California apples are affected. This fungus is not well controlled by chlorination of water systems or fungicide treatments, and it continues to grow even at 32°F (0°C). The rot that develops is very wet, and the disease can spread from fruit to fruit in storage by leakage from rotting fruit above. The propagules of *M. piriformis* are found in the orchard soil. To help control the disease, prevent fruit or bins from contacting orchard soil.

POSTHARVEST HANDLING OF PEAR FRUIT

Pear fruit are harvested by hand into picking bags. Canvas bags are often used; however, fruit injury would likely be reduced by the use of padded picking buckets. These gentler picking containers are often used for more sensitive pear varieties such as Starkrimson. Fruit are transferred into ½-ton field bins. While the majority of field bins are wooden at this time, plastic bins have increased in number each year and will likely replace wooden bins in the near future. Data indicate that plastic bins can be less damaging to fruit than wooden bins, but the plastic bins must be kept clean. Dirt or other grit in a plastic bin can cause considerable fruit abrasion. Plastic liners in wooden bins significantly reduce abrasion injury.

Dump Tank Solutions

Fruit are dumped from field bins onto the packing line either into water or onto dry tables. Immersion dumps are also used. Because pear fruit do not float, water jets or flotation salts can increase floatation if water is used. Flotation salts include sodium and calcium lignin sulfonate, sodium silicate, sodium sulfate, and sodium carbonate. Each has advantages and disadvantages. Lignin sulfonates can ferment if no disinfectant is included in the solution, rapidly creating unpleasant odors. Sodium silicate becomes a gel at low pH and should never be mixed with acids or lignin sulfonates. Sodium sulfate is difficult to dissolve unless heated and agitated water is used. Sodium carbonate solutions are somewhat caustic and corrosive and can irritate workers' skin. Many of these salts are incompatible with chlorine; however, all of them, but particularly lignin sulfonate, have been shown to be effective for controlling *Botrytis*, *Penicillium*, and *Mucor* spores, the main postharvest pathogens of pear fruit. After flotation of pears from the bins, fruit should be washed with chlorinated water (sodium hypochlorite) or other oxidizing compounds. Fruit burn is possible from the use of flotation salts or chlorine in water systems. Fruit should be rinsed with fresh water immediately upon removal from the water system. Fungicides should be applied in water or in a water dilution of a fruit coating after the fresh water rinse. Fruit can be dried if a drier is available, and then they should be boxed or packaged.

Packing

Most California European pears are packed immediately after harvest, with the exception of those stored in CA, where fruit may be sorted and placed in bins before storage. Fruit are sorted for defects, separated into fresh and canning grades, sized, and packaged. Sizing with rope sizers (based on fruit diameter) had been the most common method in California, but weight sizers, which are more accurate and allow for application of PLU (Price Look Up) stickers, are now the standard practice.

There are four basic methods of packing for fresh market: tight-fill, tissue-wrapped, tray-packed, and bagged. Tight-fill is a modification of volume filling in which the fruit are settled into the box and tightly packed to minimize fruit movement. (See chapter 23, "Reducing Fruit Injury During Postharvest Handling and Transportation.") Tissue wrapping makes up 30 to 40 percent of the fresh pear volume in California and is the most common packaging method in Oregon and Washington. Unvented, plastic-film box liners are used in Oregon and Washington but not in California, where forced-air cooling is used. Tray packs are increasing in popularity. Single-layer trays packed in open-topped boxes are sold in warehouse food stores. They are often used for ripened fruit because the fruit is better protected from mechanical damage than with tight-fill or wrapped fruit displayed in bulk. Bagging has also increased in volume in recent years, especially for warehouse food stores but also for local grocery chains. Immediately following packaging, fruit are palletized and moved to the cooler. If SmartFresh is applied, it should be applied at warm or cold temperatures at the start of storage. Most fresh-market fruit is forced-air cooled to near 32°F (0°C) and stored at 30°F (–1.1°C).

Forced-air Cooling

Forced-air cooling is the method of choice for rapid cooling of pears. Pallets or bins of fruit are placed on opposite sides of a large fan (fig. 22.1). The top of the

Table 22.4. Estimates of typical ⅞ cooling times for pears in selected package types and cooled with selected cooling systems

Package type	Cooling system	Hours to ⅞ cool
Field bin, ½ ton, with 5% Side venting and bottom vents	room cooling, 75 ft/min air velocity	48
	forced-air cooling, 0.75 cfm/lb of fruit, 0.7 inch static pressure	6
	hydrocooling, 10 gpm/ft² water flow	0.55
Wrapped 50-lb	room cooling, 75 ft/min air velocity	96
Package, 5% side area vent, no plastic liner	forced-air cooling, 0.5 cfm/lb, 0.5 inch static pressure	18
Tight-fill 36-lb box, 5% side venting	room cooling, 75 ft/min air velocity	40
	forced-air cooling, 0.4 cfm/lb of fruit, 0.6 inch static pressure	6

Note: Metric conversions: 1 lb = 0.45 kg; ½ ton = 450 kg; 1 ft/min = 0.0051 m/sec; 1 cfm/lb = 0.00104 m³/sec-kg; 1 inch static pressure (1 inch water column) = 0.25 kPa; 1 gpm/ft² = 42.2 l/m²

tunnel is covered with canvas. When the fan is running, a negative pressure develops in the tunnel that pulls the cold air from the room through the box or bin vents and past the pear fruit, providing for rapid cooling. Forced-air cooling cools pears 5 to 8 times faster than room cooling and can be used for packed boxes or field bins. However, as seen in table 22.4, forced-air cooling is slowed considerably when packaging materials, such as wraps, retard air flow. For solid-sided field bins, serpentine cooling, a modification of forced-air cooling, can be used. In this case, cold air flows either up or down through the bin rather than from side to side. Fruit temperatures should be monitored during forced-air cooling to avoid leaving fruit in the cooler after final temperatures have been reached, which can cause excessive water loss. Although hydrocooling is very effective and rapid, a limited number of studies suggests that hydrocooling may increase postharvest decay.

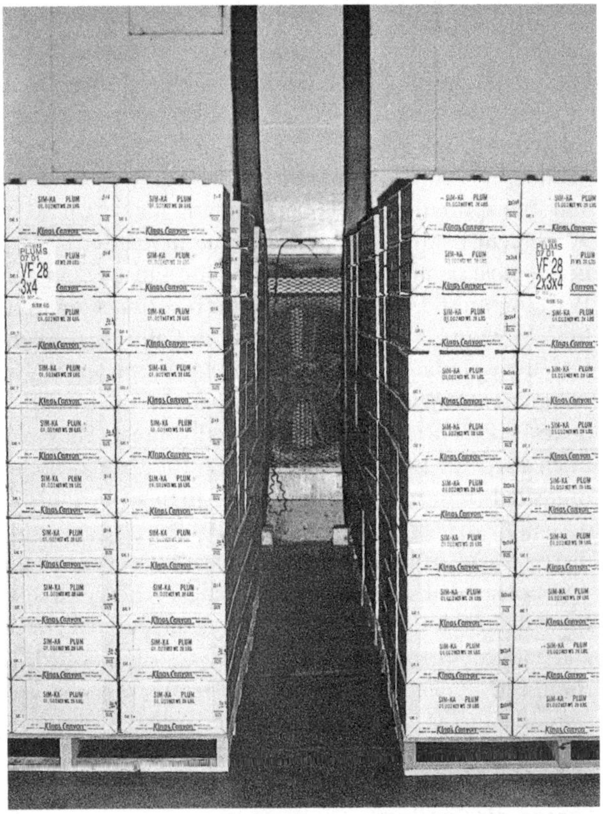

Figure 22.1. Forced-air cooling of boxed fruit. Pallets of fruit are stacked on either side of a fan mounted in a plenum. A tarp is drawn over the tunnel created by the pallets, and the air is pulled by the fan through the ventilation openings in the fruit boxes. *Photo:* E. Mitcham.

Freezing

Should freezing occur in storage, carefully warm the fruit to minimize injury. Do not move the fruit while frozen. Leave the fruit in place and slowly warm it to near 32°F (0°C). If only slight freezing has occurred, recovery of most of the fruit may be possible using this procedure.

Transportation

It is critical that pear fruit be cooled to 32°F (0°C) prior to shipment to prevent ripening en route and ensure good quality upon arrival. Transportation vehicles do not have the capability to cool fruit and can at best maintain the initial fruit temperature. Vehicles should be inspected for the condition of the refrigeration system and care should be taken to load the vehicle in a manner so as not to disrupt the flow of cold air. For more information, see chapter 23, "Reducing Fruit Injury during Postharvest Handling and Transportation."

METHODS FOR DETERMINING QUALITY OF PEAR FRUIT

The quality of fruit arriving at the packing facility or the wholesale or retail distribution center should be monitored. Quality assurance programs can greatly improve the final quality of product delivered to the consumer and can increase profits for all segments of the industry. Color, shape and size, absence of defects, and firmness are the most important quality factors The appearance of fresh pear fruit is one of the first quality determinants made by the wholesaler, retailer, or consumer. Often the

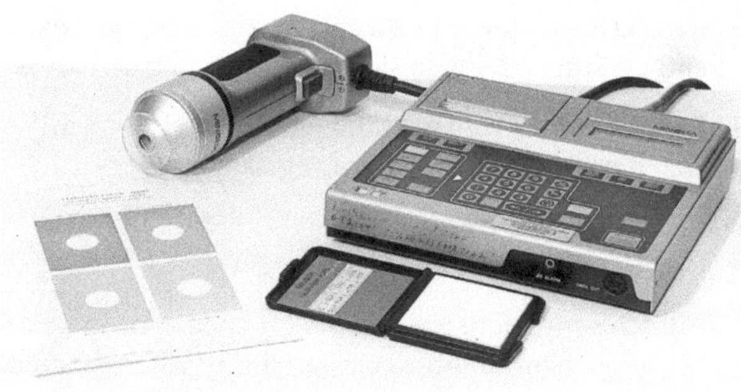

Figure 22.2. The color of fruit can be evaluated subjectively by using comparative color charts, or objectively using an instrument such as a Minolta colorimeter. Color charts can be very effective and useful if the colors truly match the color change in the fruit. The Minolta colorimeter can detect small differences in color and provides separate values for lightness to darkness, green to red and blue to yellow scales as well as hue angle. *Photo:* D. Edwards.

appearance of the commodity is the most critical factor in the initial purchase (in addition to price), while subsequent purchases may be more related to texture and flavor.

Color

Color perception depends on the type and intensity of light, chemical and physical characteristics of the commodity, and the viewer's ability to characterize color. Evaluating color can be subjective or objective (fig. 22.2).

Subjective

The human eye is used to evaluate color.

Advantages:

- Faster and easier than objective measures.
- Requires no specialized equipment.
- Color charts or guides can be used for matching and describing colors.

Disadvantages:

- Results can vary considerably due to viewers' differences in color perception and human error.
- The quantity and quality of available light can influence color perception.

Objective

An instrument provides a color value based on the light reflected from the commodity surface or transmitted through the commodity.

Advantages:

- Less variability in color measurement.
- Can accurately measure small differences in color.
- Can be automated on the packing line.
- Portable hand-held units are available.

Disadvantages:

- Requires specialized equipment at a significant cost.

Shape and Size

Uniform and characteristic shape are important quality characteristics in pear fruit. Misshapen fruit may be more susceptible to mechanical injury, may be unsuitable for some processing uses, and are generally avoided by consumers. The size of the fruit can also be important, depending on the intended use. Consumers tend to associate large size with higher quality and view larger fruit as more mature.

Absence of Defects

The fruit should be evaluated for the presence of defects. The level of tolerance for each type of defect such as cuts, scratches, russeting, scuffing, bruises, and disease should be determined. During quality evaluation, the percentage of fruit with each class of defect and the severity of defect can be determined as a guide to overall product quality.

Firmness

Firmness, or the degree of softness or crispness, is measured using objective instruments (fig. 22.3). There are several types of firmness testers available, including

- Manual penetrometers: Hand-held probe with dial gauge for pounds-force (lbf) or kilogram-force (kgf) (Effegi, Ametek). Probes may be mounted on a drill press stand for increased accuracy.
- Automatic penetrometers: Probe is mechanically inserted into fruit and data stored electronically. Test at standard depth and speed of penetration for increased accuracy.
- Nondestructive firmness testers: A number of nondestructive firmness testing devices have been tested with varying success. A durometer can give accurate information. An impact sensor in commercially available for pears from Sinclair International as benchtop and online models (see www.sinclair-international.com).

For further information on types of firmness testers and suppliers, see the Postharvest Resources Directory at the UC Davis Postharvest Technology Web site, http://postharvest.ucdavis.edu.

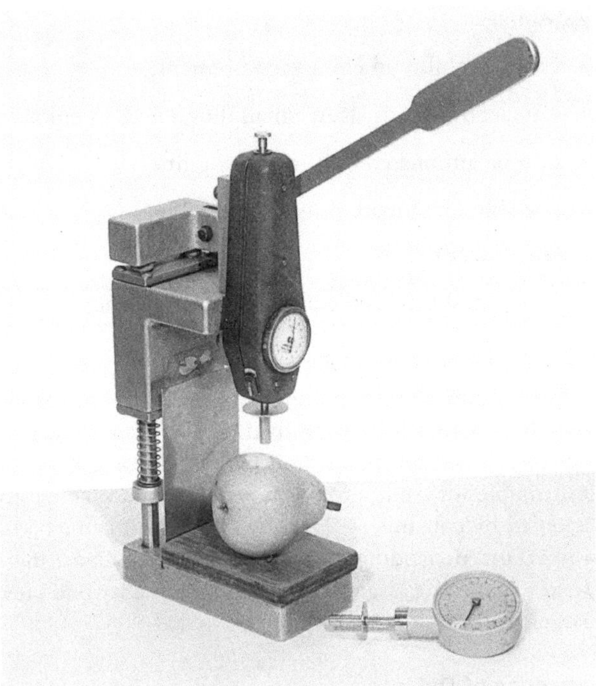

Figure 22.3. The most common way to measure firmness is resistance to compression measured in pounds-force (lbf) or kilogram-force (kgf) with a penetrometer. Two basic types of penetrometer are used. The Effegi penetrometer is lightweight and easy to carry and has an easy-to-read dial. Mounting a hand-held penetrometer, like this Ametek gauge, on a drill-press stand increases the potential accuracy of results. The drill press can be automated for increased accuracy and can provide electronic data collection. Remove a thin disc of skin approximately 1 inch (2.5 cm) in diameter with a vegetable peeler or sharp knife before testing for compression. *Photo:* D. Edwards.

Sample size and selection for firmness

- Select a random sample of fruit from several representative boxes or bins including at least 15 to 25 fruit, or 3 percent of the total.
- Select fruit of a uniform size to avoid variation in firmness due to size (large fruit are usually softer than smaller fruit).
- Make sure all fruit tested are comparable in temperature, since warm fruit are usually softer than cold fruit.

Units for measuring firmness

It is inappropriate to use the term "pressure" in association with firmness measurements using a penetrometer. Pounds-force or kilogram-force are the units preferred in the pear industry; Newton (N) is the unit used in scientific writing. The conversion factors are as follows:

$$\text{pounds-force (lbf)} \times 4.448 = \text{Newton (N)}$$

$$\text{kilogram-force (kgf)} \times 9.807 = \text{Newton (N)}$$

Using penetrometer firmness testers

- Make 2 puncture tests per fruit, once on each opposite cheek, midway between the stem and blossom end on sun and shade sides; avoid sunburned areas.
- Remove a thin disc of skin approximately 1 inch (2.5 cm) in diameter with a vegetable peeler or sharp knife.
- Use a 5⁄16-inch (8-mm) tip (plunger).
- All determinations for a given lot should be made by one person to minimize variability when manual penetrometers are used.
- Hold the fruit against a stationary hard surface or place the fruit on the platform of the drill-press stand and force the tip into the fruit at a uniform speed (take about 2 seconds).
- Penetration should consistently be to the scribed line on the tip.
- Record reading to the nearest 0.5 pounds-force (lbf) or 0.25 kilogram-force (kgf).

Maintaining penetrometer firmness testers

- Before use each day, work the plunger in and out for 10 seconds to loosen up the springs inside the instrument.
- Clean the tips after use to prevent clogging of the mechanism with juice.
- Follow the manufacturer's maintenance instructions for periodic servicing.

Calibrating firmness testers

- Steady the firmness tester in an upside-down position and place a known weight over the probe (a weight that has a hole to fit over the probe on the gauge is ideal).
- If the reading on the gauge matches the weight applied, the penetrometer is ready to use.
- If reading on the gauge does not match the weight applied, find the magnitude and direction of the differences and proceed as follows.

Effegi fruit penetrometer

1. Unscrew the chrome guide nut to remove the plunger assembly.
2. To make the instrument read lower, insert washers between the spring and the stationary brass guide.

3. To make the instrument read higher, insert washers between the chrome guide nut and the stationary brass guide on the plunger shaft.
4. Reassemble and check for calibration.
5. If the indicator needle does not stop or does not release button hold, remove the plunger assembly and lubricate the inside of the instrument with an aerosol lubricant.

Ametek penetrometer

The Ametek penetrometer must be returned to the manufacturer for calibration. Follow the manufacturer's instructions for other firmness tests.

SOLUBLE SOLIDS CONTENT (SSC)

Because sugars comprise the bulk of soluble solids in fruit juice, a measurement of soluble solids content (SSC) can be used to estimate sugar content. Organic acids, amino acids, and soluble pectins also contribute to soluble solids. The soluble solids content can be determined in a small sample of fruit juice using a hand refractometer (fig. 22.4). The refractometer measures the refractive index, which indicates how much a light beam is slowed down when it passes through the fruit juice. The refractometer has one scale for refractive index and another for equivalent degrees Brix (°Bx) or SSC percentage, which can be read directly. The term "degrees Brix" should not be used to describe fruit juices, since the term refers only to the refractive index of a pure sucrose solution. Soluble solids content is correctly used for fruit juices and other mixtures. Sample wedges should be cut from stem to blossom end and to the center of the fruit to account for variability in SSC from the top to bottom and from the inside to the outside of the fruit. A garlic press works well for juicing small samples. Cheesecloth may be necessary to remove pulp from the juice.

The temperature of the juice is a critical factor for obtaining accurate readings because juice expands when heated and becomes less dense. The percentage of sugar in a sugar solution decreases about 0.5 percent for every 10°F increase in temperature. Good-quality refractometers are able to compensate for temperature or at least include a thermometer so that the operator can make the necessary corrections. Clean refractometers between each reading and standardize them at the start of use by measuring distilled water, which has a refractive index of 0.

TITRATABLE ACIDITY (TA)

Titratable acidity (TA) can be determined by titrating a known volume of fruit juice with 0.1N NaOH (sodium hydroxide) to an end point of pH = 8.2 as indicated by

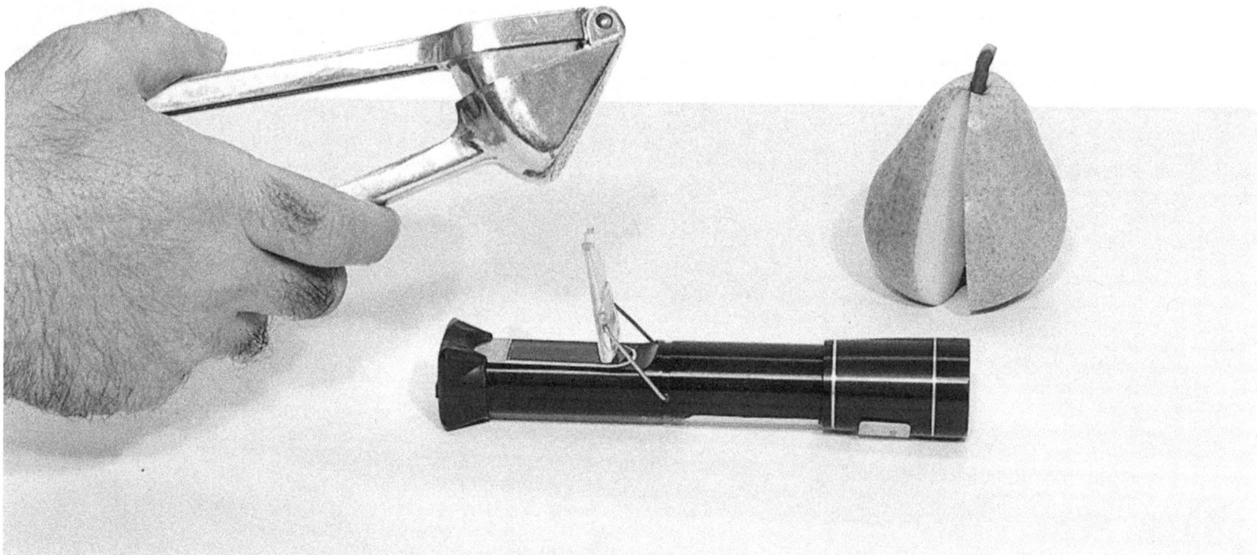

Figure 22.4. To test for soluble solids content (SSC), cut a wedge from the pear from stem to blossom end and from the outside to the center. Extract the juice using a garlic press and place a few drops onto the glass of a refractometer. Close the refractometer, hold it up to the light, and look through the eyepiece. The internal scale shows the SSC of the juice. *Photo:* D. Edwards.

a phenolphthalein indicator or by using a pH meter. NaOH is added to the juice until the pH changes to 8.2. The volume in milliliters of NaOH needed is used to calculate the TA. The percentage TA, expressed as the percentage of malic acid, can be calculated as follows:

$$\% \text{ acid} = (\text{ml NaOH} \times \text{N (NaOH)} \times 0.067) \div (\text{ml juice titrated}) \times 100$$

(0.067 is the acid milliequivalent factor for malic acid, the predominant organic acid in pears.)

REFERENCES

Apel, G. 1993. Flotation materials for pears. Tree Fruit Postharvest Journal 4(1): 19–21.

Meheriuk, M, R. K. Prange, P. D. Lidster, and S. W. Porritt. 1994. Postharvest disorders of apples and pears. Ottawa: Agriculture and Agri-Food Canada, Publication 1737/E.

Sommer, N. F., R. J. Fortlage, and D. C. Edwards. 2002. Postharvest diseases of selected commodities. In A. A. Kader, ed., Postharvest technology of horticultural crops. 3rd ed. Oakland: University of California Division of Agriculture and Natural Resources Publication 3311.

23

Reducing Fruit Injury during Postharvest Handling and Transportation

JAMES F. THOMPSON

Pears are very sensitive to mechanical damage even though they are firm and unripe when they are handled at harvest, packing, and in distribution. The damage is not usually noticeable immediately after it occurs, except for stem punctures. But after several hours, and especially as the fruit ripens, injuries become apparent, particularly with varieties like Bartlett that have very little natural skin russeting and skin that is especially sensitive to damage. Consumers often avoid fruit that has considerable surface marking caused by accumulated damage the fruit has undergone. Mechanical damage provides points of entry for decay organisms that cause rots and also tends to increase water loss and shriveling near the injury. Wounded tissue produces ethylene that may stimulate premature ripening.

SCUFFING

Scuffing is caused when fruit are rubbed or pressed against a rough surface. The susceptibility to scuffing injury increases as fruit ripen and get softer. The damage appears as dark brown to black marks on the skin. Sometimes the rough surface leaves a characteristic pattern on the fruit, allowing the cause of damage to be identified. For example, fruit pushed against a rough picking bag will show the pattern of the fabric on its surface. The common cause for this is pickers leaning against filled picking bags. Fruit moving against rough-walled wood bins or damaged and dirty plastic pallet bins will be scuffed. Wood bins should be lined with plastic sheeting to minimize damage (fig. 23.1). Plastic bins should be cleaned regularly to prevent dirt from sticking to interior walls. Packinghouse machinery can also cause scuffing. Fruit that rubs against rough conveyor walls or catches on a machine part and rubs against a moving belt will be damaged. Considerable scuffing can occur during retail handling by workers and consumers.

IMPACT BRUISING

Fruit falling against a hard surface is subject to flesh damage at the point of impact. The damage is often not immediately visible on the skin, but within a few hours at room temperature the flesh will turn brown (fig. 23.2). After transit to market, consumers see external damage on the fruit. Impact bruising is also a serious defect in canned pears.

The drop height and the nature of the surface on which the fruit lands control the bruising severity.

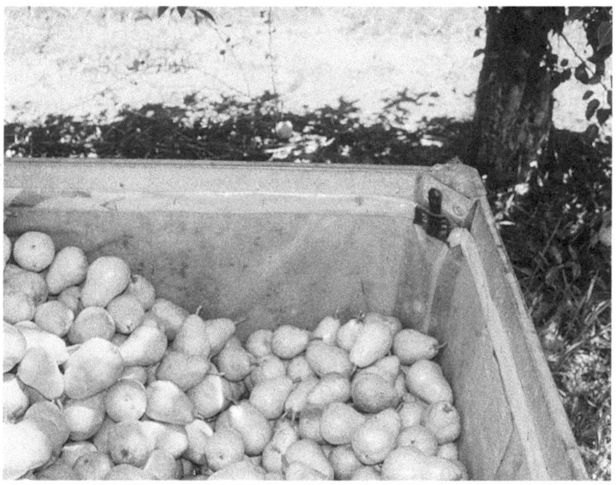

Figure 23.1. Plastic film bin liner used to reduce scuffing damage. *Photo:* E. Mitcham.

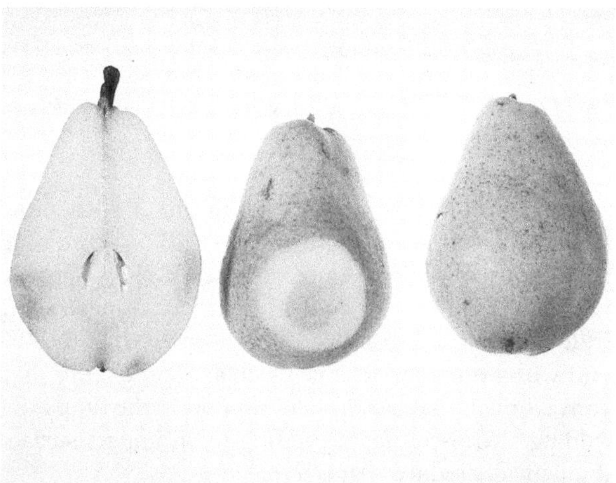

Figure 23.2. Flesh browning caused by an impact. No discoloration of skin is noticeable. *Photo:* D. Edwards.

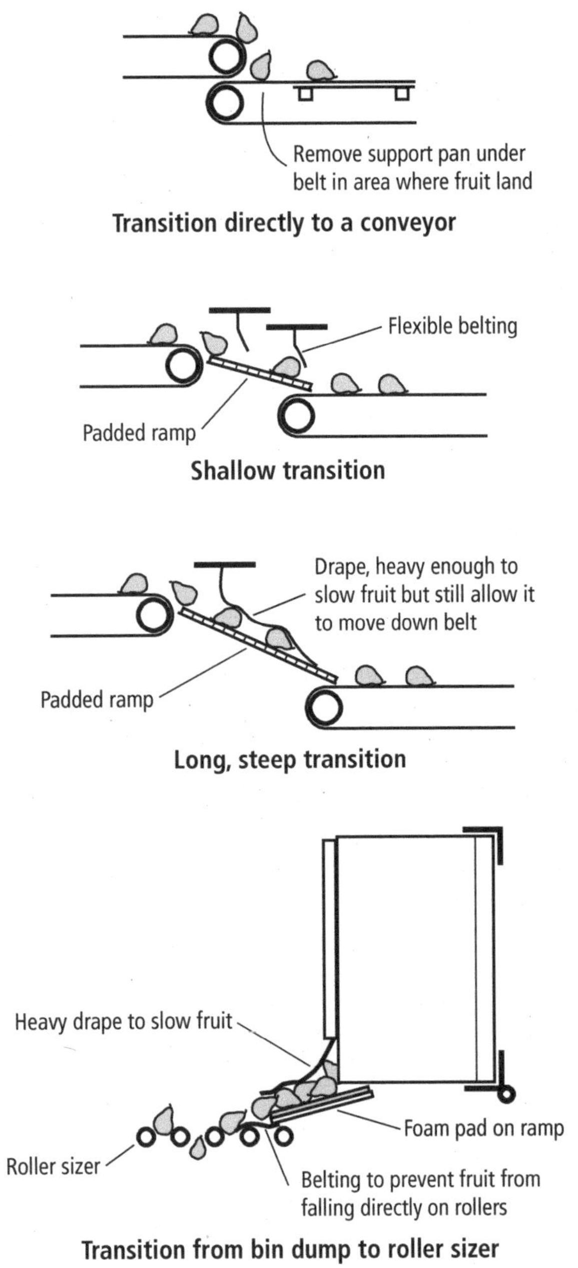

Figure 23.3. Methods of reducing bruising in packing line transfers.

Studies with freshly harvested California Bartlett pears showed that drops of 1½ inches (4 cm) or more onto metal rollers usually produce a bruise. Drops of 3 inches (7.5 cm) or more onto a hard flat surface damage some fruit. A pear can be dropped 6 to 12 inches (15 to 30.5 cm) with little injury if it lands on another fruit or on a conveyor belt that has no solid support in the drop area. Adding closed-cell foam padding to an impact surface dramatically reduces damage.

A survey of several Bartlett packinghouses showed that 20 to 30 percent of the fruit arrived with at least one bruise. Much more bruising occurred in packing: 60 to 90 percent of the fruit in packed boxes had at least one detectable bruise. Impact damage is likely to occur by careless transfer of fruit from picking bags to pallet bins and at fruit transfers in the packinghouse. Pickers should be required to unload their picking bags gently into the pallet bin. As shown in figure 23.3, packinghouse damage can be reduced by:

- installing drapes to slow fruit and padded ramps for fruit to land on at bin dumps and bin fillers
- never allowing fruit to land directly on metal rollers
- minimizing drop heights at conveyor transfers
- cutting out the steel support pan beneath the area where fruit lands on a conveyor belt
- using closed-cell foam padding at unavoidably high drops

Foam should be at least 0.25 inch (6.5 mm) thick and have a density of 5.5 to 15 pounds per cubic foot (88 to 240 kg•m^{-3}). Because foam deteriorates, losing its ability to absorb impact energy, it should be replaced regularly.

STEM PUNCTURE

Skin puncturing occurs when the fruit skin is broken because it is pushed against the stem of a neighboring fruit (fig. 23.4). Methods of reducing puncturing have not been studied, but damage is likely to occur whenever fruit is rehandled in bulk. For example, some damage may occur in the picking bag as random fruit orientation occasionally causes a fruit to come in contact with the stem of another fruit. When the picking bag is dumped, fruit is repositioned, and there is another opportunity for fruit to be punctured by a stem. Minimizing handling should reduce stem puncturing. Cutting the stem near the fruit surface would also reduce this damage, but it is a costly (labor-intensive) solution to the problem.

TRANSPORT VIBRATION DAMAGE

This is sometimes called roller bruising because the vibrating fruit slowly turn, causing a dark blemish ring around the fruit. The flesh is not damaged, but the dark brown to black mark is a major cause of rejection at the market.

The damage occurs when fruit vibrate with an acceleration greater than the acceleration of gravity (1 g).

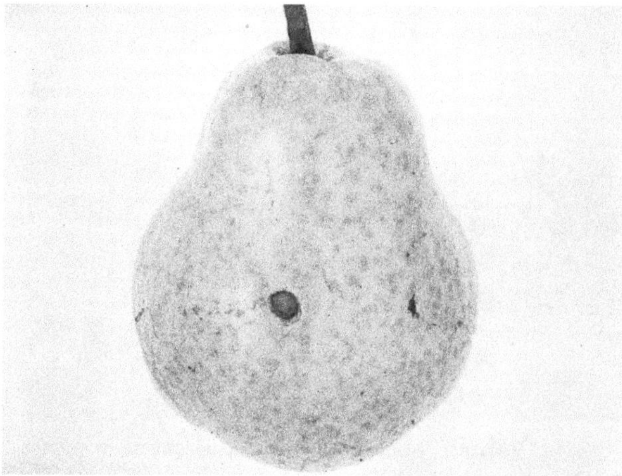

Figure 23.4. Stem puncture damage to a Bartlett pear. *Photo:* D. Edwards.

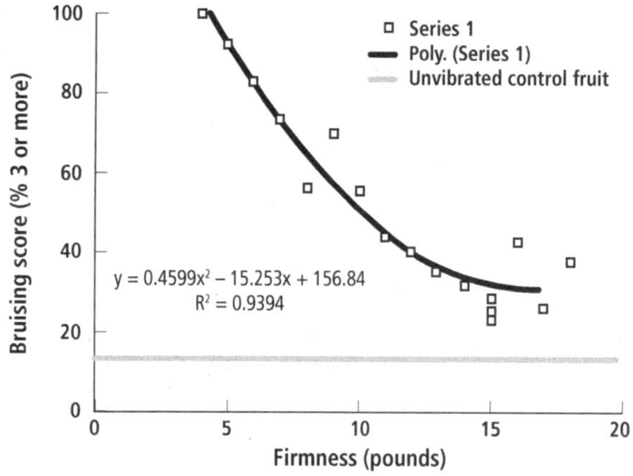

Figure 23.5. Effect of Bartlett pear firmness on susceptibility to vibration bruising. Pears were tight-fill packed in full-telescope boxes subjected to simulated long-distance highway transport. A score of 3 is enough damage to cause a consumer to consider not buying the fruit.

At this level fruit can move in relation to neighboring fruit, causing surface damage. Below 1 g the fruit mass vibrates as a whole, with no relative movement between individual fruit. Thirty minutes of exposure to 1 g acceleration can severely damage fruit. However, 60 hours of 0.5 g acceleration causes no damage. California Bartlett pears vary in their susceptibility to vibration bruising. Fruit harvested early or in the middle of the season for a particular growing region are not very susceptible. Late-season fruit are more susceptible, and stored fruit is very susceptible to damage (table 23.1). Ripe fruit is much more susceptible to vibration bruising than firm fruit (fig. 23.5). This type of damage can be eliminated by ensuring that vibration never exceeds 1 g, or by using packaging that prevents fruit from moving against neighboring fruit or packaging material.

Table 23.1. Laboratory evaluation of the effect of harvest date and growing region on the susceptibility of Bartlett pears to vibration injury

Pears with moderate* or greater damage (%)	Harvest date	Relative harvest period in region	Growing region
23	9/20/93	late	Mendocino
11	7/28/93	late	River
5	7/21/93	mid	River
3	7/14/93	early	River
4	8/25/93	mid	Mendocino
1	8/11/93	early	Mendocino
1	Nonvibrated fruit from each harvest		

Note: *Moderate damage is severe enough that a consumer might reject the fruit.

Table 23.2. Laboratory simulation of vibration damage to Bartlett pears in a cross-country highway shipment

Internal packing method	Pears with moderate* or greater damage (%)
3-lb polyethylene bags	12
5-lb polyethylene bags	19
Paper wrapped	25
Tight-fill	60

Note: *Moderate damage is severe enough that a consumer might reject the fruit.

Vibration damage can be managed by minimizing forklift transfer of fruit in the orchard, grading orchard roads, and selecting routes from the orchard to the packinghouse that have smooth pavement. Plywood bin covers can be used to immobilize fruit and prevent damage during longer trips to the packinghouse. The cover fits inside a bin and is lined on the underside with a foam pad. Stiff rubber cords at each corner pull the cover down tight against the fruit. Air ride suspension reduces the vibration reaching the bed of a truck and is very effective in reducing vibration damage.

Air ride suspension systems have also proven to dramatically reduce damaging vibration in highway vehicles. Air ride equipment costs more than steel spring suspensions, but it is becoming commonly available on trailers. It also increases tire life and may reduce vibration damage to the trailer itself. Nearly all long-haul tractors have air ride suspension for driver comfort, and semi trailers usually have air ride suspension for the axle under the front.

Vibration is greatest directly over axles, and vibration acceleration increases as it is transmitted through boxes on a pallet. If damage occurs, it is usually to fruit in top boxes near the rear of a highway trailer. Keeping fruit that is susceptible to damage (late-harvested or stored fruit) away from the rear of a trailer with steel spring suspension greatly reduces vibration damage. For more information, see Thompson, Brecht, and Hinsch 2002.

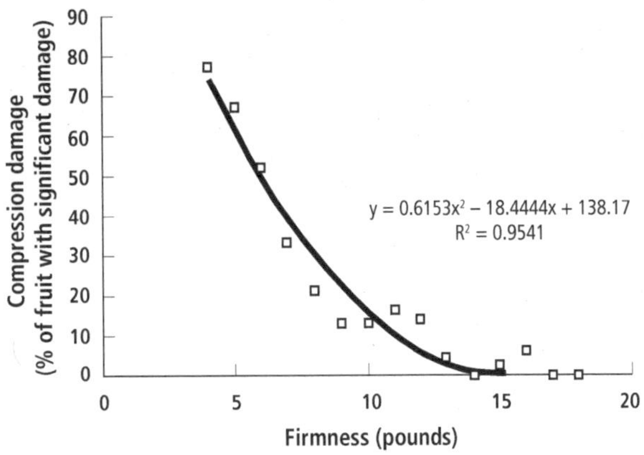

Figure 23.6. Effect of Bartlett pear firmness on susceptibility to compression bruising. Pears were tight-fill packed in full-telescope boxes subjected to simulated long-distance highway transport.

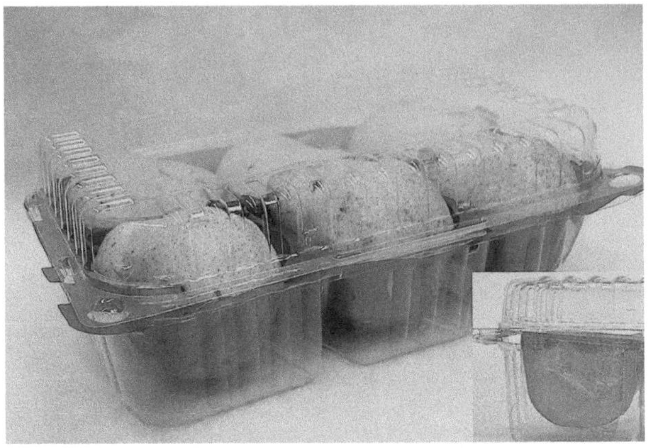

Figure 23.7. Suspended tray clamshell package for Bartlett pears (Hammock pack). *Photo:* J. Thompson.

The choice of packing method can reduce vibration damage. Tight-fill packing was developed in the late 1960s as a method of immobilizing fruit during transit. Corrugated fiberboard boxes are slightly over-filled and passed through a vibrator-sealer that settles the fruit and closes the box. Individual fruit cannot move in a properly packed tight-fill box. Proper tight-fill packing requires that

- boxes are formed to a consistent size.
- a consistent volume of fruit is packed in each box. The fruit weight in the boxes should not vary by more than ¼ pound (113 g). Net fruit weight increases as soluble solids concentration increases during the season, so the average fruit weight in the box must increase to keep a constant fruit volume.
- an excelsior (wood fiber) top pad is used that maintains a slight pressure on the pack even as fruit loses weight and boxes deform because of high humidity.
- the box is tightly closed with staples or bands.

Proper tight-fill packing has proven difficult to accomplish in commercial practice. Recent studies have shown that packing pears in polyethylene bags protects pears from vibration damage. Table 23.2 shows that pears shipped in the rear of a highway trailer with steel spring suspension are subjected to much less damage if they are packaged in loosely filled polyethylene bags placed in a conventional, full-telescope, corrugated fiberboard box. Bartlett pears packed in single-layer plastic trays are also less susceptible to vibration bruising.

A newly developed suspended tray package known as the Hammock pack almost completely protects ripe Bartlett pears from bruising damage (fig. 23.7). The packaging can be designed for a consumer clamshell package or for a larger corrugated master container for food service applications.

COMPRESSION BRUISING

Slowly squeezing pears causes flesh browning that starts at the surface and can extend deep within the fruit. It is usually caused by overfilling boxes, causing fruit to support the boxes stacked above, rather than allowing the box to provide this support. Pears become much more susceptible to compression bruising as their firmness drops during ripening (fig. 23.6). Failure of corrugated fiberboard boxes is another common cause of compression damage. Boxes lose strength because of

- improper assembly
- cross stacking on a pallet
- placing too much weight on boxes by double-stacking pallets
- gaining moisture during exposure to high relative humidity

Special treatments can be used during fiberboard manufacture to slow the rate of moisture gain, but even these will eventually fail under prolonged exposure to high relative humidity. Long-term storage is better done in plastic or wood containers.

REFERENCES

Thompson, J. F., P. E. Brecht, and T. Hinsch. 2002. Refrigerated trailer transport of perishable products. Oakland: University of California Agriculture and Natural Resources Publication 21614.

24

Postharvest Handling of Winter Pears

DAVID SUGAR

Winter pears are varieties that are usually harvested in the fall, require postharvest chilling in order to ripen properly, and have sufficient storage life to be marketed during the winter months. Presently, the leading winter pear cultivars produced in the United States are Anjou, Bosc, and Comice, while Packham's Triumph is the leading winter pear cultivar of the Southern Hemisphere. Winter pears can usually be maintained with good quality in storage for a minimum of 4 months and may keep up to 10 months, depending on the cultivar and storage conditions. Improvements in the technology of refrigeration, use of controlled atmosphere (CA) (see chapter 25, "Controlled Atmospheres"), and controlling diseases and disorders have allowed the rapid expansion of winter pear production in recent years. However, as increasing volumes of winter pears are harvested and the limits of storability are pressed, difficulties in maintaining fruit free from decay and physiological disorders have been experienced. In order to successfully maintain the quality of winter pears in long-term storage, it is critical that harvest and handling techniques focus on the unique requirements of winter pears and on factors that influence storability.

POSTHARVEST CHILL REQUIREMENT FOR RIPENING

Postharvest chilling induces the capacity to ripen in winter pears, and it promotes uniform ripening throughout the flesh of individual fruit as well as synchronizes ripening among fruit in a lot. The amount of postharvest chilling required differs among winter pear cultivars and is influenced by fruit maturity. Late-harvested Bosc pears may ripen to a buttery texture at room temperature without exposure to cold, but the uniformity of ripening is improved with some chilling. Bosc harvested in the early half of the maturity range may need 2 to 4 weeks of postharvest chill for proper ripening. Anjou pears may need from 60 days of chilling when harvested early to 30 days when harvested late in the maturity range. Comice pears typically require 30 days of chilling.

During cold storage, pears develop the capacity for producing ethylene at a sufficient rate to activate the ripening process when exposed to warmer temperatures. External treatment of insufficiently chilled pears with 100 ppm ethylene gas for 2 to 3 days at room temperature can reduce or replace the chilling requirement for stimulating uniform ripening. The precise length of chilling required also varies with the internal condition of the fruit. For example, fruit low in calcium require less postharvest chilling than fruit higher in calcium. Since high calcium content may retard fruit maturity relative to low-calcium pears, it may be that the effect of calcium on fruit maturity influences the fruit requirement for postharvest chilling. Other cultural factors that affect fruit maturity can likewise be expected to indirectly affect the amount of postharvest chilling required for proper ripening.

FRUIT MATURITY

Proper management of fruit maturity is probably the single most important factor affecting the storage quality of winter pears that can be manipulated by the grower. Fruit harvested early in the maturity range have the greatest potential to maintain good quality in long-term storage. The later fruit are harvested, the faster quality deteriorates during storage and the greater the fruit susceptibility to fungal decay during storage. Flesh texture is generally coarser in later-harvested pears. However, earlier-harvested pears have a greater susceptibility to shrivel, scald, and friction discoloration. Practices that reduce incidence of these disorders are described in subsequent sections of this chapter.

The firmness test is currently the best indicator of harvest timing for winter pears and has proven to be a reasonable indicator of the time to begin harvest (see chapter 22, "Postharvest Biology and Technology"). However, several factors may affect the continued decline of fruit firmness on the tree, and the value of firmness test data diminishes the longer that fruit are on the tree. A useful method is to begin harvest when the firmness test indicates the top of the range for a given variety in the growing district, and then set a goal of completing harvest within 10 to 14 days of that point. Under high temperature conditions during maturity and harvest, Anjou fruit tend to deteriorate in quality more rapidly than they do under cool conditions, and the harvest period should be shortened accordingly. In contrast, Bosc is relatively heat-tolerant and can sustain fruit quality during hot periods prior to and during harvest. In general, late-harvested fruit should be clearly identified as having diminished storage potential and sold as quickly as possible after harvest and chilling. Controlled atmosphere storage cannot "repair" the condition of late-harvested fruit; only top-quality fruit harvested early in the maturity range can dependably be kept in long-term CA.

Most winter pear varieties show a change in the surface color when mature. While much less pronounced than in Bartlett, winter pears generally increase in yellow and decrease in green at the onset of maturity on the tree. The waxiness of the fruit surface also becomes more apparent, especially in Anjou.

Tree age and vigor can affect fruit maturity. Generally, fruit on young or vigorously growing trees are slower to mature than fruit on older or low-vigor trees. In order to accurately identify the storage potential of fruit lots from orchards with interplanted trees, it may be advisable to harvest fruit from older or less-vigorous trees separately from the fruit from younger or more-vigorous trees.

STORAGE CONDITIONS

Fruit begin the process of aging and quality loss the moment they are removed from the tree. This physiological process is greatly slowed by low temperatures and by limiting the oxygen supply to the fruit, both of which lower the rate of fruit respiration. Accordingly, the storage life of fruit is maximized by cooling fruit as soon as possible after harvest, maintaining a low temperature during packing and storage, and, if CA is used, establishing a low-oxygen environment as promptly as possible after cooling.

Fruit may be safely maintained at temperatures slightly above their freezing point. The minimum temperature that the fruit can tolerate should be consistently maintained during the storage period. Typically, winter pears are stored at temperatures between 29.5° and 32°F (−1.4 and 0°C). The actual freezing temperature is influenced by the soluble solids content of the fruit; at soluble solids levels of 10 to 12%, the freezing point would be approximately 28.9° to 28.0°F (−1.7 to −2.2°C).

Winter pears are susceptible to moisture loss and consequent shrivel during long-term storage. This is particularly true of Bosc, where the narrow neck and long, thin stem show shrivel before the thicker portions of the fruit. Shrivel is largely prevented by maintaining relative humidity levels of 90 to 95 percent within the storage container. Fruit stored in polyethylene box liners perforated with ⅛-inch (3.2-mm) holes are generally free of shrivel. However, fruit stored in bins are at risk of shrivel, particularly in storage rooms where large amounts of moisture are drawn from the air by refrigeration units.

SUPERFICIAL SCALD

The Anjou pear is susceptible to a skin browning known as superficial scald, which is distinct from surface browning of several varieties as they near the end of their storage life (senescent scald). The condition occurs as certain constituents of the peel change over time through contact with oxygen. Earlier-harvested Anjou pears are more susceptible to superficial scald than later-harvested fruit. Warm temperatures before harvest also promote scald development. Low-oxygen storage slows the appearance of the disorder.

Treatment of the fruit prior to storage with the antioxidant compound ethoxyquin, either as a drench, line-spray, or in treated paper wraps, has been highly effective in preventing superficial scald in Anjou pears. The current recommendation for bulk-stored Anjou pears in the Pacific Northwest is for 2,700 ppm of ethoxyquin, which may be applied as two treatments, bin drench and line spray, each delivering one-half of the dosage. In the Southern Hemisphere, diphenylamine (DPA) is often applied postharvest to prevent scald development in Packham's Triumph pears. Neither of these materials is registered for use on pears in California. Alternative antioxidant compounds and low-oxygen storage strategies for scald prevention are currently being investigated.

FRICTION DISCOLORATION

The tender-skinned winter pear varieties Anjou and Comice are subject to surface marking from contact between the fruit and postharvest handling equipment. This is known as friction discoloration, or belt burn (see chapter 23, "Reducing Fruit Injury during Postharvest Handling and Transportation"). The problem is reduced by use of equipment that handles the fruit slowly with a minimum of friction, but it is difficult to avoid completely. Anjou pears become increasingly susceptible as the duration of storage increases. Anjou pears held in bins for long-term storage then packed for shipment during the winter may suffer from this problem; fruit packed within approximately 10 to 12 weeks of harvest are rarely affected. The problem is most pronounced in fruit harvested early and following relatively cool growing seasons. Pears handled while cold are less prone to discoloration than those that have warmed during packing.

Comice pears are highly prone to show surface marks where contact was made with branches or neighboring fruit on the tree, or from friction during postharvest handling, as with Anjou. Some packers have indicated that sucrose ester coatings applied to fruit as a line spray give protection from friction discoloration occurring further down the packing line. Sucrose ester coatings, which are currently available as commercial products, may also prolong the shelf life of the fruit. Proper rates must be used to avoid adversely affecting the ability of the fruit to ripen.

DECAY CONTROL

Since winter pears are usually held in storage for long periods of time, they are vulnerable to various types of fungal decay. Decay is rarely a problem in fruit stored for less than 3 months, but it becomes an increasing risk with longer storage. Bosc pears are especially prone to postharvest decay, both because of inherent susceptibility and because their shape makes smooth movement difficult and their long, woody stems easily puncture the flesh of neighboring fruit during bulk handling. Comice pears are also easily punctured by jagged stems that result when the fruit do not cleanly separate from spurs at the abscission layer. In addition to decay that develops at small punctures, Anjou pears may decay through the fleshy stems, which remain susceptible to fungus penetration long after harvest.

The most common decays that develop in winter pears can be grouped into three types: blue mold and gray mold, side rot, and bull's eye rot (see chapter 22, "Postharvest Biology and Technology").

Blue mold (*Penicillium expansum*) and gray mold (*Botrytis cinerea*) are rapidly developing decays that are problematic worldwide. Spores develop on decaying organic matter in the orchard and are common in air and soil. The currently registered fungicide thiabendazole (TBZ) is primarily targeted at these diseases.

Side rot is caused by several slow-growing fungi, including *Phialophora malorum*, *Cladosporium herbarum*, and *Alternaria* spp. They typically develop after 4 or more months of storage, when fruit have begun to "weaken" with age. Although *Cladosporium* is sensitive to TBZ, fungicide treatment generally gives little control of side rot. However, this group of diseases responds well to factors that promote keeping quality in the fruit: early harvest, calcium sprays in the orchard, reduction of tree vigor and nitrogen fertilization, and CA storage. The spores of these fungi are brought into the packinghouse in soil on bin skids, with debris, and on the fruit surfaces.

Bull's-eye rot (*Neofabraea* spp.) begins with fruit infection in the orchard following spore dispersal with rain or over-tree irrigation. The fungus grows beneath the fruit cuticle and is unreachable by fungicide treatments after infection, even though the rot does not appear until 4 or more months after harvest. Control is based on applications of fungicide in the orchard during the petal fall–first cover spray period, and immediately before or after preharvest rains. When specific orchards are known to be infected with the bull's-eye rot fungus, fruit from them can be targeted for early sales.

CONTROLLED ATMOSPHERE STORAGE

Several quality aspects of winter pears are better maintained in controlled atmosphere (CA) storage than in normal air (see chapter 25, "Controlled Atmospheres"). Appropriate oxygen and carbon dioxide levels vary with growing district and season and among different pear cultivars. Generally, the lowest oxygen level and highest carbon dioxide level that the fruit will tolerate without adverse effects is desirable. Carbon dioxide injury is influenced by oxygen level: as the oxygen level drops below 5 percent, the risk of injury from carbon dioxide levels above 1 percent increases. Typical firmness values

Table 24.1. Harvest firmness, storage life, and CA conditions for selected winter pear cultivars

Cultivar	Harvest flesh firmness* (lb)	Typical maximum storage life* (mo)	
		Air	CA†
Abate Fetel	13–15	3–4	4–6
Anjou	13–15	6–7	8–10
Bosc	14–16	4	5–7
Comice	11–13	4	5–6
Concorde	11–13	4–5	6–7
Conference	11–13	5–6	7–8
Forelle	13–15	4–5	6–7
Packham	14–16	6	7–8
Winter Nelis	13–15	7–8	8–10

Notes: *Data have been collected and generalized from various sources. Appropriate values for specific growing regions and conditions may vary from those presented.

†Refer to chapter 25, "Controlled Atmospheres," for recommended CA conditions.

indicating the range of harvestable maturity for winter pear varieties and expected storage life in regular air storage and in CA are listed in table 24.1.

Fruit selected for long-term storage in CA should have all possible advantages to enhance keeping quality. Fruit harvested at the top of the maturity range from orchards with good storage histories have the greatest potential for maintaining good quality in CA. Although some pathogens of the side rot type are suppressed by CA, most postharvest diseases are relatively unaffected. Fruit to be stored in CA benefit from being selected for freedom from punctures and treated to minimize decay risk.

25 Controlled Atmospheres

ADEL A. KADER

Pears are best enjoyed by consumers when harvested at the mature-green stage and ripened after harvest to optimal eating quality. Between harvest and ripening initiation, pears can be stored for various durations to extend their marketing season. Maintaining the fruit at 30° to 32°F (–1.1° to 0°C) and a relative humidity of 90 to 95 percent is the most important tool for preserving the quality and extending the postharvest life of pears.

An important supplemental treatment is atmospheric modification, that is, lowering the oxygen concentration, elevating the carbon dioxide concentration, and/or removing ethylene from the storage environment. This is known as modified atmosphere (MA) or controlled atmosphere (CA) storage and can be used during transport or storage. In general, CA requires greater precision in control of oxygen and carbon dioxide concentrations than MA. Modified atmosphere packaging (MAP) can provide the desired atmospheric composition during postharvest handling and marketing of pears.

Pear cultivars differ widely in the length of time they can be stored while maintaining the ability to ripen and attain good flavor and texture. These differences exist in air storage and in CA storage. Generally it is possible to extend the postharvest life of pear cultivars in CA by about 50 to 100 percent relative to air storage at the same temperature and relative humidity. For example, the storage potential of Bartlett pears is up to 3 months in air and up to 6 months in CA (2% oxygen + 0.5% carbon dioxide) (plate 25.1). Longer storage life may be possible on the basis of appearance quality, but not on the basis of flavor and textural quality when ripe. Beyond about 2 months of storage, CA-stored pears usually have better texture and flavor upon ripening than do air-stored pears. Winter pear cultivars can be stored in CA at 30°F (–1.1°C) for up to 4 months (Bosc and Comice) or 6 months (Anjou) while maintaining their capacity to ripen and retain good texture and flavor. These durations are shorter than those included in table 25.1 because they are based on flavor quality while those in the table are based on appearance and textural quality.

The optimal CA range for most pear cultivars is 1 to 3 percent oxygen and 0 to 3 percent carbon dioxide at 30° to 32°F (–1.1 to 0°C) and 90 to 95% relative humidity (table 25.1). The lower oxygen levels (1 percent or lower) must be combined with scrubbing of carbon dioxide to levels below 0.5 percent to minimize

Table 25.1. Optimal ranges of CA conditions and storage life for major cultivars of pears kept at 32°F ± 2°F (0°C ± 1°C) and 90–95% relative humidity

Cultivar*	O_2 (%)	CO_2 (%)	Storage life (months)
Abate Fetel	1.0–2.0	0–1.0	4–6
Anjou	1.0–2.5	0–0.5	8–10
Bartlett	1.0–2.0	0–0.5†	3–6
Bosc	1.0–2.5	0.5–1.5	5–7
Comice	1.5–4.0	0.5–4.0	5–6
Concorde	0.5–2.0	0–1.0	6–7
Conference	1.0–2.5	0.5–1.5	7–8
Forelle	1.0–2.0	0–1.5	6–7
Packham's Triumph	1.5–2.0	1.5–2.5	7–8
Winter Nelis	1.0–3.0	0–1.0	8–10

Notes: *For information about other cultivars, see Richardson and Kupferman 1997.
†Early- to mid-season California fruit can be stored with up to 5% CO_2.

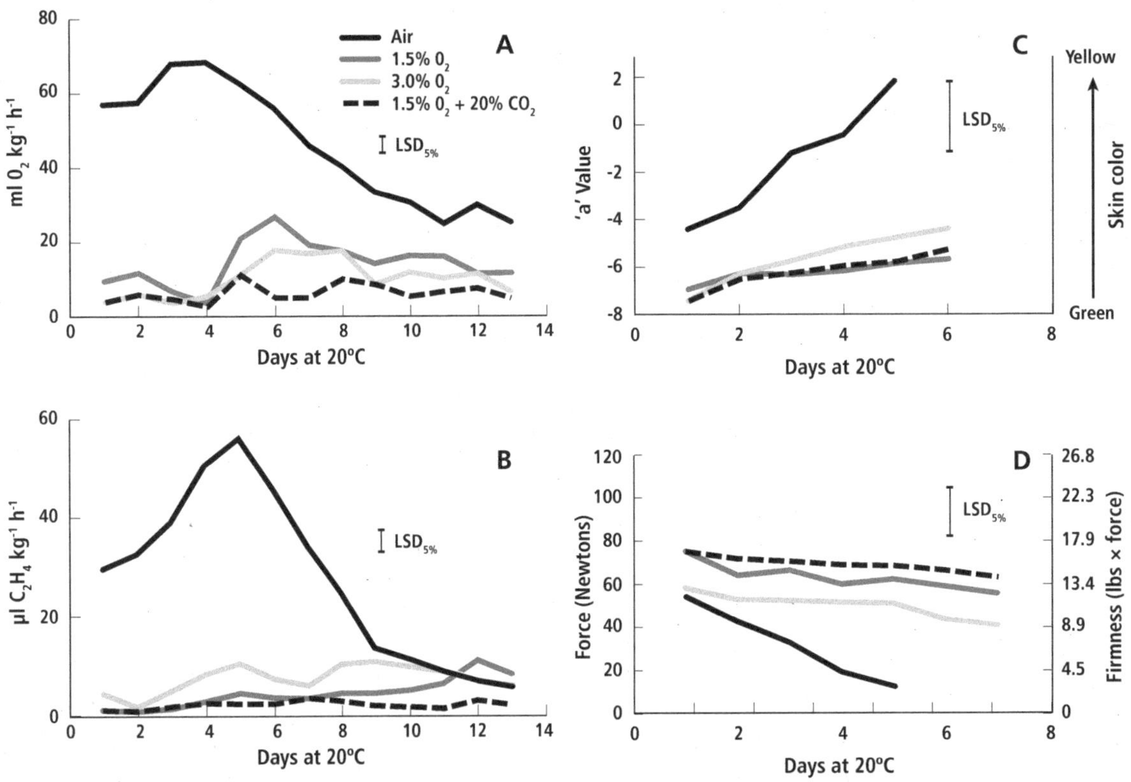

Figure 25.1. Effects of CA on rates of respiration (oxygen uptake in ml/kg·hr) (A), ethylene production (µlC_2H_4/kg·hr) (B), skin color changes (lower value = more green) (C), and flesh softening (penetration force) (D) of Bartlett pears kept at 68°F (20°C) and 90–95% relative humidity. *Source:* Ong 1987, pp. 22, 27, 31, 33.

physiological disorders. For California-grown Bartlett pears, the optimal CA condition for long-term storage of early- and mid-season-harvested fruit is 1.5 to 2 percent oxygen and 1 to 5 percent carbon dioxide for up to 6 months at 30°F (–1.1°C). For late-season California-grown Bartlett pears, carbon dioxide should be kept below 1 percent because of their greater sensitivity to carbon dioxide injury (core and flesh browning), especially after 3 months of storage. Bartlett pears grown in other regions of the United States do not tolerate carbon dioxide levels above 1% regardless of harvest date.

Reduced oxygen levels (1 to 3%) slow the rates of respiration, ethylene production, color change from green to yellow, conversion of starch to sugar, softening, and loss of acidity in pears (fig. 25.1). Elevated carbon dioxide levels (up to 5%), if tolerated, as in the case of early- and mid-season California Bartlett pears, have a synergistic effect with the reduced oxygen levels in delaying senescence. Controlled atmosphere conditions render pears less sensitive to ethylene action and also suppress scald development and decay incidence and severity.

The presence of ethylene in storage environments of Bartlett pears advances ripening, as indicated by loss of green color, flesh softening, and increased rates of respiration and ethylene production. However, these effects are greatly reduced as fruit temperature is lowered to 30°F (–1.1°C) under CA conditions. Removal of ethylene from CA storage may not be cost-effective since most studies have shown no advantage to ethylene removal unless the concentration is maintained below 1 ppm, which is very difficult to achieve.

Controlled atmosphere conditions can influence postharvest decay-causing pathogens such as *Botrytis cinerea* and *Penicillium expansum* directly by inhibiting their spore germination and slowing their growth or indirectly by retarding fruit senescence and maintaining fruit health. Generally, oxygen levels below 1 percent or carbon dioxide levels above 10 percent are needed to significantly suppress fungal growth. Pears will not tolerate these atmospheres during long-term storage, but these fungistatic atmospheres can be used during transport or short-term storage of less than 1 month at 30° to 41°F (–1.1° to 5°C).

Insecticidal CA (below 1 percent oxygen or 30 to 60 percent carbon dioxide) can be tolerated by Bartlett pears for up to 10 days at temperatures between 32° and 41°F (0° and 5°C) or up to 4 days at 50°F (10°C). Such atmospheres, alone or in combination with heat treatments, have been shown to kill some insects and may have potential as quarantine treatments.

Exposure of pears to unfavorable atmospheric composition (levels of oxygen below or levels of carbon dioxide above those tolerated by the pears at a given temperature and exposure time) can induce physiological disorders and increase the incidence and severity of pathological diseases. The severity of the effect increases with lower oxygen levels, higher carbon dioxide levels, higher temperatures, and longer storage duration. Pears stressed by CA become unable to ripen upon removal from CA conditions, exhibit flesh browning, develop cavities in damaged tissues (plate 25.2), and undergo anaerobic respiration that cause the accumulation of acetaldehyde, ethanol, and ethyl acetate, and the development of off-flavors. The oxygen concentration at which aerobic respiration of mature-green Bartlett pears shifts to anaerobic respiration is about 0.3 percent oxygen at 32°F (0°C); it increases with higher temperatures. Symptoms of carbon dioxide injury include browning of interior walls of carpels and adjacent core tissue. In severe cases, cortex tissue turns light brown and cavities may develop in damaged tissue. The susceptibility of pears to carbon dioxide injury increases with advanced maturity, delayed cooling, prolonged storage, and low oxygen concentrations in the storage atmosphere. More-mature and partially ripe pears are more susceptible to low oxygen injury (core browning).

Refrigerated, gas-tight rooms are required for CA storage of pears. Various devices are used to compensate for fluctuations in atmospheric pressure within the CA rooms. Oxygen concentrations are reduced by flushing the room with nitrogen gas that is generated from liquid nitrogen or from compressed air using a membrane or molecular sieve system to separate the nitrogen from the oxygen and carbon dioxide. Carbon dioxide scrubbing can be done using lime, charcoal, or molecular sieve absorbers. Concentrations of oxygen and carbon dioxide should be monitored and adjusted as needed using automated systems to assure that the optimal atmospheric composition is maintained.

For further information on CA and MA, see Kader 2002.

REFERENCES

Blanpied, G. D. 1990. Controlled atmosphere storage of apples and pears. In M. Calderon and R. Barkai-Golan, eds., Food preservation by modified atmospheres. Boca Raton: CRC Press. 265–299.

Carlson, S. A. 1984. Effect of controlled atmospheres on 'Bartlett' pear storage and handling. MS thesis, Department of Pomology, University of California, Davis.

Chavez-Franco, S. H., and A. A. Kader. 1993. Effects of CO_2 on ethylene biosynthesis on 'Bartlett' pears. Postharvest Biology and Technology 3:183–190.

Chen, P. M., R. A. Spotts, D. M. Varga, and L. A. Cervantes. 1995. Ripening behavior and combined fungicide and prestorage heat effects on decay control of 'Bosc' pears in air or step-wise low oxygen storage. Postharvest Biology and Technology 6:235–248.

Claypool, L. L. 1973. Further studies on controlled atmosphere storage of 'Bartlett' pears. Journal of the American Society for Horticultural Science 98:289–293.

Kader, A. A. 1989. Mode of action of oxygen and carbon dioxide on postharvest physiology of 'Bartlett' pears. Acta Horticulturae 258:161–167.

——. 2002. Modified atmospheres during transport and storage. In A. A. Kader, ed., Postharvest technology of horticultural crops. 3rd ed. Oakland: University of California Division of Agriculture and Natural Resources Publication 3311. 135–144.

Ke, D., H. Van Gorsel, and A. A. Kader. 1990. Physiological and quality responses of 'Bartlett' pears to reduced O_2 and enhanced CO_2 levels and storage temperature. Journal of the American Society for Horticultural Science 115:435–439.

Meheriuk, M., and P. Sholberg. 1990. Postharvest treatments of pears. Postharvest News and Information 1:441–446.

Ong, B. T. 1987. The effect of controlled atmospheres on phenolic metabolism in 'Bartlett' pears. MS thesis, Department of Pomology, University of California, Davis.

Richardson, D. G., and E. Kupferman. 1997. Controlled atmosphere storage of pears. In E. J. Mitcham, ed., CA '97 Proceedings, vol. 2: Apples and pears. Postharvest Horticulture Series No. 16. University of California, Davis, Department of Pomology. 31–35.

26

Conditioning and Ripening of Bartlett Pears

ELIZABETH J. MITCHAM AND F. GORDON MITCHELL

CONDITIONING PEARS FOR UNIFORM RIPENING

European pears do not ripen properly without a period of cold storage to condition the fruit. For Bartlett pears grown in California, this is particularly true of fruit harvested in the first week or two of the season from each growing location, that is, Bartletts harvested at generally greater than 15 pounds-force (lbf) (7.5 kgf) firmness. Unconditioned fruit cannot produce much ethylene and tend to ripen slowly and nonuniformly (fig. 26.1). In addition, they tend to have poor texture and flavor compared with fruit conditioned in cold storage.

During cold storage, ethylene precursors are formed within the pear tissue so that ethylene production and uniform ripening occur when the fruit are placed at ripening temperatures. Bartlett pears require 2 to 3 weeks of cold storage at 30° to 32°F (–1.1° to 0°C) for the maximum effect in promoting uniform ripening. Without prior cold storage, Bartlett pears can remain hard and unripe for several weeks after transfer to ripening temperatures.

Ethylene treatment after harvest can substitute for cold storage conditioning (see fig. 26.1). During ethylene treatment, which has been called "quality conditioning" by the California pear industry, pears are exposed to ethylene gas immediately after harvest to ensure prompt, uniform ripening. Fruit are placed in air containing 100 ppm ethylene (although 10 ppm should be adequate) at 68° to 77°F (20° to 25°C) for 24 to 48 hours. It is important that the fruit be maintained at a uniform temperature and that the temperature not vary within the load of fruit or during the time in conditioning. Fruit response to ethylene is slower at lower temperatures: for example, it takes twice as long to condition fruit at 50°F (10°C) than at 68°F (20°C).

Conditioning does not ripen fruit, and the storage life of conditioned fruit is not affected if the fruit are promptly cooled to near 32°F (0°C) following treatment. Pears that receive conditioning followed by prompt cooling are not significantly softer than cold-stored fruit. Late-harvested, highly mature fruit that are already starting to soften should not be conditioned with ethylene. Also, there is no advantage to treating fruit with ethylene if the fruit will be subsequently cold-stored for 3 weeks or more, since the cold storage will substitute for ethylene treatment; further, the delay in cooling could exaggerate the development of core breakdown problems following prolonged storage (see chapter 22, "Postharvest Biology and Technology"). Postharvest treatment of pears with SmartFresh (1-methylcyclopropane, 1-MCP) at harvest, which inhibits the effects of ethylene gas, can be used to improve fruit quality after extended cold storage. However, it is important to be aware that this treatment could affect the fruit's ability to ripen, even after cold storage, despite exposure to warm temperatures and ethylene, if not properly applied.

PEAR RIPENING

Studies have shown that consumers prefer ripened pears by a margin of 3 to 1 over unripe pears. Fruit should be ripened at or near the retail market because partially ripe fruit are much more susceptible to bruising injury during transportation. Partially ripened fruit can be transported to distant markets, but softer pears will be much more susceptible to bruising. (See chapter 23, "Reducing Fruit Injury during Postharvest Handling and Transportation," for more details on the relationship between fruit firmness and bruising susceptibility.) Store personnel should be trained to handle partially ripened fruit carefully. Fresh fruit are best ripened in the original shipping container; cannery fruit are best ripened in highly ventilated plastic field bins. The fresh fruit container should be well ventilated and without a plastic liner. Ethylene gas at 100 ppm should be provided during the first 24 to 48 hours of ripening unless the fruit has already been conditioned with cold storage or ethylene treatment at harvest (quality conditioned; see above). Without ethylene or cold storage, pears ripen very slowly and nonuniformly, and the eating quality may be poor.

Rates of pear ripening at various temperatures are shown in figure 26.2. Temperatures can be adjusted

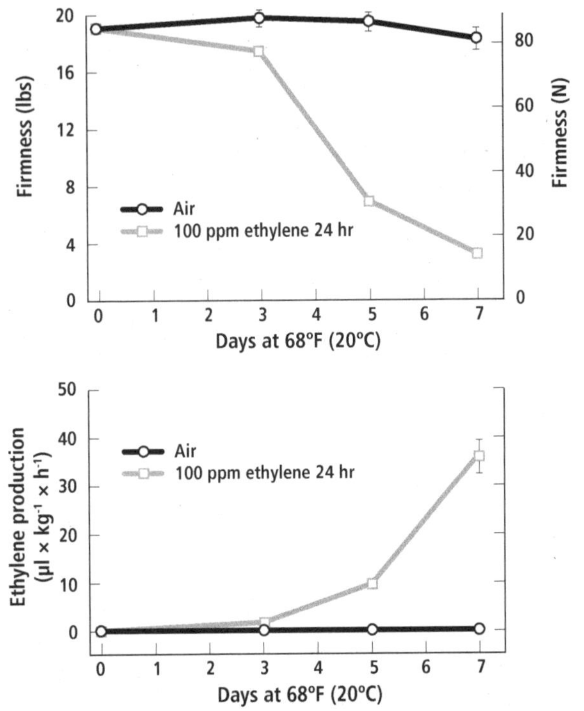

Figure 26.1. Changes in firmness and ethylene production of Bartlett pears ripened at 68°F (20°C). Fruit were treated with 100 ppm ethylene or air at 68°F (20°C) for the first 24 hours of ripening.

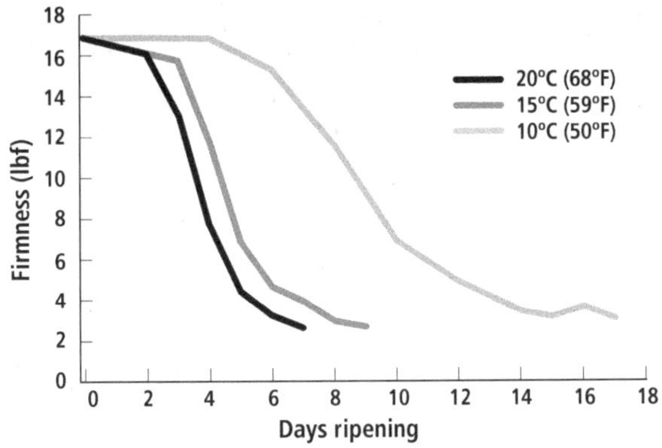

Figure 26.2. Rate of softening of Bartlett pears at 68°F (20°C), 59°F (15°C) and 50°F (10°C). Fruit were treated with 100 ppm ethylene at 68°F (20°C) for the first 24 hours of ripening.

slightly to change the speed of ripening to better meet marketing or processing windows. Temperatures between 64° and 72°F (17.8° and 22.2°C) provide for optimal ripened quality; the temperature must be 55°F (12.8°C) or greater for adequate ripening. At temperatures above 77°F (25°C), full yellow color and juiciness may not develop. For maximum ripening uniformity, fruit must be uniformly warmed. Positive air circulation using fans or blowers should be maintained until fruit in the center of the load are warm. Pressurized ripening rooms are ideal for this purpose. Otherwise, boxes can be spread on pallets to warm more quickly. Temperature gradients in large ripening rooms of cannery pears can result in considerable variability in fruit ripeness. Adequate air circulation and steady room temperatures yield more uniform fruit temperatures and ripening rates. The relative humidity during ripening should be 90 to 95 percent to prevent water loss and fruit shrivel.

The best eating quality is achieved for fresh pears at 3 pounds-force (1.5 kgf) or less, although some consumers prefer slightly crunchy pears at 4 to 6 pounds-force (2 to 3 kgf) (for more information on firmness testing, see chapter 22, "Postharvest Biology and Technology"). Because partially ripened fruit is very susceptible to mechanical injury, it is recommended that fruit for retail display be ripened to no less than 8 to 10 pounds-force (4 to 5 kgf). For canning, pears are processed at varying levels of firmness from 5 to 8 pounds-force (2.5 to 4 kgf), depending on the final product.

A penetrometer-type firmness tester is needed to accurately monitor ripening changes, as skin color is not always well correlated with fruit ripeness. Firmness should be measured on a minimum of 15 fruit collected from throughout the load at the start of ripening and twice a day thereafter. Nondestructive firmness testers for pears are also available. They may reduce the need to destructively sample fruit for firmness measurements and allow for constant monitoring of fruit firmness during ripening. Fruit firmness can change rapidly during the ripening period (see fig. 26.2). At 8 to 10 pounds-force (4 to 5 kgf), the color change from green to yellow has begun in Bartlett pears, but the fruit still feels fairly hard and resists bruising better than softer fruit.

Once fruit are ripened to 8 to 10 pounds-force, cold-temperature storage only slows the ripening process. Therefore, ripening should coincide with anticipated sales as much as possible. Ripe fruit can be successfully held near 32°F (0°C) for only a few days.

Nevertheless, refrigerated retail display is recommended for partially ripened pears to slow further softening and extend shelf life.

REFERENCES

Agar, I. I., W. V. Biasi, and E. J. Mitcham. 1999. Exogenous ethylene accelerates ripening responses in Bartlett pears regardless of maturity or growing region. Postharvest Biology and Technology 17:67–78.

———. 2000a. Cold storage duration influences ethylene biosynthesis and ripening of 'Bartlett' pears. HortScience 35:687–690.

———. 2000b. Temperature and exposure time during ethylene conditioning affect ripening of Bartlett pears. Journal of Agricultural and Food Chemistry 48:165–170.

Chen, P. M., W. M. Mellenthin, and D. M. Borgic. 1983. Changes in ripening behavior of 'd'Anjou' pears (*Pyrus communis* L.) after cold storage. Scientia Horticulturae 21:137–146.

Hansen, E., and W. M. Mellenthin. 1979. Commercial handling and storage practices for winter pears. Oregon State University Agricultural Experiment Station Special Report 550.

Mitchell, F. G., G. Mayer, W. W. Coates, and R. J. Snyder. 1975. Breakdown and ripening of processing 'Bartlett' pears. Sacramento: Report to the California Tree Fruit Agreement and Canners League of California.

Puig, L., D. M. Varga, P. M. Chen, and E. A. Mielke. 1996. Synchronizing ripening in individual 'Bartlett' pears with ethylene. HortTechnology 6:24–27.

Glossary

abiotic disease. A disease caused by factors other than pathogens.
abscise (abscission). To fall or cut off; especially refers to a leaf or fruit.
adventitious. Arising or occurring sporadically from unexpected tissue origin.
antibiotic. A chemical, usually of microbial origin, that inhibits or kills other organisms.
apical bud. Floral or vegetative bud located at apex (tip) of a shoot.
apical dominance. Tendency for apical vegetative buds at or near the shoot apex to grow and develop while suppressing the growth of lower buds.
asexual. Vegetative; without sex organs, sex cells, or sexual spores, as in the imperfect stage of a fungus.
asexual propagation. Production of a new plant by vegetative means: budding, grafting, or rooting of cuttings.
axillary. Pertaining to or placed within an axil.
bacteria. Large, widely distributed group of typically one-celled organisms, chiefly parasitic or saprophytic.
biological control. Disease or pest control through counterbalance by microorganisms and other natural components of the environment.
blight. A disease characterized by general and rapid killing of leaves, flowers, and branches.
budding. Art of placing the bud of a desired variety into a rootstock or other variety.
calyx. Base of the flower enclosing the ovary.
chilling requirement. Number of hours cold temperatures (below 45°F, or 7.2°C) a plant requires in the dormant season to resume normal growth and fruit development the following spring.
chlamydospore. Thick-walled or double-walled asexual resting spore formed from hyphal cells (terminal or intercalary) or by transformation of conidial cells.
chlorosis (chlorotic). Yellowing of leaves.
compatibility (propagation). Ability of a bud or graft to unite with the rootstock to form a long-lasting productive tree.
controlled atmosphere (CA). Cold storage room in which an atmosphere of low oxygen and high carbon dioxide is maintained to slow fruit respiration and extend storage life.
deciduous. Plants whose leaves fall in autumn at the end of the growing season.
diapause. Period of physiologically controlled dormancy in insects.
dormancy. Condition in a flower or shoot bud (and insects) of inactivity or prolonged rest.
epiphytotic. General and serious outbreak of a plant disease.
ET. Evapotranspiration, or water loss, from soil evaporation and canopy transpiration.
evapotranspiration. The loss of water from soil by the combination of soil surface evaporation and transpiration by plants.
fertilization. The sexual union of two protoplasts resulting in doubling of chromosome numbers.
flagellum (pl. flagella). Hairlike, whiplike, or tinsel-like appendage of a motile cell (bacteria or zoospores) that provides locomotion.
fruitwood. Vegetative portion of a tree on which fruit is produced.
fungi. Organisms having no leaves, flowers, or chlorophyll and reproducing by means of spores.
fungicide. Chemical agent used to prevent or control fungus infections.
girdle. To kill, damage, or remove a ring of bark around a branch, trunk, or root.
graft. Form of asexual reproduction in which a scion of one variety is placed in contact with the cambium layer of another variety or rootstock until they grow together.
growth hormone (plant hormone). Any naturally occurring or synthetic chemical that is transported within the plant to a location where it evokes a response that regulates plant growth.
gypsum. Hydrated sulfate of calcium, $CaSO_4{\cdot}2H_2O$.
hardpan. Hard, impervious layer in the soil, usually located within 5 feet (1.5 m) of the surface.
hormone. Plant-synthesized growth regulator.
hyphae. The threads that make up the mycelium of a fungus.
internode. Area of stem between two nodes.
leaching. Process in which water dissolves minerals or salts from near the soil surface and moves them downward out of the root zone.
lesion. Localized spot of diseased tissue.
macronutrients. Chemical elements used in large quantities by plants for growth and development, such as nitrogen, potassium, and phosphorus.
meristem (meristematic). Plant tissue that functions principally in cell division and differentiation; a mass of growing cells capable of frequent cell division.

micronutrients. Chemical elements used in small quantities by plants for growth and development, such as zinc, manganese, and iron; toxic at high concentrations.

micropropagation. Propagation of plants in tissue culture.

mycelium. Mass of interwoven hyphae that forms the vegetative portion of a fungus and is often submerged in another substance or body.

nematode. Any of a class of soil parasitic worms.

node. Branch point on a stem where leaves and buds arise.

oospore. Thick-walled resting spore that develops from an oosphere produced by fertilization or parthenogenesis.

perennials. Plants that may live three or more seasons and flower at least twice.

pH. An expression of the hydrogen ion concentration (acidity or alkalinity) of a solution or soil solution. Neutral solutions have a pH of 7. Basic solutions have a pH greater than 7 and acid solutions have a pH less than 7. It is a logarithmic scale in which pH 6.0 has 10 times the hydrogen ion concentration of pH 7.0.

pheromone. Substance secreted by an organism to affect the behavior or development of a member of the same species. Sex pheromones attract the opposite sex for mating.

photosynthetic. Tissue capable of photosynthesis, the synthesis of carbohydrates from simple molecules in the presence of light by chlorophyll in the chloroplast.

phytotoxic. Causing injury to plants.

pollination. Transfer of pollen from an anther to a stigma.

postemergence herbicide. Chemicals that control weeds only after they have germinated and emerged above the soil surface.

preemergence herbicide. Chemicals that control weeds before or after germination but before they emerge above the soil surface.

propagule. Any part of an organism capable of independent growth.

radiation frost. Cold air temperature resulting from loss of heat from surrounding objects due to radiation.

rat-tail bloom. A secondary bloom in Bartlett pears that results when terminal buds form and open on the current season's growth.

rhizomorph. Threadlike or cordlike structure made of hyphae.

ripping. Breaking hardpan by pulling a vertical bar (shank) through the soil.

rootstock. A plant having certain desirable root characteristics to which other varieties are grafted or budded to produce a commercially acceptable compound tree; the lower portion of a grafted tree that makes up the crown and root system (see **scion**).

scaffold. Larger branch arising from the central portion or trunk of the tree.

scion. Plant with certain desirable fruit characteristics that is grafted to a rootstock; the cultivar or variety.

senesce (senescent, n. senescence). Decline, as with maturation, age, or disease stress.

senescent. Growing old; preparing to abscise.

sexual. Participating in or produced as a result of a union of nuclei in which genetic recombination occurs.

sporangia. Sacs or receptacles in which fungus are produced.

spore. A reproductive body produced by certain fungi and other organisms capable of growing into a new individual under proper conditions.

spur. Small, stubby branch on which fruit is born.

stomata. Small openings in leaves through which gas exchange takes place.

stone cells. Specialized cells in the fruit with very thick cell walls that can give a gritty texture to fruit.

temperature inversion. Condition in which cold air settles near the ground beneath warm air on a windless, clear night.

topwork. Changing the variety of an established tree by budding or grafting a new variety to its trunk or branches.

translocation. Movement of water, minerals, food, or pathogens within a plant.

transpiration (transpire). Loss of water vapor from aerial parts of plants, chiefly through stomata in the leaves.

turgid (turgidity). Swollen or plump as a result of internal water pressure.

volatile. Evaporating or vaporizing readily.

water potential. A measure of the tendency of water to move from one location to another within the plant.

xylem. Woody portion of a tree located inside the cambium, containing nonliving vascular tissues that conduct water and nutrients from the roots to the top of the tree.

zoospore. Swimming asexual spore.

Index

Tables, figures, and glossary entries are indicated with italic typeface, e.g., 145*t*, 163*f*, 184*g*.

PEAR PRODUCTION AND HANDLING MANUAL

COLOR PLATES

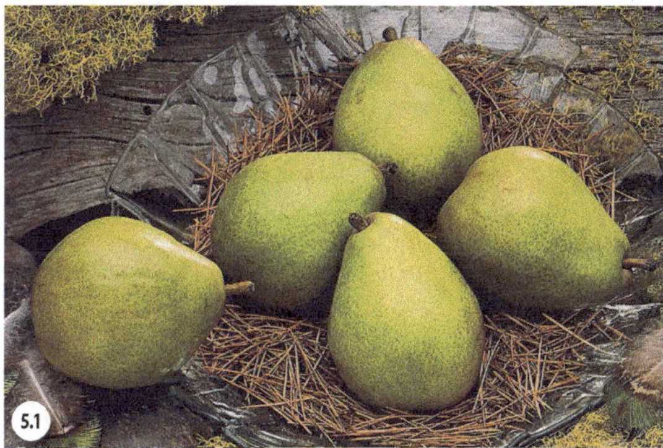

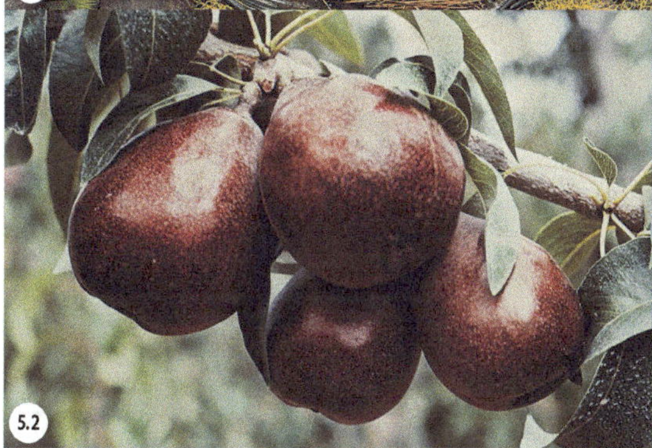

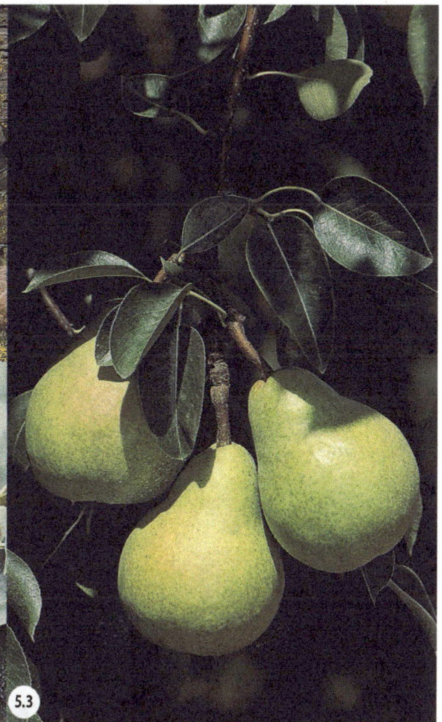

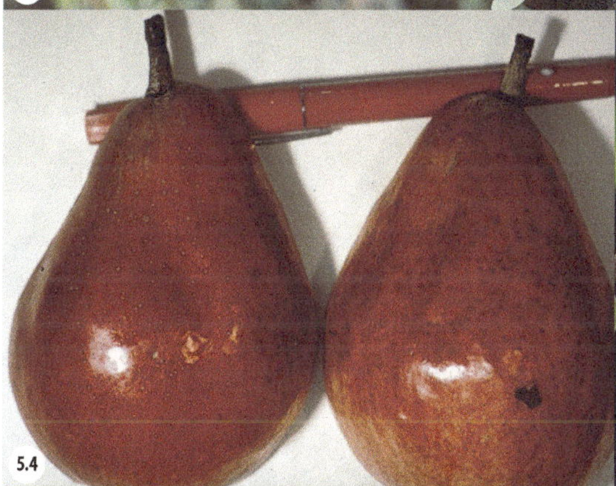

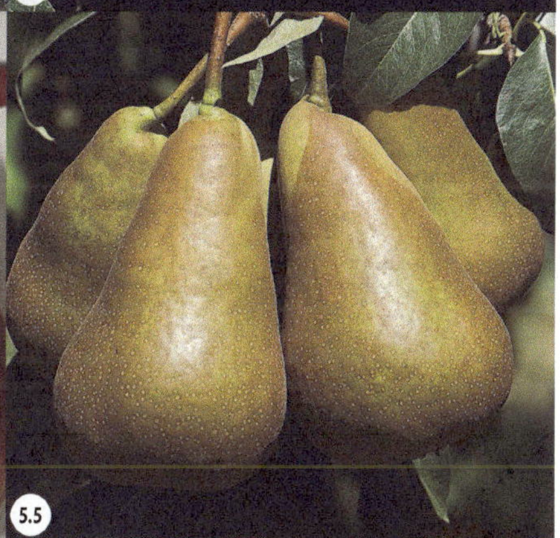

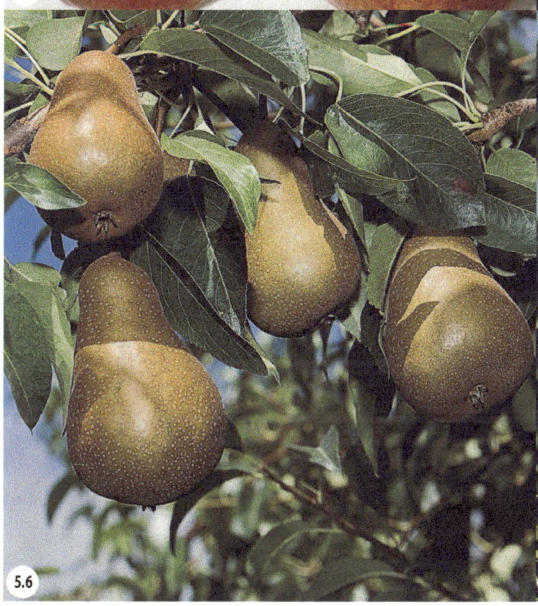

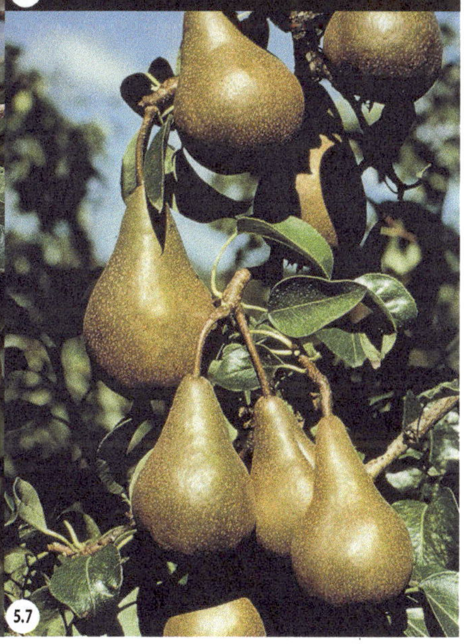

Plate 5.1. Anjou.
Photo: J. K. Clark.

Plate 5.2. Red Anjou.
Photo: R. Elkins.

Plate 5.3. Bartlett.
Photo: J. K. Clark.

Plate 5.4. Red Bartlett.
Photo: R. Elkins.

Plate 5.5. Bosc.
Photo: J. K. Clark.

Plate 5.6. Bronze Beauty Bosc.
Photo: J. K. Clark.

Plate 5.7. Golden Russet Bosc.
Photo: D. Burkhart.

Plate 5.8. Noble Russet
Photo: J. K. Clark.

Plate 5.9. Comice.
Photo: J. K. Clark.

Plate 5.10. Forelle.
Photo: C. Ingels.

Plate 5.11. Packham's Truimph.
Photo: J. K. Clark.

Plate 5.12. Red Clapp Favorite (Starkrimson).
Photo: J. K. Clark.

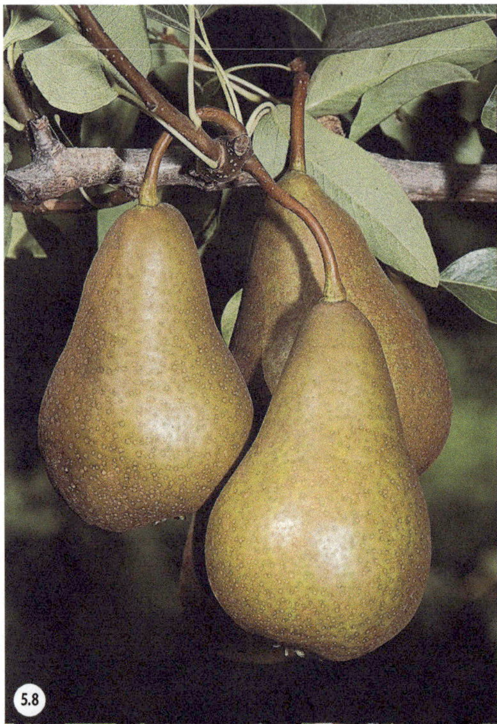

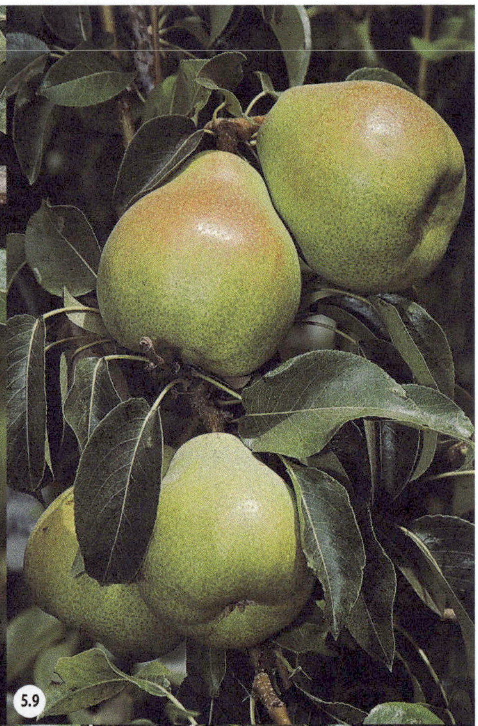

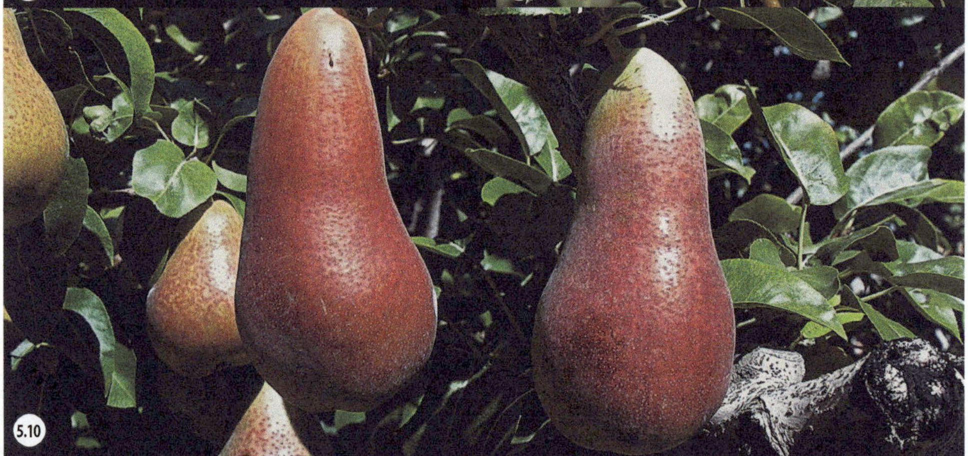

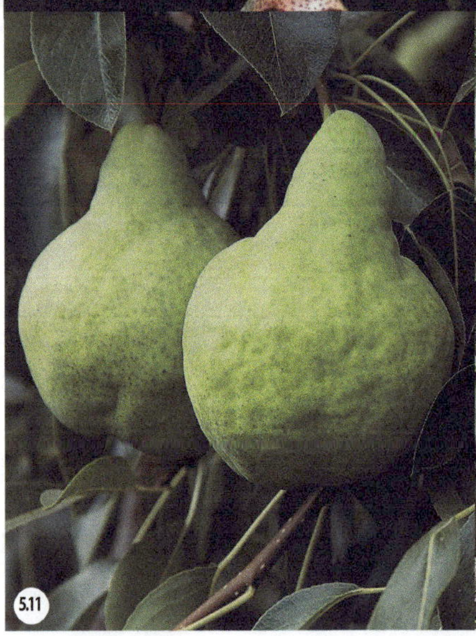

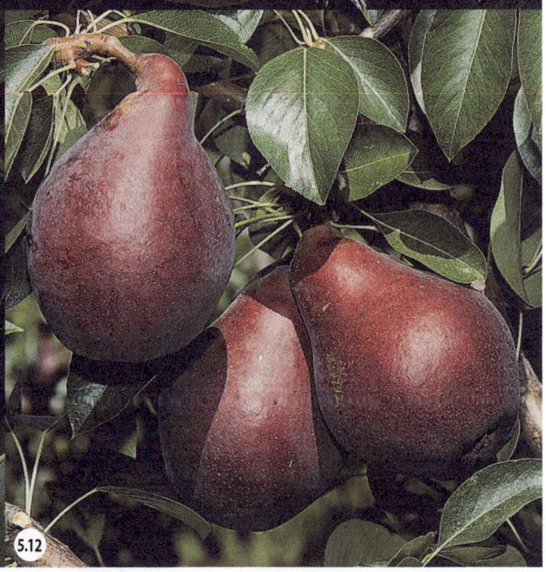

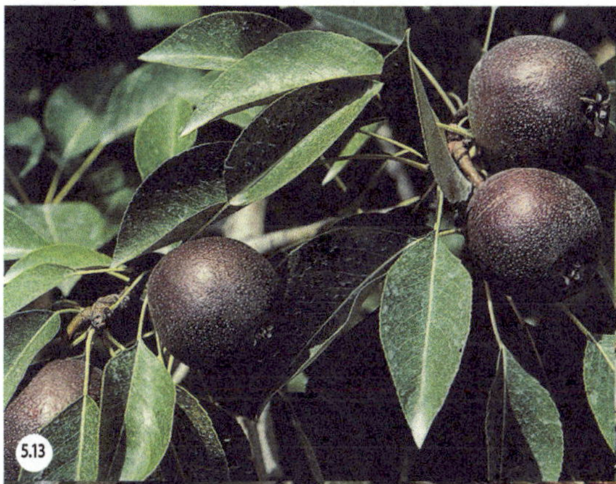

Plate 5.13. Seckel.
Photo: D. Burkhart.

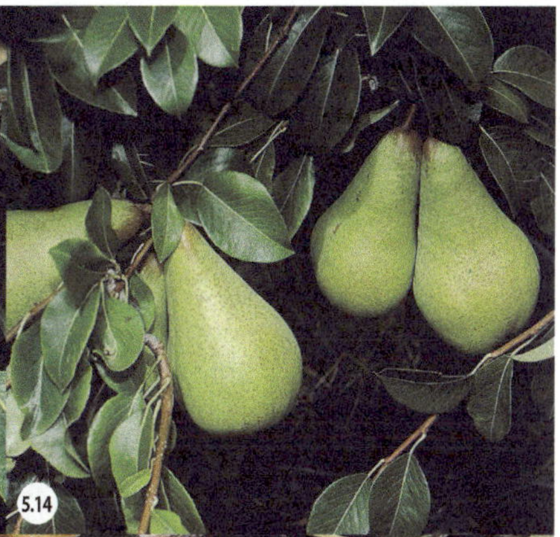

Plate 5.14. Abate Fetel.
Photo: J. K. Clark.

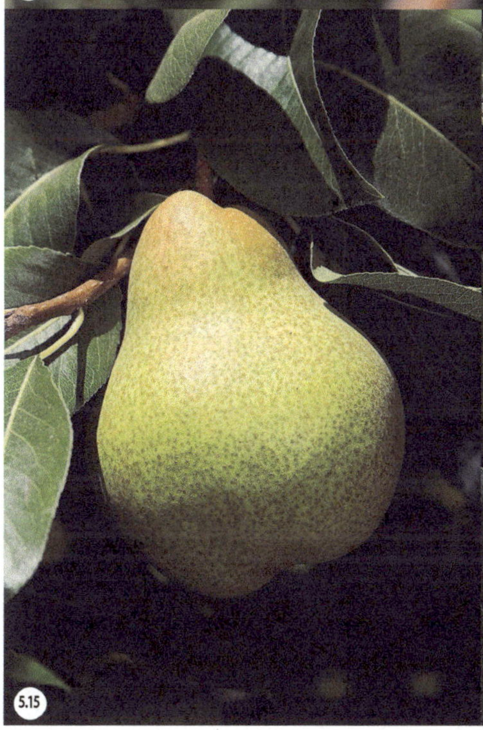

Plate 5.15. Blake's Pride.
Photo: J. K. Clark.

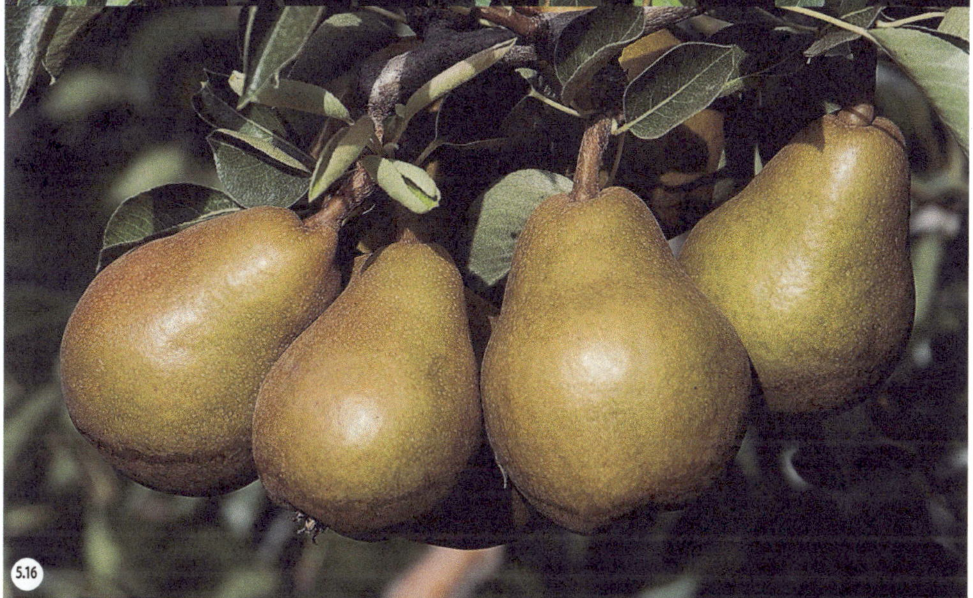

Plate 5.16. Cinnamon.
Photo: J. K. Clark.

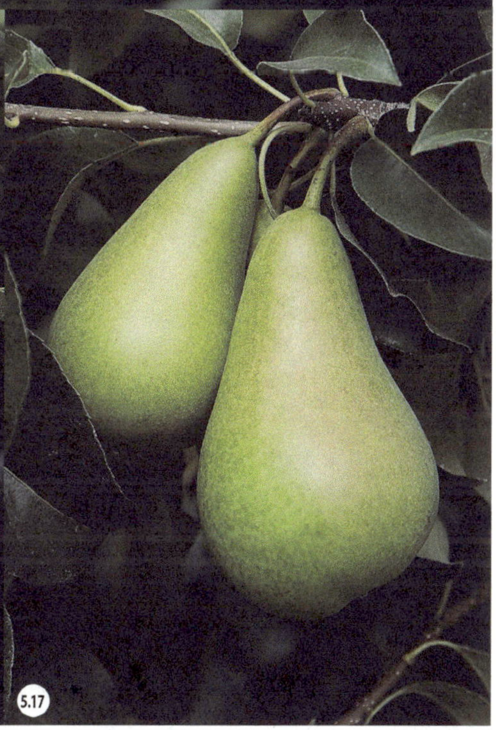

Plate 5.17. Concorde.
Photo: J. K. Clark.

Plate 5.18. Conference. *Photo:* J. K. Clark.

Plate 5.19. Potomac. *Photo:* J. K. Clark.

Plate 5.20. Rosemarie. *Photo:* J. K. Clark.

Plate 5.21. Taylor's Gold. Russeting is less consistent in drier climates. *Photos:* J. K. Clark (5.21A); K. Zielindki (5.21B).

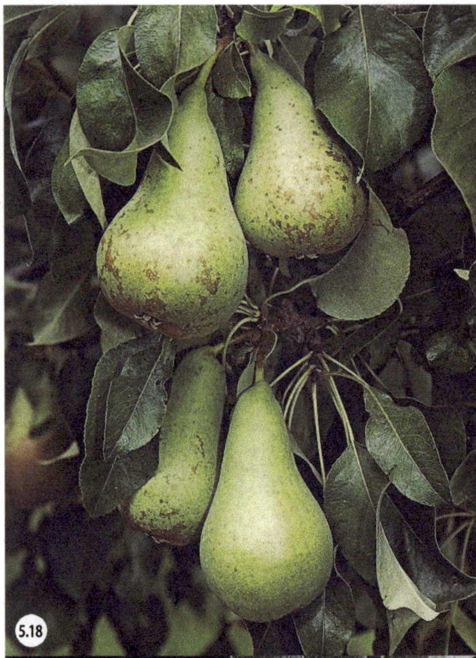

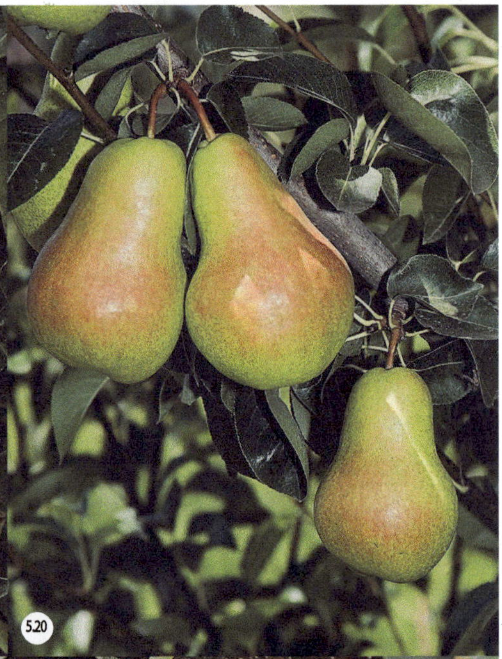

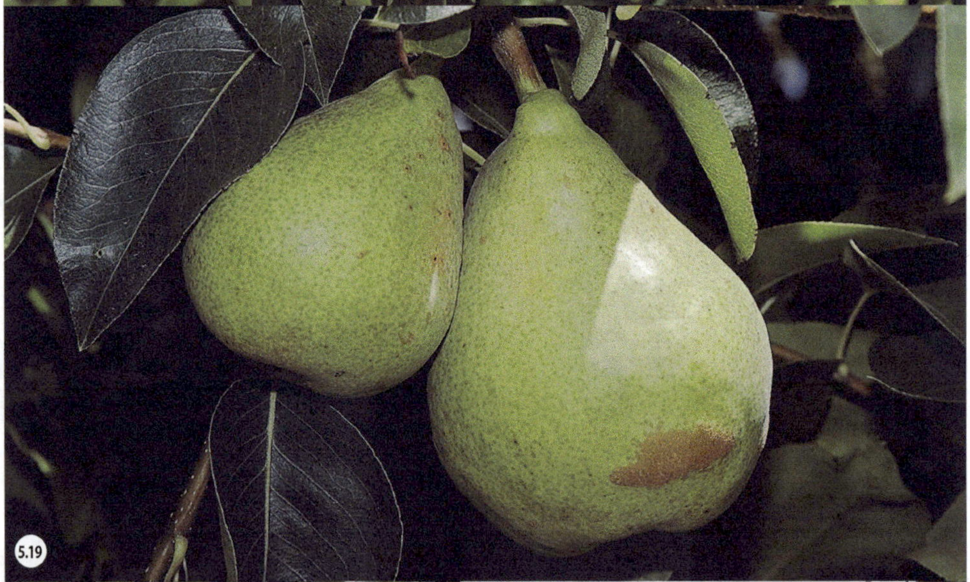

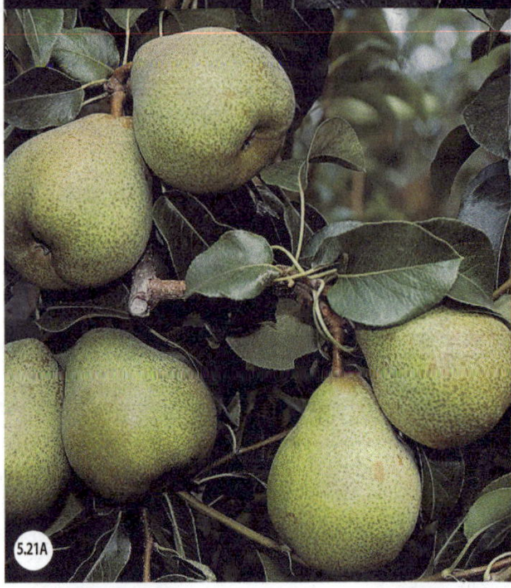

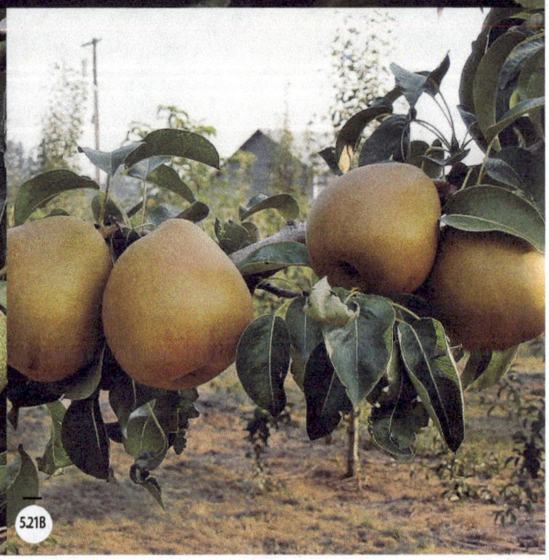

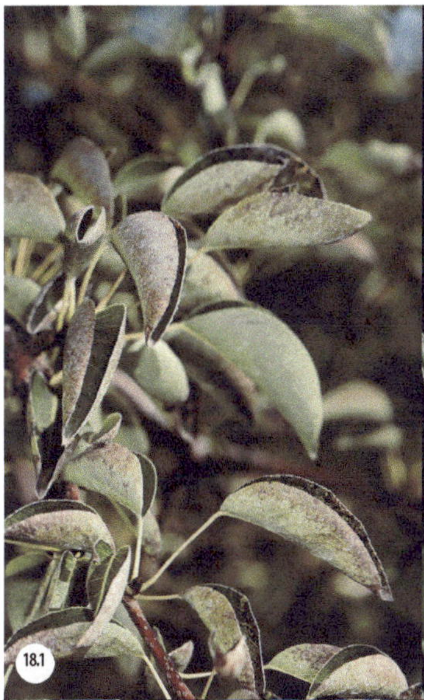

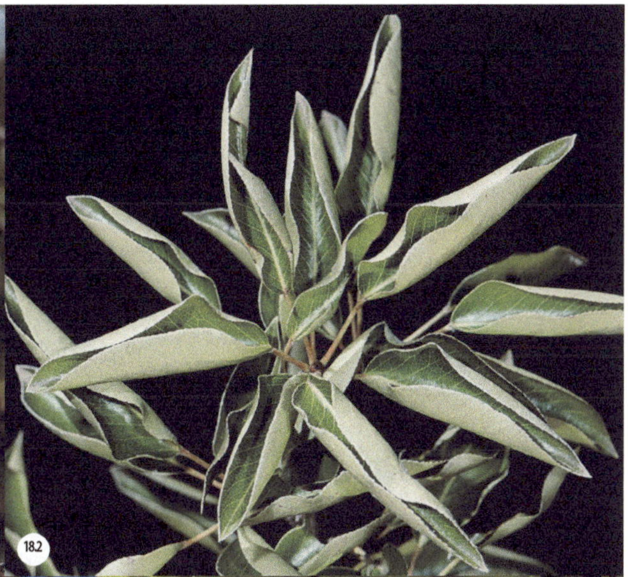

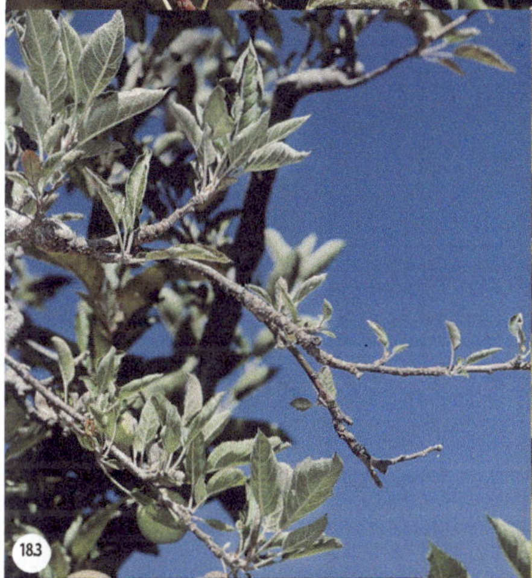

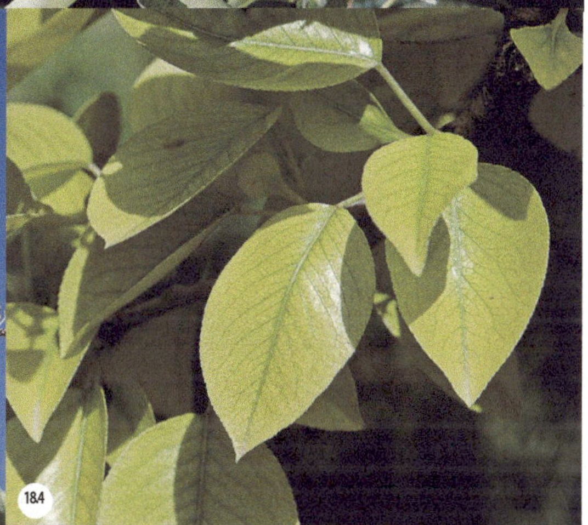

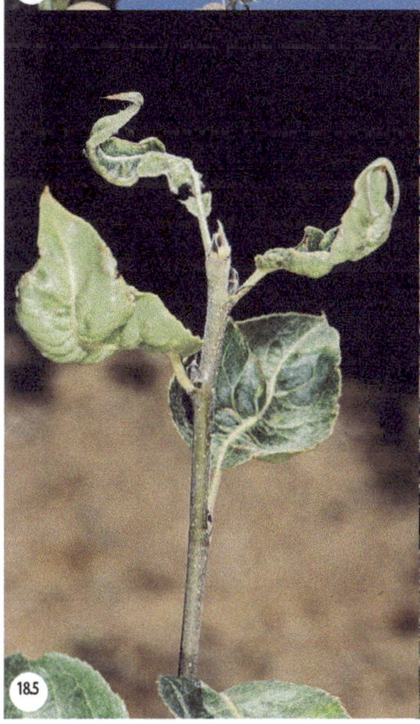

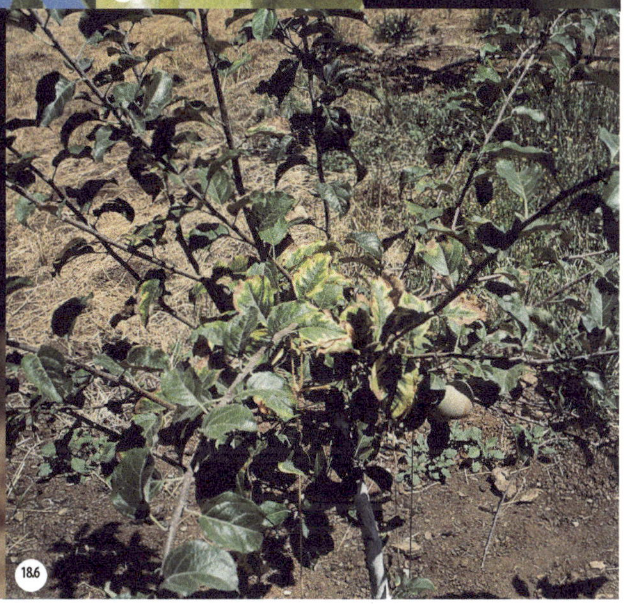

Plate 18.1. Potassium deficiency symptoms. *Photo:* J. K. Clark.

Plate 18.2. Phosphorous deficiency symptoms. *Photo:* J. K. Clark.

Plate 18.3. Zinc deficiency symptoms. *Photo:* J. K. Clark.

Plate 18.4. Iron deficiency symptoms. *Photo:* J. K. Clark.

Plate 18.5. Boron deficiency symptoms. *Photo:* J. K. Clark.

Plate 18.6. Magnesium deficiency symptoms. *Photo:* J. K. Clark.

Plate 20.1. Adult male (right) and female (left) codling moth. *Photo:* J. K. Clark.

Plate 20.2. Black head stage of codling moth egg and a newly hatched codling moth larva. *Photo:* J. K. Clark.

Plate 20.3. Mature diapausing codling moth larvae. *Photo:* J. K. Clark.

Plate 20.4. Stem, side, and calyx entries made by codling moth larvae. *Photo:* J. K. Clark.

Plate 20.5. Adult winter form pear psylla. *Photo:* J. K. Clark.

Plate 20.6. Young pear psylla nymph. *Photo:* J. K. Clark.

Plate 20.7. Hard shell, or fifth-instar, pear psylla nymph. *Photo:* J. K. Clark.

Plate 20.8. Pear psylla eggs deposited by an overwintering female. *Photo:* J. K. Clark.

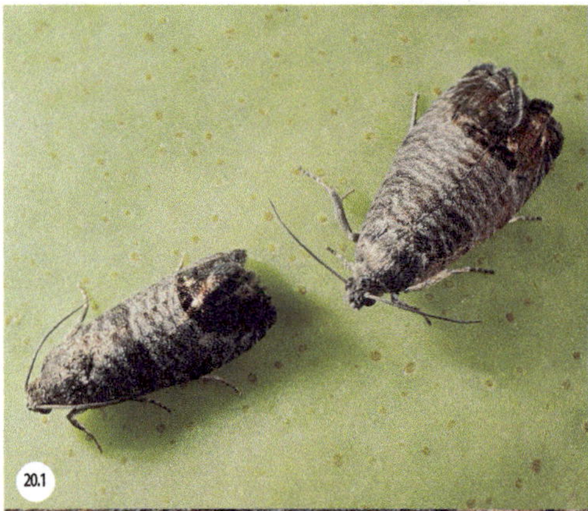

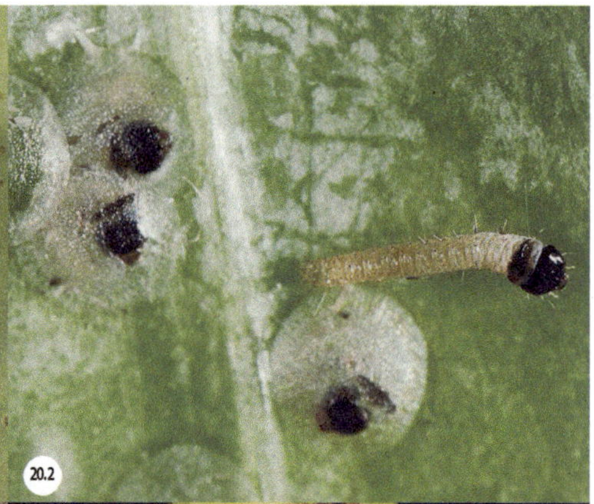

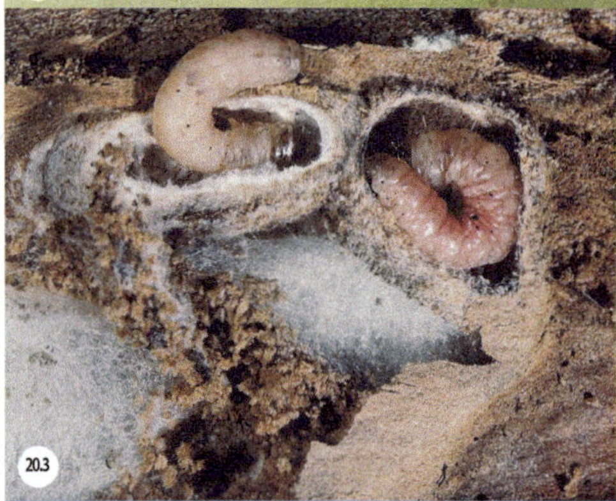

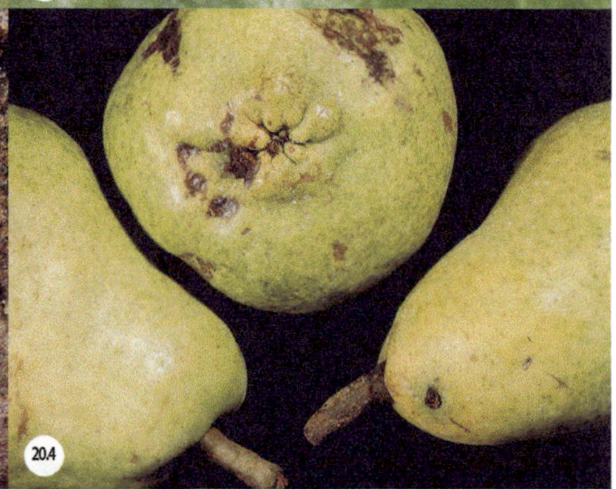

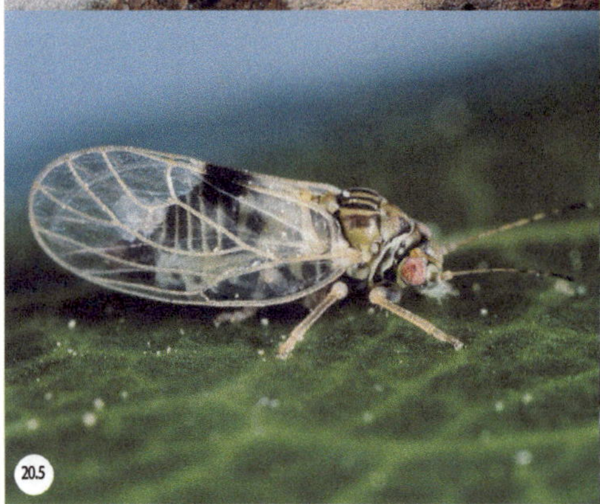

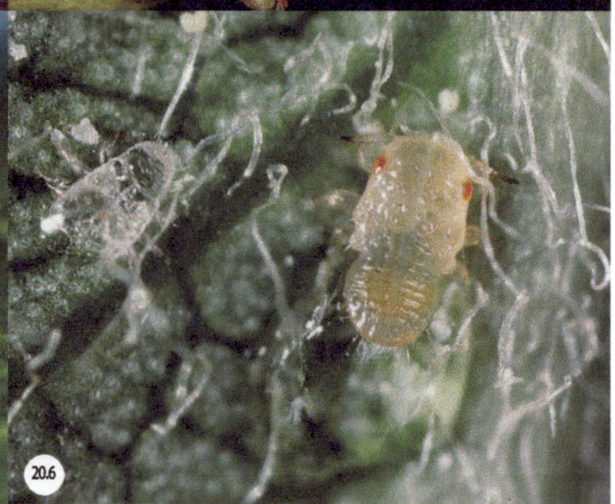

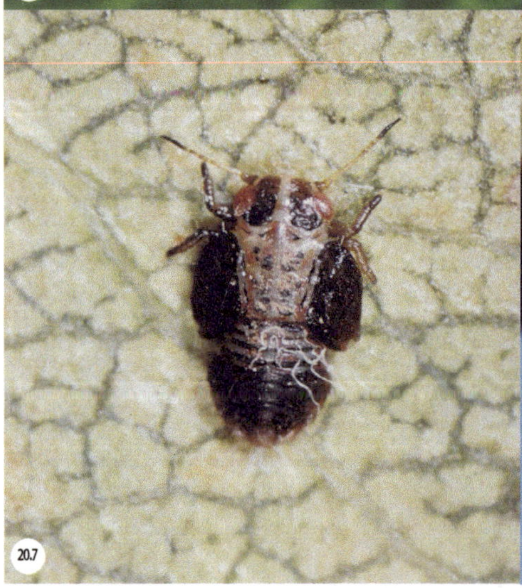

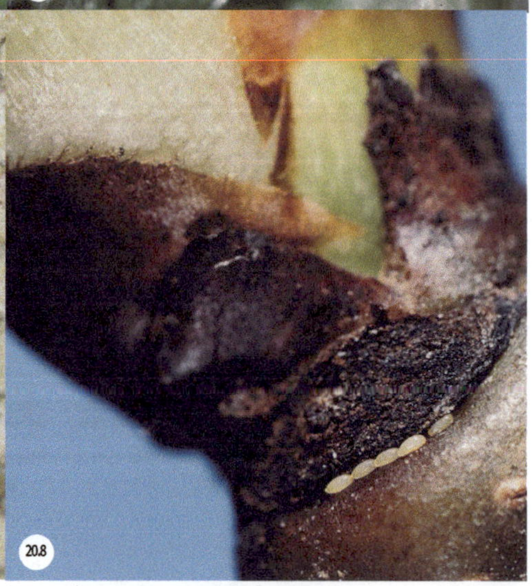

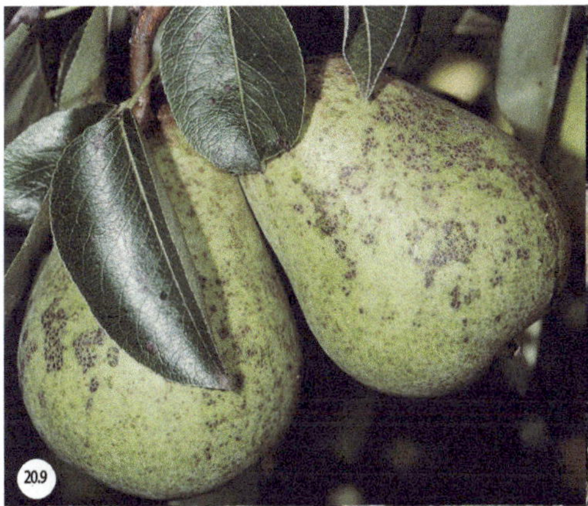

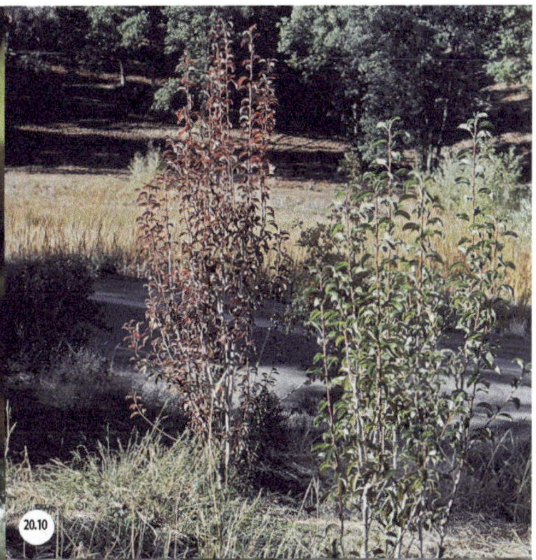

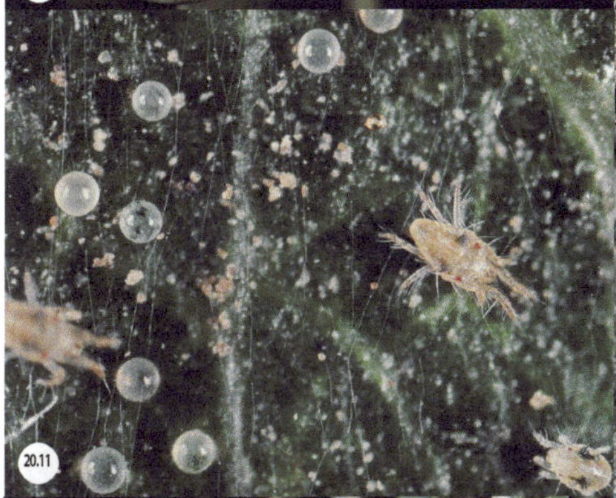

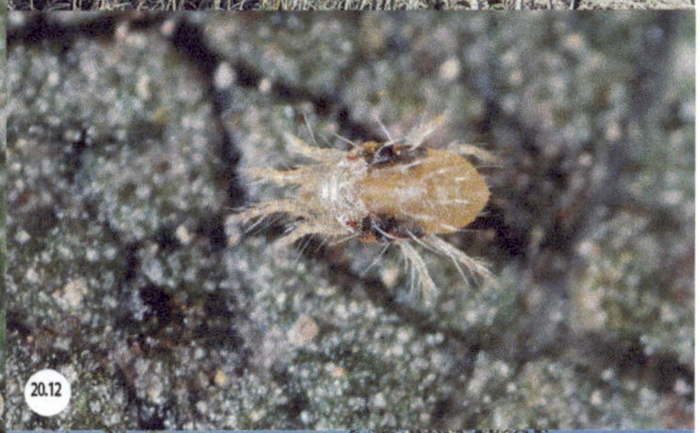

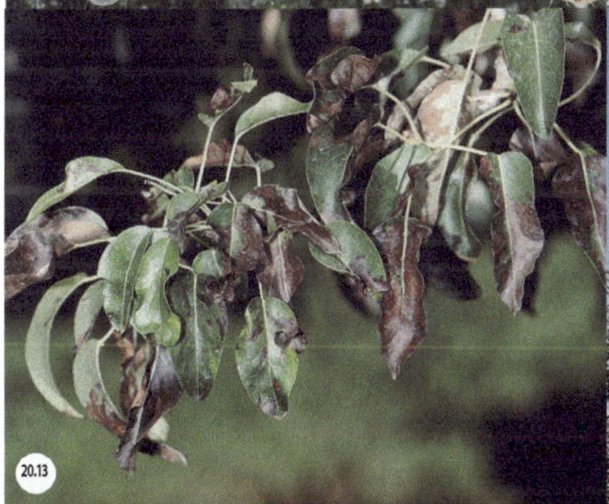

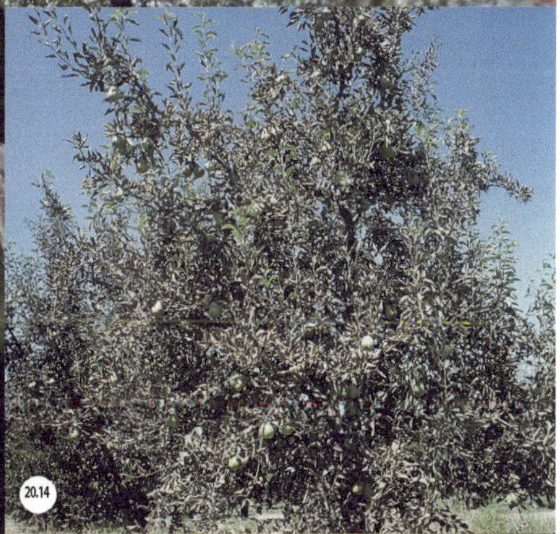

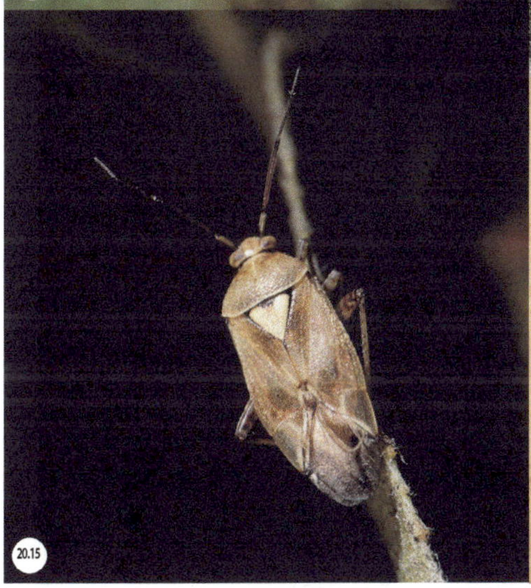

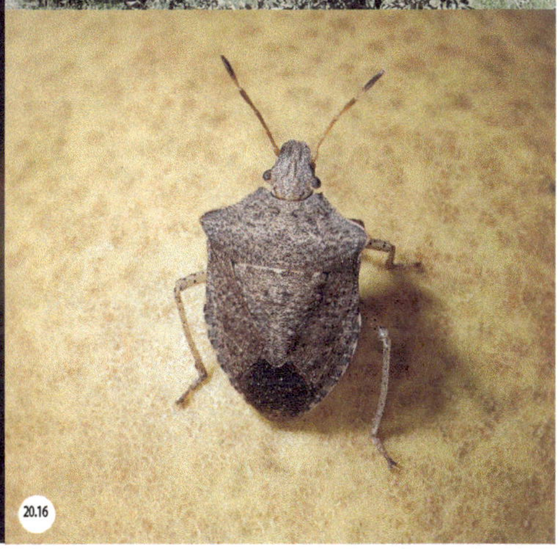

Plate 20.9. Sooty mold growth on honeydew produced by pear psylla, with resulting fruit russeting. *Photo:* J. K. Clark.

Plate 20.10. Pear trees suffering from pear decline. *Photo:* J. K. Clark.

Plate 20.11. Eggs and nymphs of twospotted spider mite. *Photo:* J. K. Clark.

Plate 20.12. Adult female twospotted spider mite. *Photo:* J. K. Clark.

Plate 20.13. Leaf burn caused by twospotted spider mite. *Photo:* J. K. Clark.

Plate 20.14. Damage to foliage caused by a high population of twospotted spider mite. *Photo:* J. K. Clark.

Plate 20.15. Adult lygus bug. *Photo:* J. K. Clark.

Plate 20.16. Adult consperse stink bug. *Photo:* J. K. Clark.

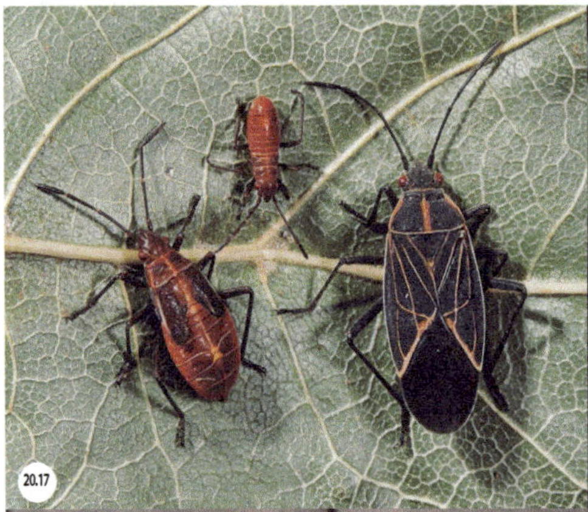

Plate 20.17. Boxelder bug adult (right) and nymphs. *Photo:* J. K. Clark.

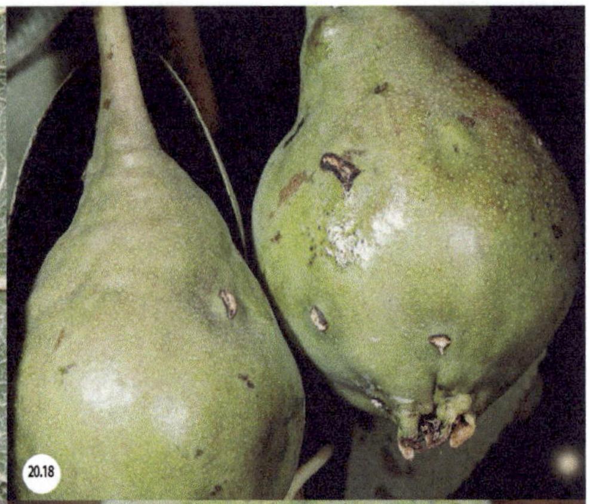

Plate 20.18. Early-season lygus bug damage. *Photo:* J. K. Clark.

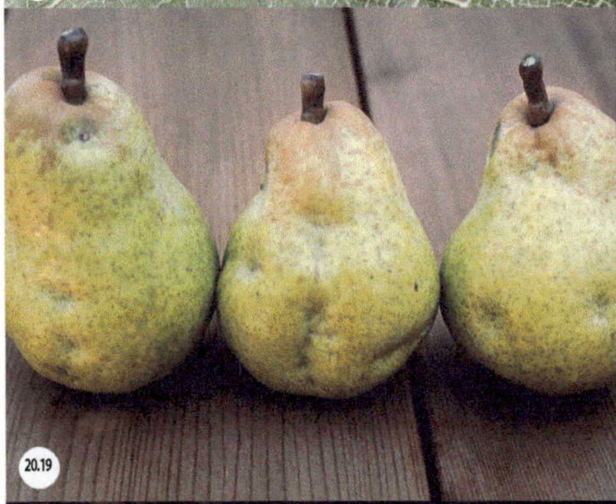

Plate 20.19. Mid- to late-season lygus bug damage. *Photo:* J. K. Clark.

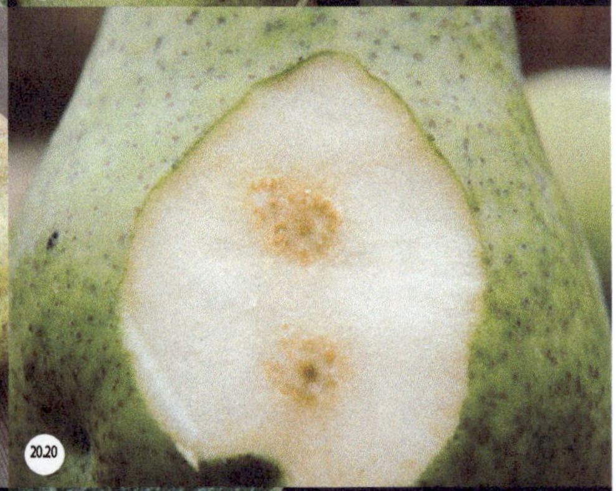

Plate 20.20. Darkened mass of stone cells caused by lygus bug feeding. *Photo:* J. K. Clark.

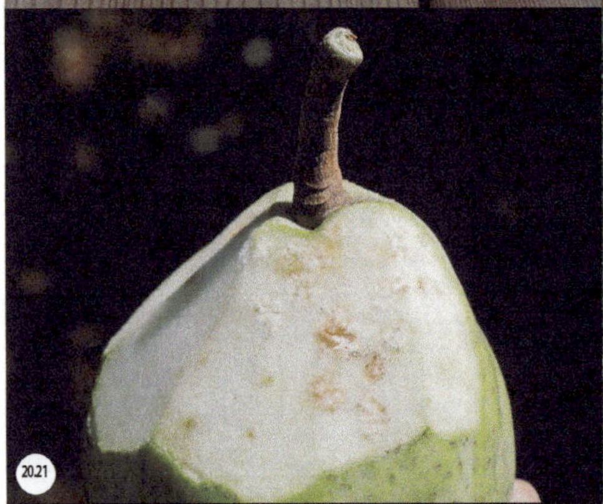

Plate 20.21. Brown, pithy areas caused by stink bug feeding. *Photo:* J. K. Clark.

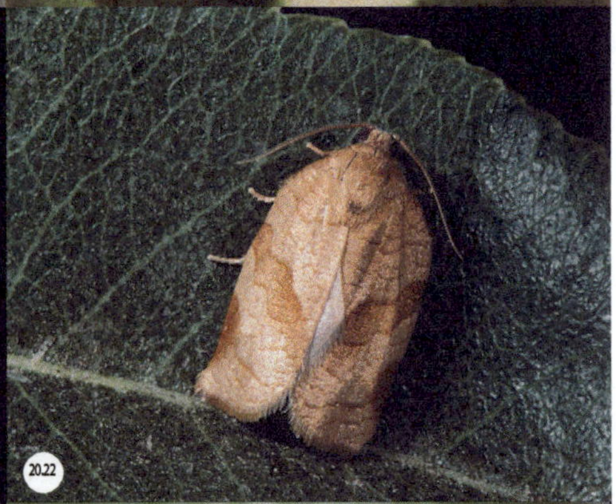

Plate 20.22. Adult obliquebanded leafroller. *Photo:* J. K. Clark.

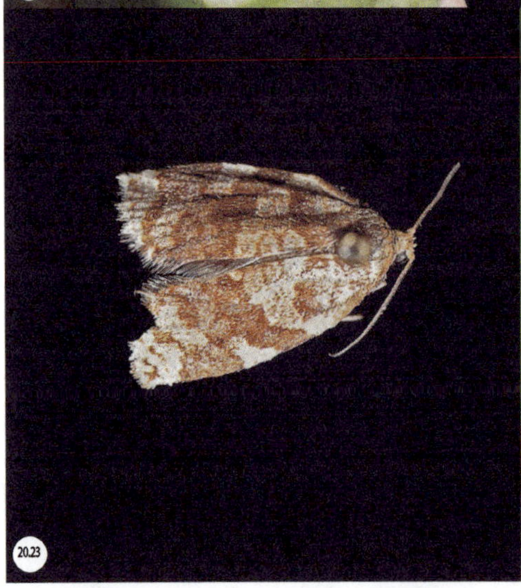

Plate 20.23. Adult fruittree leafroller. *Photo:* J. K. Clark.

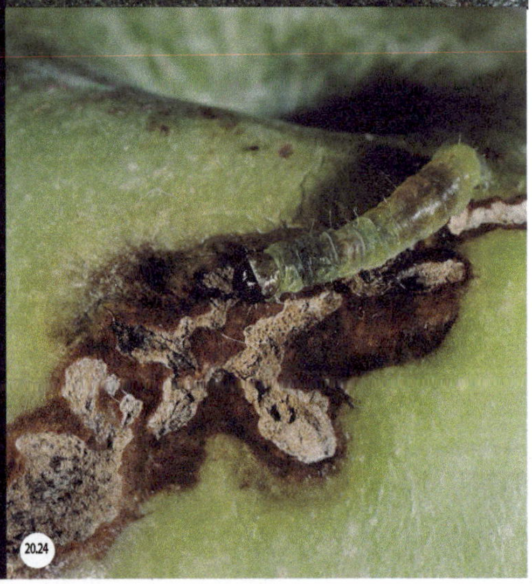

Plate 20.24. Obliquebanded leafroller larva feeding on a pear fruit. *Photo:* J. K. Clark.

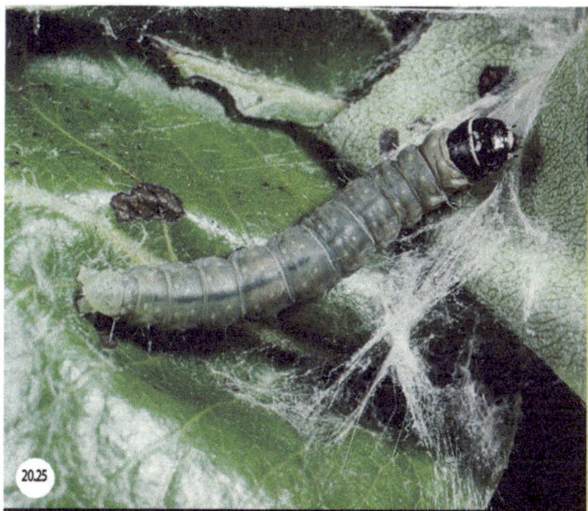
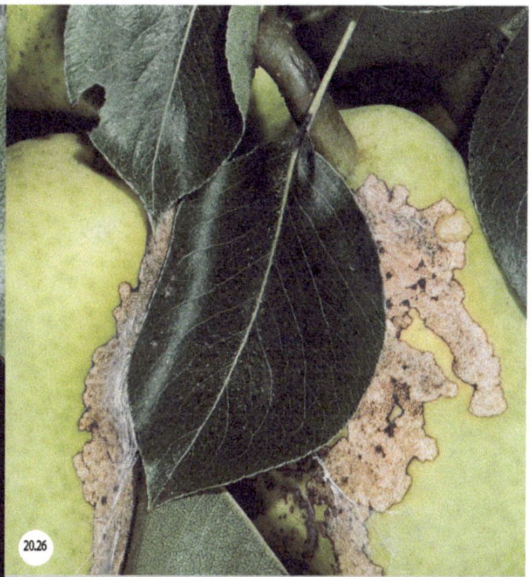
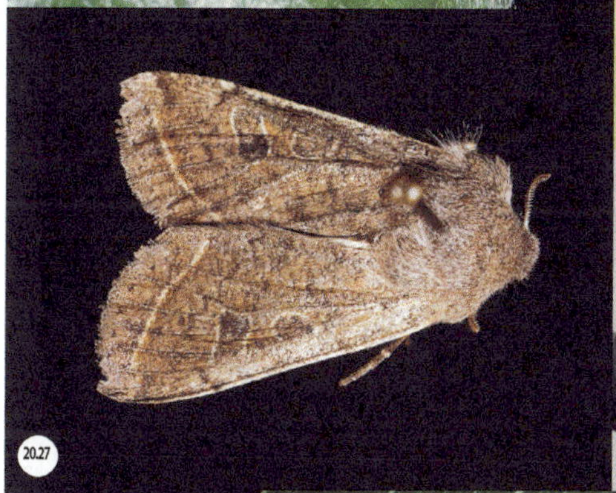
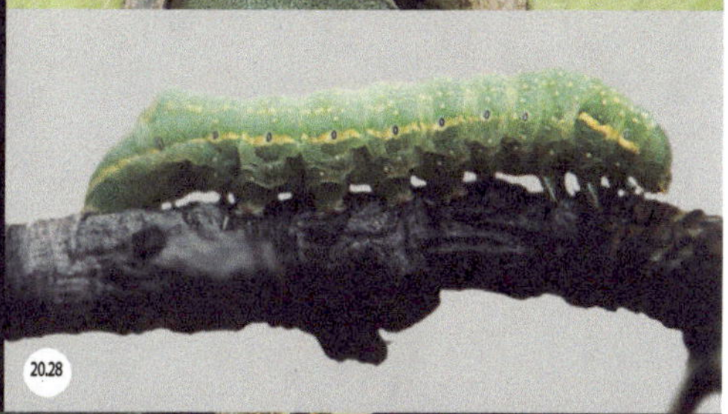
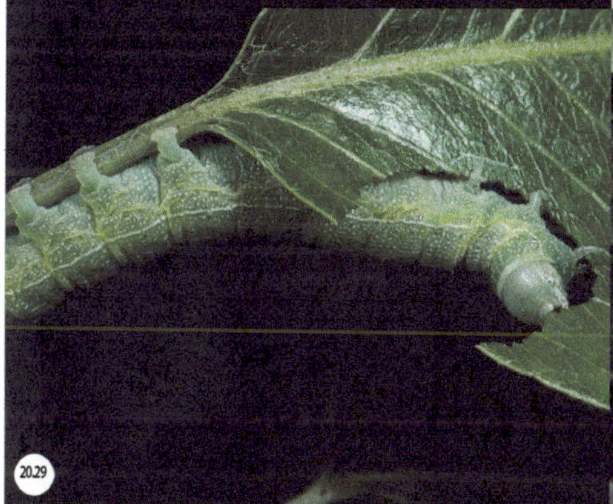
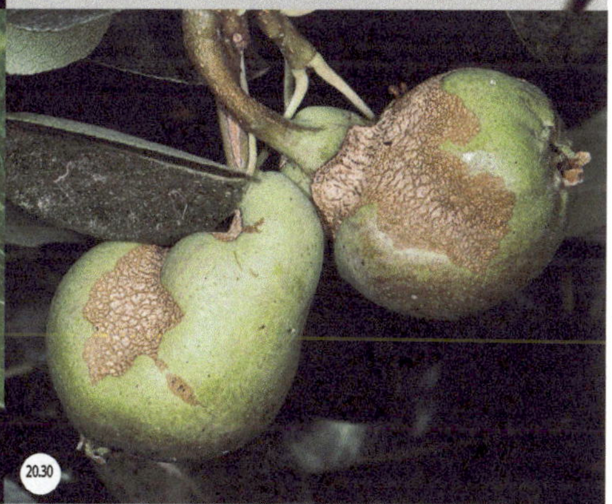
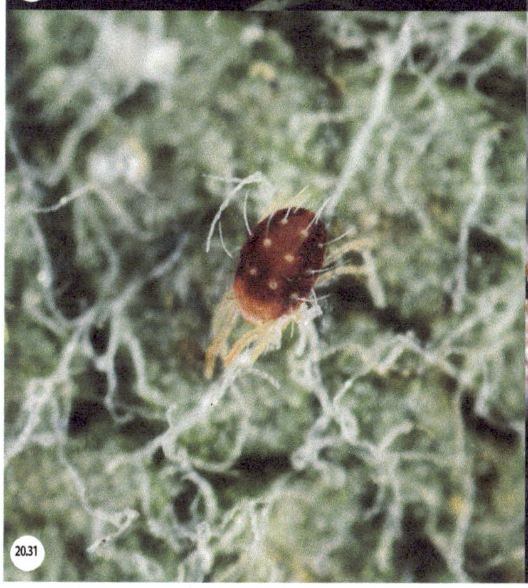
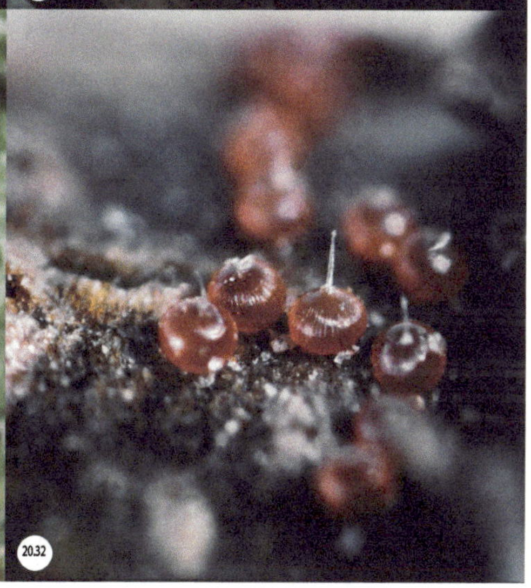

Plate 20.25. Fruittree leafroller larva tying leaves together with silk to form a leaf roll. *Photo:* J. K. Clark.

Plate 20.26. Characteristic leafroller feeding. *Photo:* J. K. Clark.

Plate 20.27. Adult speckled green fruitworm. *Photo:* J. K. Clark.

Plate 20.28. Humped green fruitworm larva. *Photo:* J. K. Clark.

Plate 20.29. Speckled green fruitworm larva. *Photo:* J. K. Clark.

Plate 20.30. Damage caused by green fruitworm. *Photo:* J. K. Clark.

Plate 20.31. Adult European red mite. *Photo:* J. K. Clark.

Plate 20.32. European red mite eggs. *Photo:* J. K. Clark.

Plate 20.33. San Jose scale crawlers. *Photo:* J. K. Clark.

Plate 20.34. Adult female San Jose scale with cap. *Photo:* J. K. Clark.

Plate 20.35. San Jose scale infested fruit with red halo. *Photo:* J. K. Clark.

Plate 20.36. Mature mealybug in the calyx end of fruit. *Photo:* J. K. Clark.

Plate 20.37. Black sooty fungus growing on honeydew produced by pear psylla. *Photo:* J. K. Clark.

Plate 20.38. Russeting caused by pear rust mite. *Photo:* J. K. Clark.

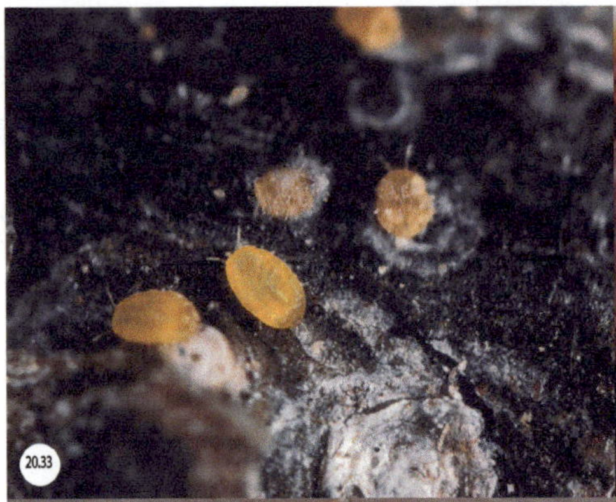

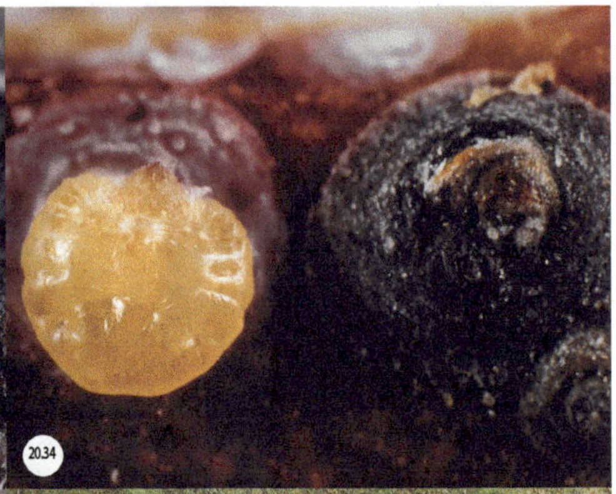

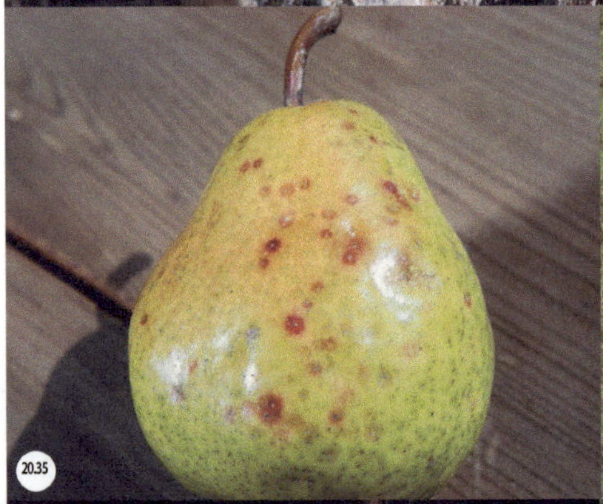

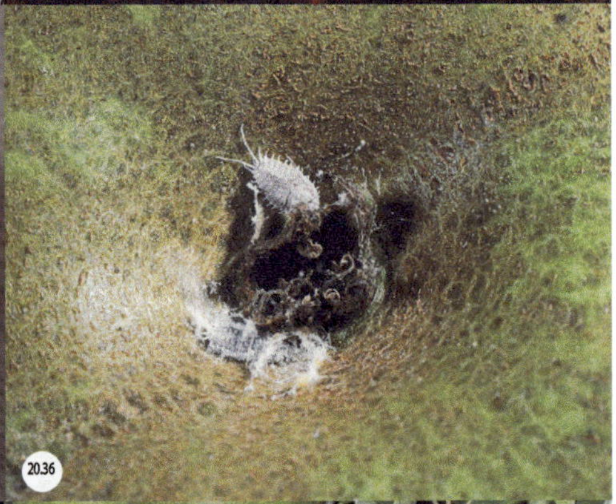

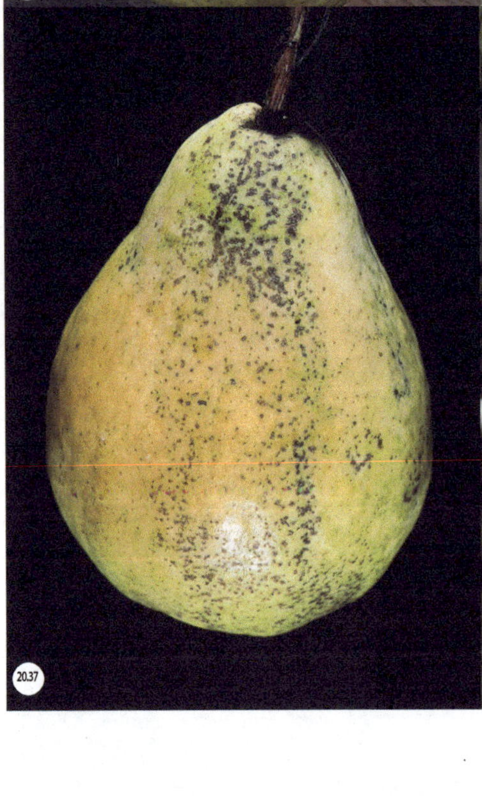

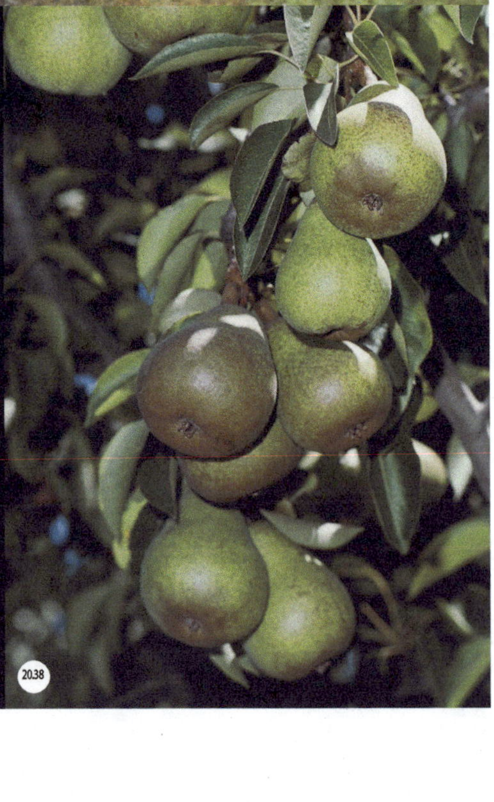

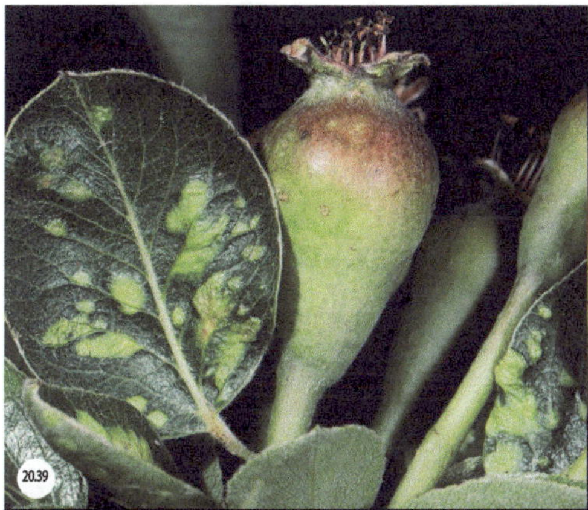

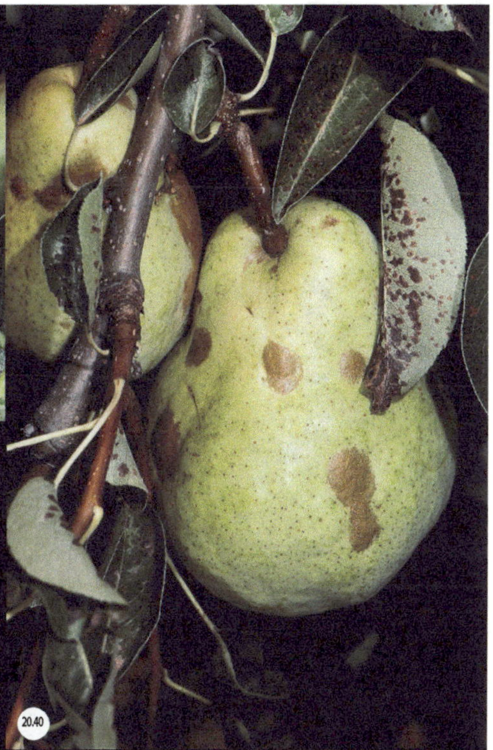

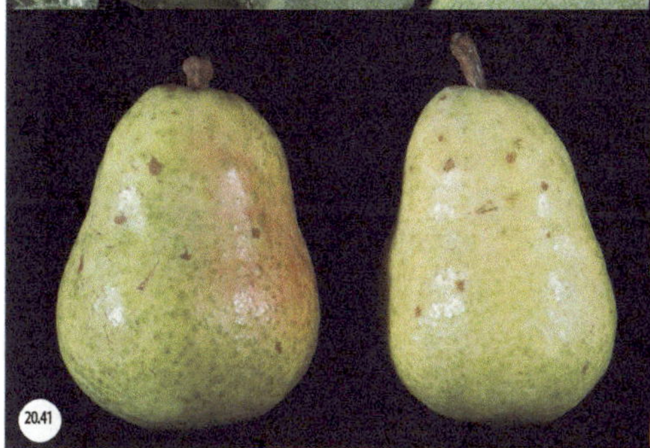

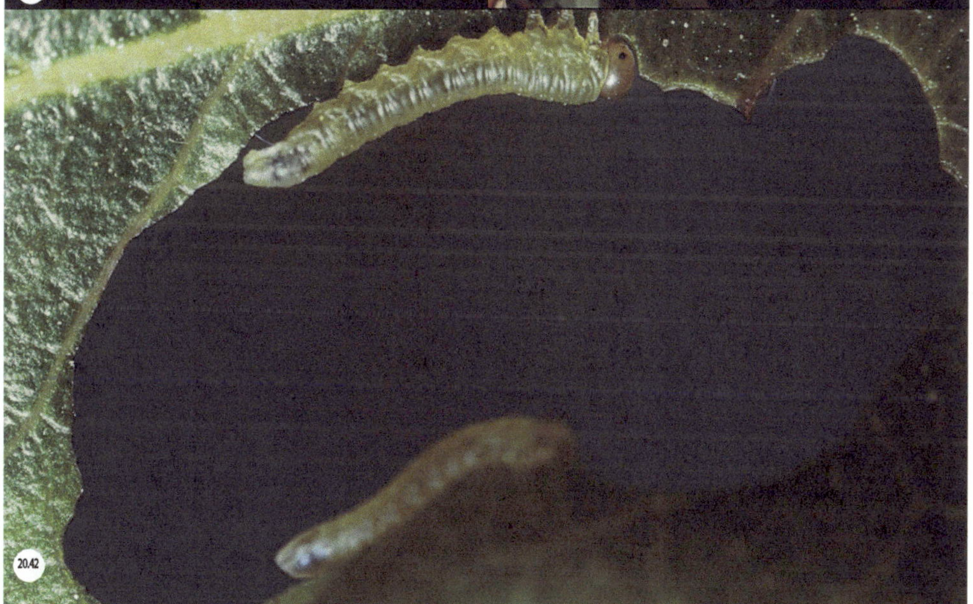

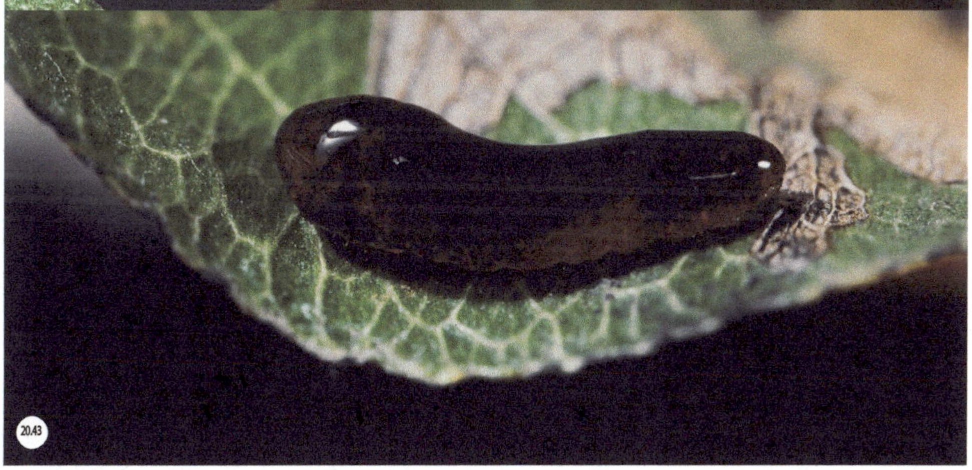

Plate 20.39. Pearleaf blister mite damage on foliage. *Photo:* J. K. Clark.

Plate 20.40. Pearleaf blister mite damage on fruit. *Photo:* J. K. Clark.

Plate 20.41. Russet spots caused by thrips feeding. *Photo:* J. K. Clark.

Plate 20.42. California pear sawfly feeding in circular patterns, leaving oval holes in the leaves. *Photo:* J. K. Clark.

Plate 20.43. Pear sawfly feeding on leaf tissue, leaving only a fine network of veins. *Photo:* J. K. Clark.

Plate 21.1. Flower clusters infected by fire blight wilt and turn dark. *Photo:* J. K. Clark.

Plate 21.2. Branches infected with fire blight show a typical terminal crook. The bacteria probably entered the branch through the rat-tail bloom near the terminal. *Photo:* J. K. Clark.

Plate 21.3. Fire blight bacterial ooze typically occurs during humid periods. *Photo:* J. K. Clark.

Plate 21.4. Cut away the bark from a fire blight canker and look at the color of the wood underneath. Where the wood is reddish, the fire blight bacteria are still active – as seen in the lower left part of cut. The dark brown area in the upper part of the cut is where the wood has died and the bacteria are no longer active. *Photo:* J. K. Clark.

Plate 21.5. Dead leaves clinging to pear trees testify to an earlier fire blight epidemic. *Photo:* J. K. Clark.

Plate 21.6. Brown, cracked bark on this pear limb covers an overwintering fire blight canker. *Photo:* J. K. Clark.

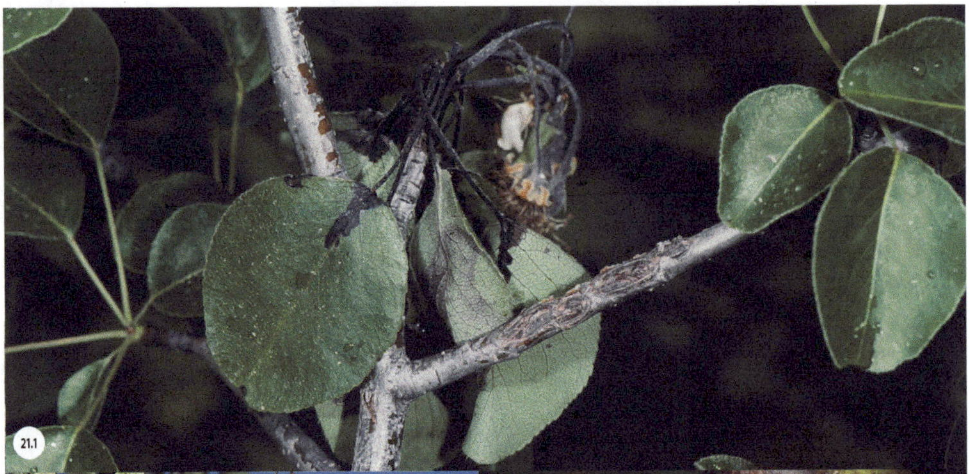

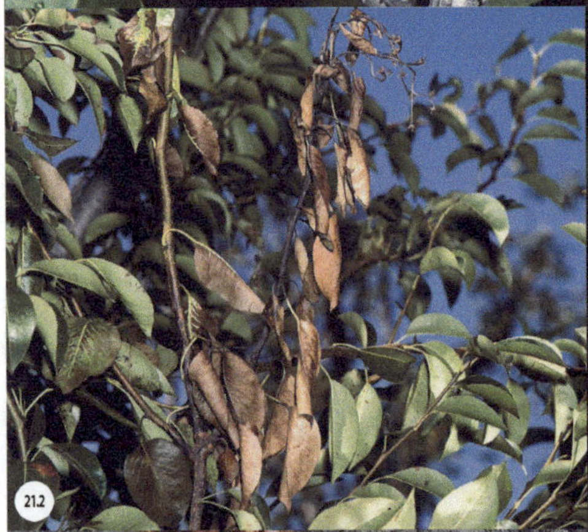

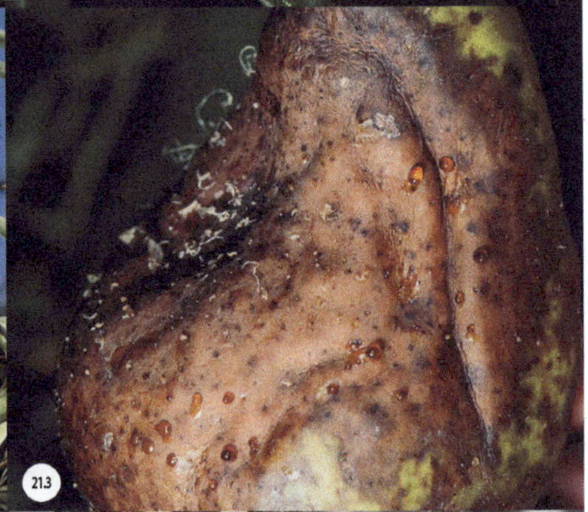

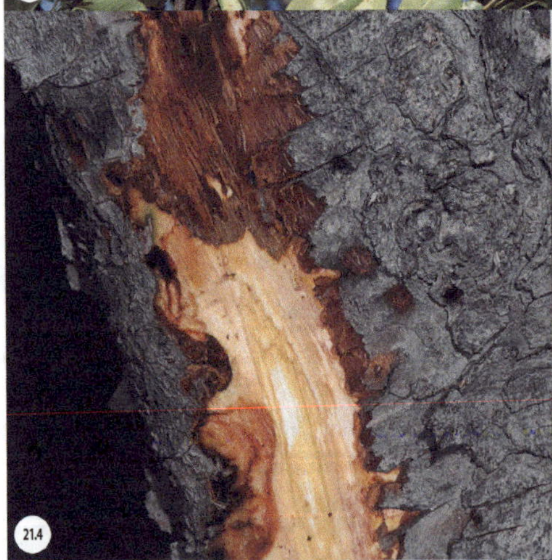

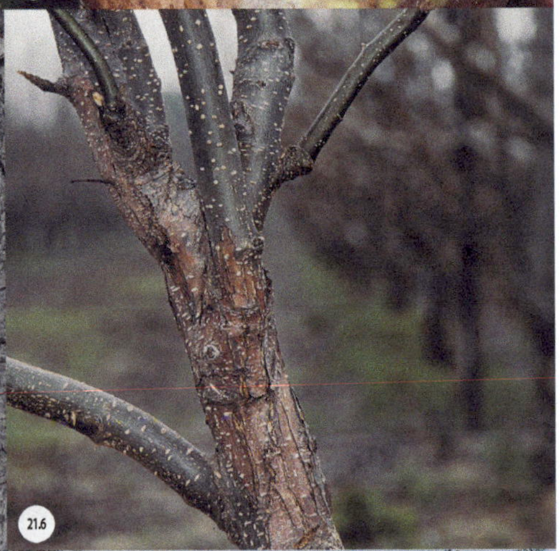

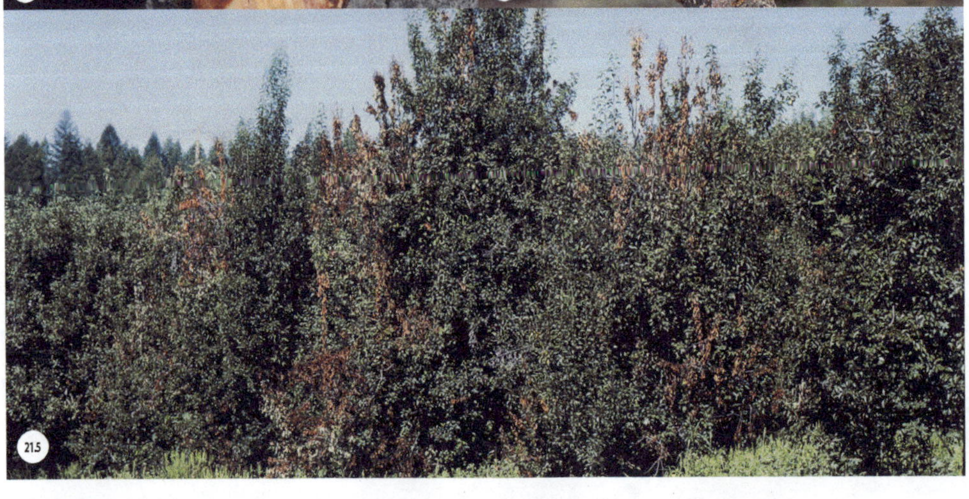

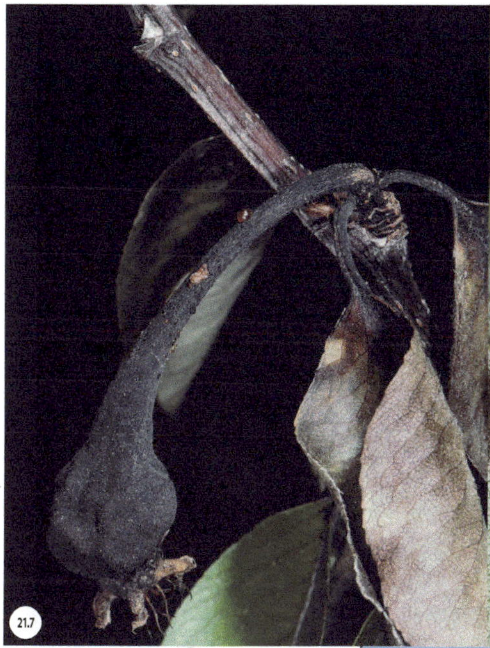

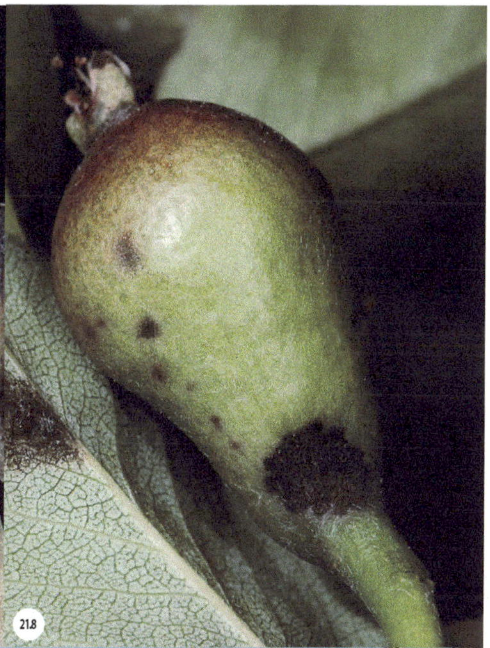

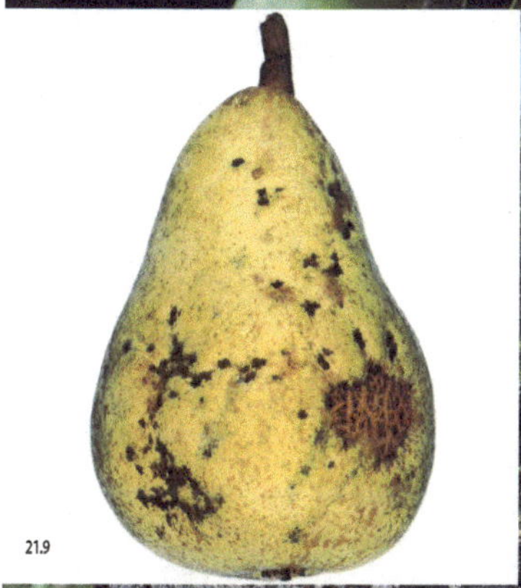

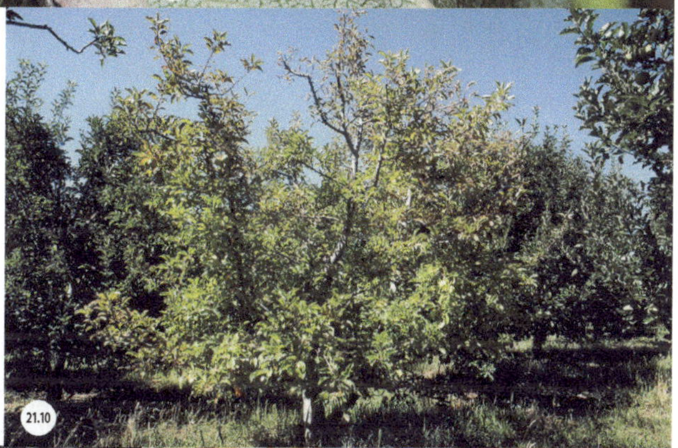

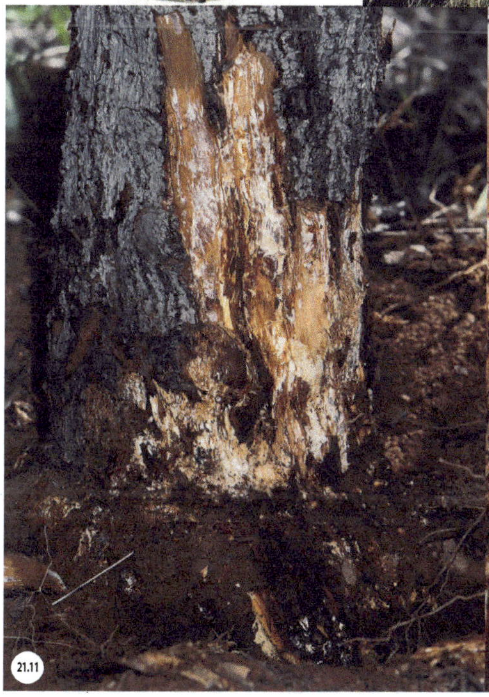

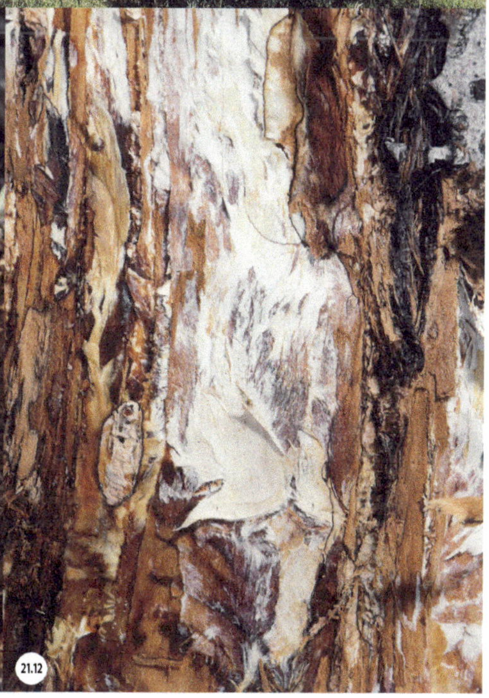

Plate 21.7. Blackening of the young fruit and tan droplets of ooze leading to shriveled fruit are typical of a fire blight infection. *Photo:* J. K. Clark.

Plate 21.8. Pear scab first appears as velvety black spots on young fruit. Leaf lesions generally follow and appear as dark spots on the undersurface of the leaf. *Photo:* J. K. Clark.

Plate 21.9. Large scab lesion on the right side of this pear was probably the source of the secondary, pinpoint lesions that cover the surface of the fruit. *Photo:* J. K. Clark.

Plate 21.10. Small, chlorotic leaves and shoot dieback indicate an *Armillaria* infection. *Photo:* J. K. Clark.

Plate 21.11. Remove soil from the crown area of an infected tree and peel the bark off to check for thin, white leathery patches of *Armillaria* fungus. *Photo:* J. K. Clark.

Plate 21.12. Close-up of *Armillaria* fungus reveals white, leathery patches typical of the fungus. *Photo:* J. K. Clark.

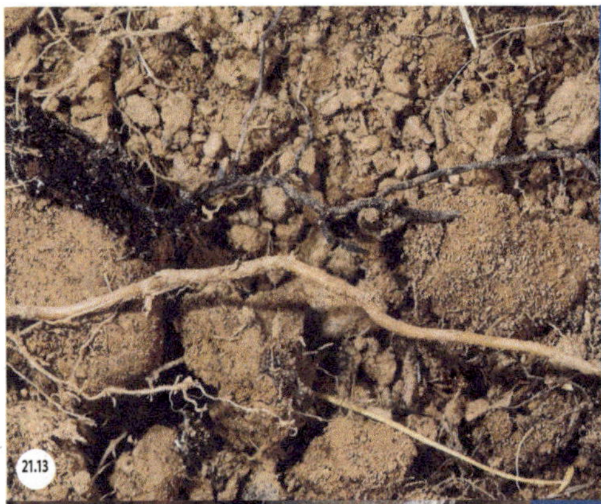

Plate 21.13. The top structure in this photo is an *Armillaria* rhizomorph, characterized by its dark, shoestringlike appearance. Beneath it is a healthy tree root. *Photo:* J. K. Clark.

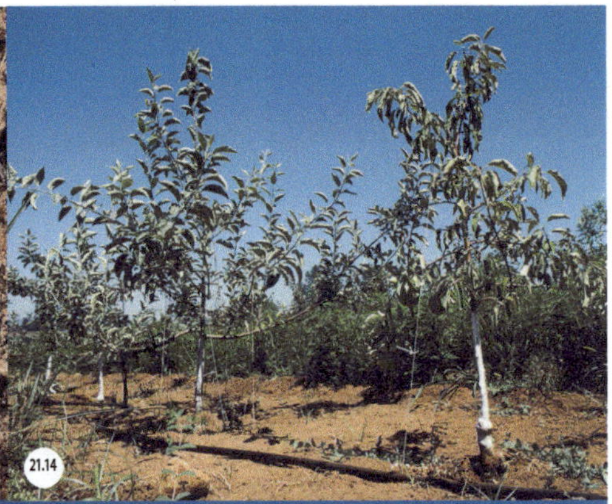

Plate 21.14. Wilted leaves and branches on these young apple trees are a result of a *Phytophthora* infection. *Photo:* J. K. Clark.

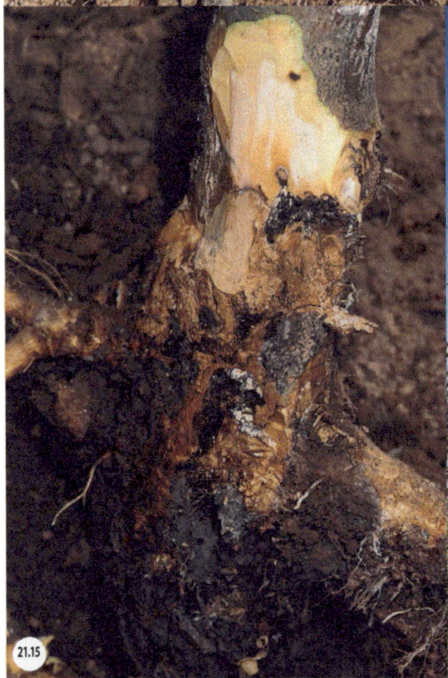

Plate 21.15. By removing dirt from around the crown and peeling off the bark, the dark, discolored wood of a *Phytophthora* infection is evident. *Photo:* J. K. Clark.

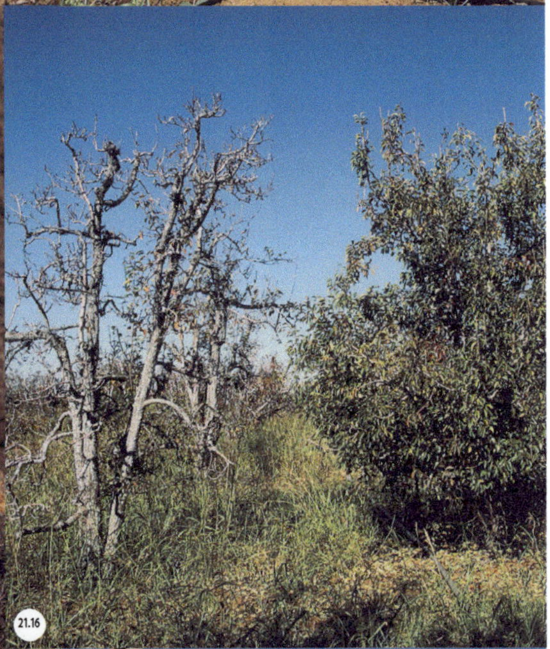

Plate 21.16. Slow decline is characterized by reddish foliage, early defoliation, and twig dieback. *Photo:* J. K. Clark.

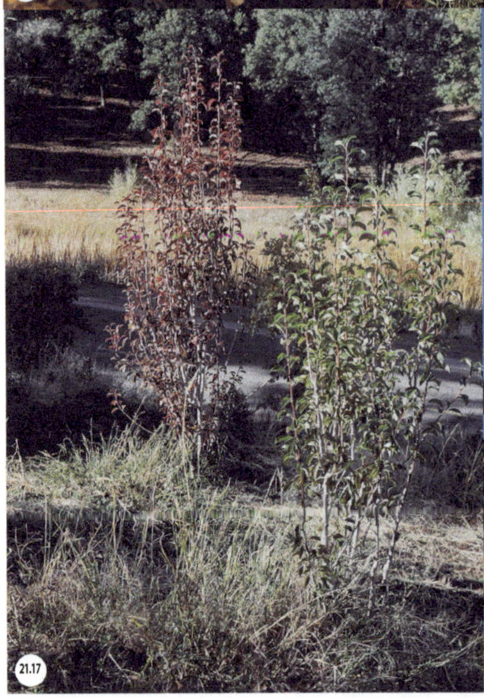

Plate 21.17. Pear decline in young trees, as evidenced by the red foliage on the tree (left), is primarily a result of the use of decline-susceptible rootstocks. *Photo:* J. K. Clark.

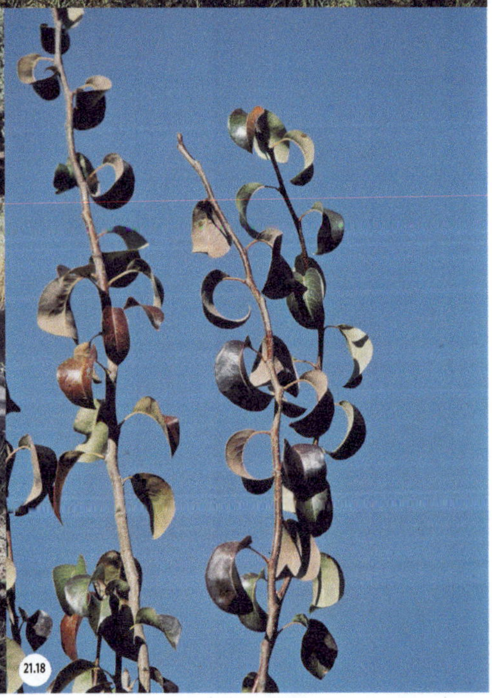

Plate 21.18. Purple leaf curl symptoms are characteristic of pear decline on the Bosc variety. *Photo:* J. K. Clark.

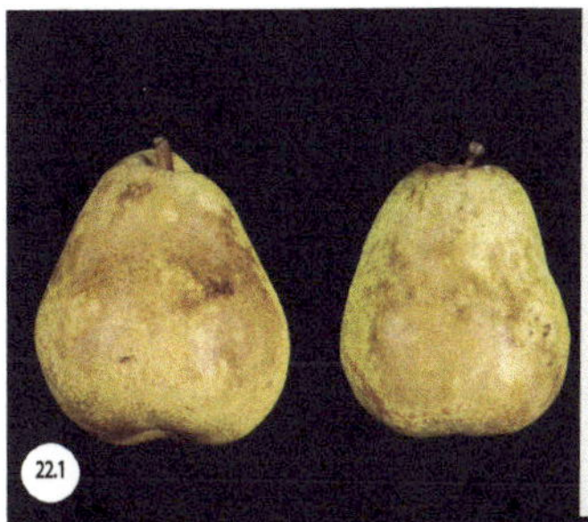

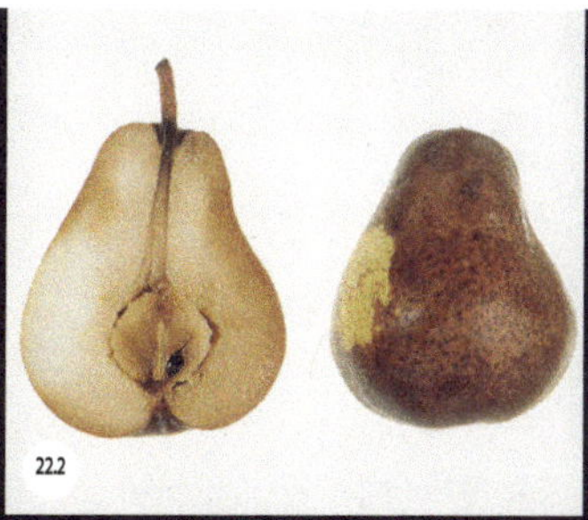

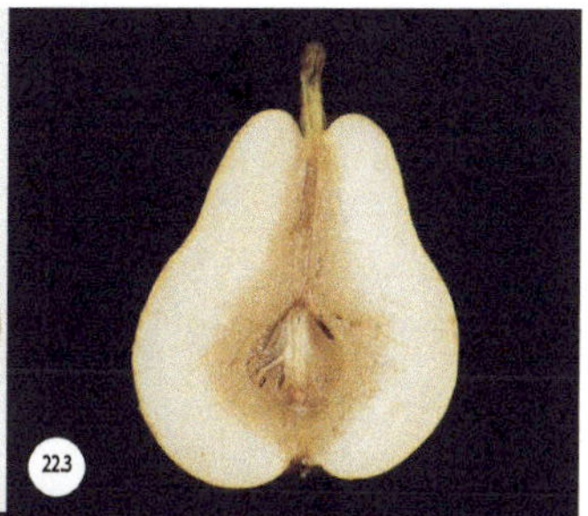

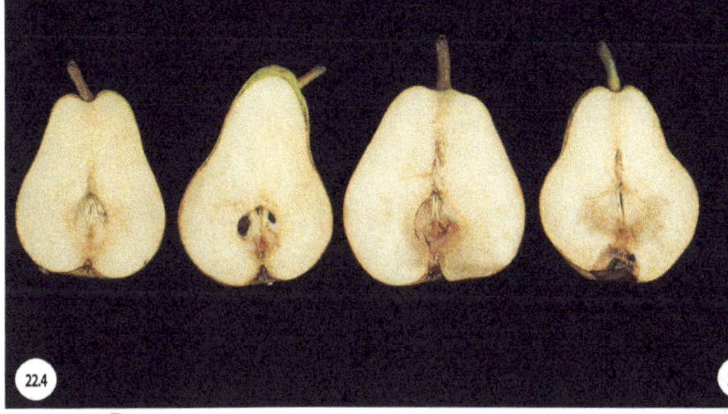

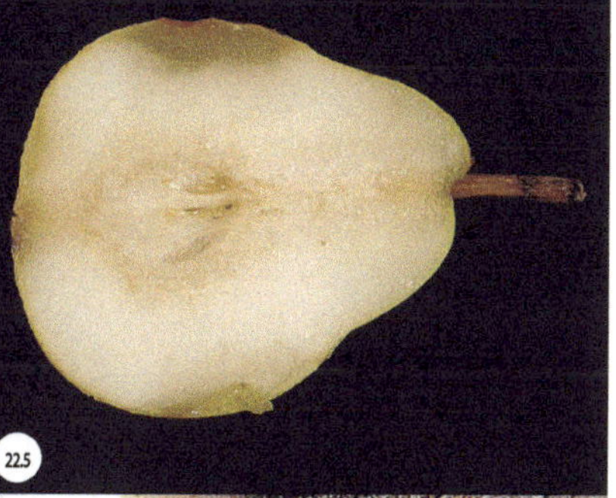

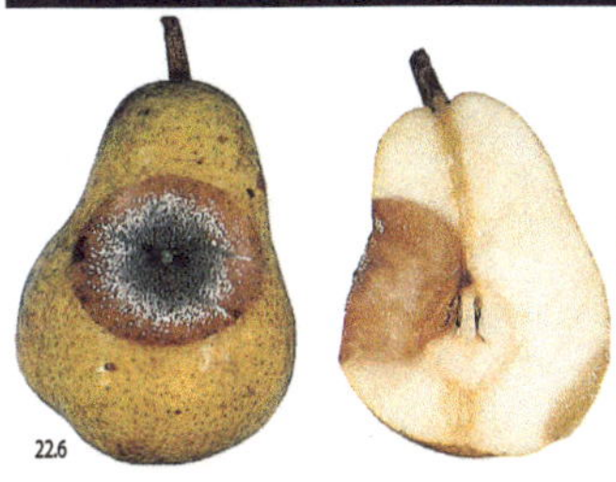

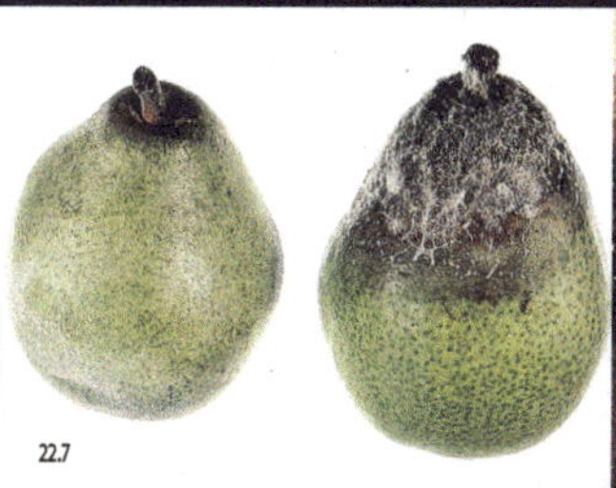

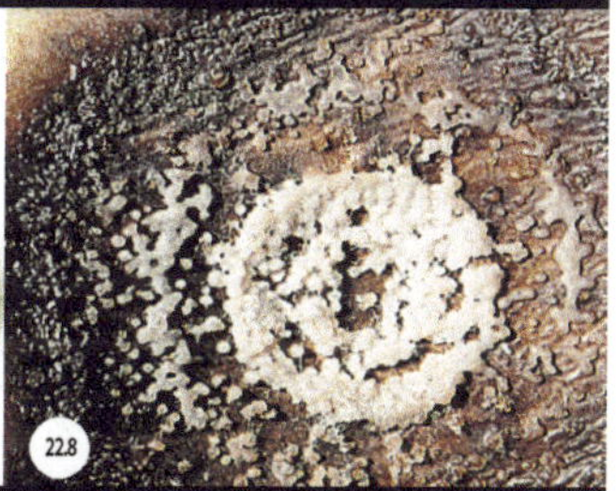

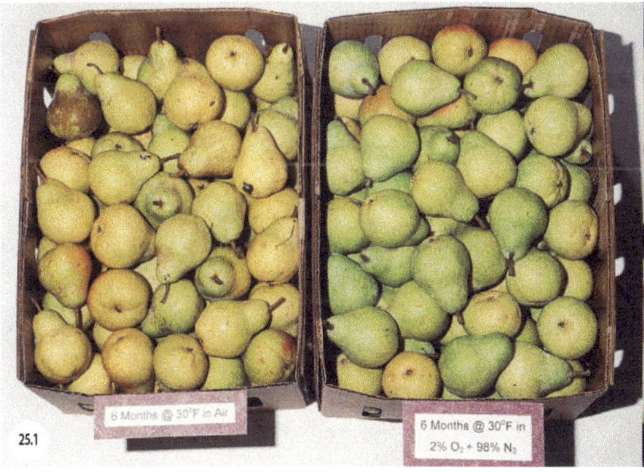

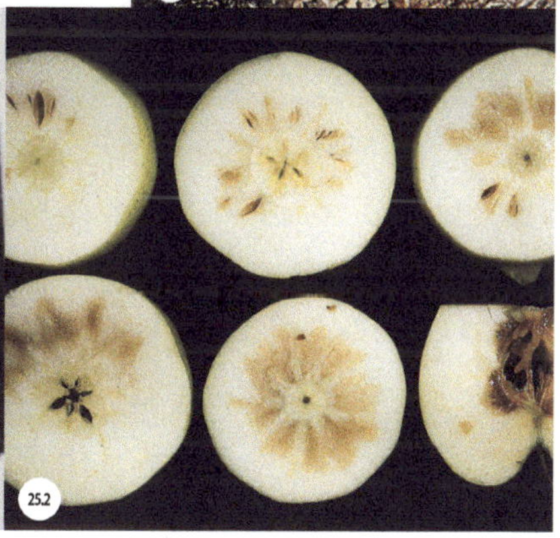

Plate 22.1. Storage scald on Bartlett pear. *Photo:* D. Edwards.

Plate 22.2. Senescent scald of Bartlett pear. *Photo:* D. Edwards.

Plate 22.3. Core breakdown of Bartlett pear. *Photo:* D. Edwards.

Plate 22.4. Pink end (premature ripening) of Bartlett pear. *Photo:* D. Edwards.

Plate 22.5. Watery breakdown of Bartlett pear. *Photo:* D. Edwards.

Plate 22.6. Blue mold rot *(Penicillium expansum)* in ripe Bartlett pear. *Photo:* J. K. Clark.

Plate 22.7. Gray mold rot *(Botrytis cinerea)* of Beurré d'Anjou pear. *Photo:* J. K. Clark.

Plate 22.8. Bull's-eye rot *(Pezicula malicorticis)* showing sporulating lesion. *Photo:* J. K. Clark.

Plate 25.1. A 2% oxygen atmosphere delays ripening and deterioration of Bartlett pears kept at 30°F (–1.1°C) for 6 months. *Photo:* A. Kader.

Plate 25.2. CO_2 injury in Bartlett pears. *Photo:* A. Kader.